COMPOUNDING MATERIALS
FOR THE POLYMER INDUSTRIES

To My Mother and Father

COMPOUNDING MATERIALS
FOR THE
POLYMER INDUSTRIES

A Concise Guide
to Polymers, Rubbers,
Adhesives, and Coatings

by

John S. Dick

np | **NOYES PUBLICATIONS**
Park Ridge, New Jersey, U.S.A.

Copyright © 1987 by John S. Dick
   No part of this book may be reproduced in any form
   without permission in writing from the Publisher.
Library of Congress Catalog Card Number: 87-12220
ISBN: 0-8155-1135-3
Printed in the United States

Published in the United States of America by
Noyes Publications
Mill Road, Park Ridge, New Jersey 07656

10 9 8 7 6 5 4 3 2 1

Library of Congress Cataloging-in-Publication Data

Dick, John S.
   Compounding materials for the polymer industries.

   Bibliography: p.
   Includes index.
   1. Polymers and polymerization--Additives. I. Title.
TP1142.D53 1987   668.9   87-12220
ISBN 0-8155-1135-3

Preface

Over one-half of the chemist population works with polymers. A large portion of these chemists either directly work as compounders and formulators or must have some understanding of these disciplines to carry out other peripheral activities such as organic synthesis of new compounding ingredients or laboratory chemical analysis of existing polymeric compounds or formulated products.

Most of the formal chemistry or polymer science curricula today are devoid of courses relating to polymer compounding or formulating. Yet practically every product in the plastics, rubber, adhesives, and coatings industries is to some extent the result of the compounding of a polymer (or polymers) with other materials. This book is designed to instruct the reader in the nature of these compounding materials, their end use, and their role in producing the end product.

Even though we cover the functions of these materials separately in order of the four industries discussed (i.e., plastics, rubber, adhesives, and coatings), still one can see the commonality in use of these materials between different industries. In fact, the interindustrial similarities among these classes of materials are just as important to study as the differences. Many times important principles that have been learned in one industry have been transferred to another. Technology transfer among these allied polymer industries is common.

The purpose of this book is to provide a broad overview of the materials and principles used in compounding and formulating in the polymer industries. It is not intended to make the reader a master compounder in any of these industries. This can only come through experience. Rather it should serve as a guide to understanding the complexities of this science and art. Also it can be a useful reference to the experienced compounder, especially in helping him in finding new ideas to apply to his specific industrial needs from other referenced areas outside his home industry. In other words, the solution to a specific problem may already exist in an allied industry; so why reinvent the wheel?

Caution: Every chemical or compounding material discussed in this book can be a health and safety hazard if not properly used. Proper handling and safety precautions are beyond the scope of this book. One should never work with any chemical or compounding material without first thoroughly reviewing and understanding the Material Safety Data Sheet (MSDS) which is available from the manufacturer (supplier). The Supplier's MSDS is required reading and should alert you to potential hazards such as acute and chronic toxicity, carcinogenicity, sensitization, aspiration dangers, flammability, and explosivity. Only one who has been professionally trained in proper chemical handling techniques and chemical safety procedures can be allowed to work with these chemicals and compounding materials. Also, one must research other relevant chemical literature to anticipate possible special hazards (perhaps not mentioned in the MSDS) that can result from a given chemical reacting with another chemical substance(s). One should abide by all Occupational Safety and Health Administration (OSHA) and Environmental Protection Agency (EPA) rules and regulations.

August, 1987

John S. Dick

Acknowledgments

The author is grateful to the following individuals for help received in completing this venture: Howard L. Stephens of the University of Akron; Robert H. Gerster of Goodyear Tire and Rubber Co. (retired); Jim Duddley of Goodyear Tire and Rubber Co.; Earl Melby of GenCorp; N.J. Maraschin of Union Carbide Corporation; James P. Walton of Monsanto Chemical Co.; George Wagner of Exxon Chemical Co.; Robert Carpenter of Ashland Chemical Co.; Vincent George of P.D. George Co.; Ralph Schupp of American Cyanamid Co.; Fred Watts of Ashland Chemical Co.; Carl J. Knauss of Kent State University; Ronald L. Pastorino of Witco Chemical Corporation; R.M. Friedel of Aristech Chemical Corporation; Douglas J. Bolton of Pennwalt Corporation; Russell H. Tobias and Terry Sprow of Owens/Corning Fiberglass; Robert S. Miller of Occidental; Mike Kallaur of Freeman Chemical Co.; Sam Sumner of Interez Inc.; Allan R. Meath of Dow Chemical Corporation; Donald L. Christman and Tony O'Driscoll of BASF Corporation; Jim Prom of Freeman Chemical Co.; Dennis J. Olszanski of Mooney Chemical, Inc.; Ron Eritano of Mobay Corporation; G.L. Wilkes of Virginia Polytechnic Institute; and Steve Velten, of Northern Adhesives Co. Also, special thanks to John McCool (B.F. Goodrich, retired) for his help and advice.

To all those whom I may have overlooked, my sincere apologies.

NOTICE

To the best of our knowledge the information in this publication is accurate; however the Publisher and Author do not assume any responsibility or liability for the accuracy or completeness of, or consequences arising from, such information. This guide does not purport to contain detailed user instructions, and by its range and scope could not possibly do so.

Compounding raw materials can be toxic, and therefore due caution should always be exercised in the use of these hazardous materials. Final determination of the suitability of any information or product for use contemplated by any user, and the manner of that use, is the sole responsibility of the user. We strongly recommend that users seek and adhere to a manufacturer's or supplier's current instructions for handling each material they use.

Contents

1

The Plastics Industry

Of the four industries to be discussed in this work, the plastics industry is by far the largest. Plastics are involved in every aspect of our daily lives. This very minute, while you are reading this page, you can probably identify several objects within a few feet of you that are made of plastic. These objects might include an ink pen, a TV cabinet, a chair, a desk, a light fixture, a telephone, eyeglasses, etc. At the beginning of the 1980s, the United States produced 14 billion dollars worth of plastic resins to be fabricated. Of course the value added to these plastics after molding into products amounts to additional billions of dollars.

MARKETS

The plastics industry has historically been an industry of rapid growth. Although synthetic plastics have been around since the late nineteenth century, it was not until after World War II with the availability of relatively cheap petrochemical feedstocks that growth took off and launched us into what is called the 'plastics age'. In the 1967 motion picture *The Graduate*, the protagonist is advised that the best career opportunities lie in plastics. Indeed, before the 1973 Arab oil embargo, the annual growth rate in the plastics industry was typically 12 percent. Growth has now slowed down somewhat because of higher oil prices; however the growth rate is still higher than many other industries. This is because new technical and economic advantages are constantly being found for substituting plastics for traditional materials such as metal, glass, wood, ceramics, etc. Examples of these market substitutions are easily found. Plastic pipe has replaced metal pipe because of cost and the ease with which plastic pipe can be installed and connected. Plastic parts have replaced many metal parts in automobiles in order to reduce their weight and thus improve fuel efficiency. Plastic house siding is replacing aluminum siding because it requires less energy to manufacture. Plastic bottles have replaced glass bottles because plastic bottles don't break. Plastics possess some unique qualities which make them superior substitutes as floor tiles, in furniture fabrication, and in electrical cable insulation. These are just a few of the many areas in which plastics have achieved a great deal of market penetration.

Today there are many different plastic polymers available commercially; however, approximately 70 percent of the total tonnage of plastic production is represented by the four commodity plastic groups, i.e. polyvinyl chloride (PVC), polyethylene (PE), polypropylene (PP), and polystyrene (PS). In each of these four commodity areas there is a great deal of competition. Generally there are 10 to 20 manufacturers (usually either chemical or oil firms) producing each of these commodity plastics. The four-firm concentration in each of these commodity areas is usually between 40 and 60 percent and the largest single firm usually has between 15 and 30 percent of the total productive capacity in each commodity area. The eight largest U.S. plastic companies (based on total world-wide pounds of productive capacity) are given below in decreasing order of size:

(1) Dow Chemical

(2) Shell

(3) Union Carbide

(4) Dupont

(5) Monsanto

(6) B.F. Goodrich

(7) Exxon

(8) Phillips

Overall, price competition in the commodity plastics markets can be quite active. On the other hand, specialty plastics, which may only have one or two manufacturers, may not be as price competitive.

If there is competition among commodity resin producers, then there is intense competition among plastic fabricators which buy the resins and mold them into the final plastic products. These fabricating firms number about six thousand. Many times these manufacturers are quite small.

HISTORY

Throughout the chemical literature of the nineteenth century, organic chemists had reported the existence of many residues from various organic reactions which are now known as polymers. The problem was that chemists lacked the know-how to convert these "residues" into useful products. The quest for a substitute for ivory in the middle of the last century led John and Isaiah Hyatt to discover, in 1869, that cellulose nitrate could be effectively processed by the incorporation of camphor. Thus, cellulose nitrate represents the first commercial plastic. However, this plastic was extremely flammable and dangerous to work with. Therefore, in 1908, G. W. Miles developed cellulose acetate as a safer alternative. This new plastic did gain acceptance.

While cellulose nitrate was the first man-made commercial plastic, phenol-formaldehyde resin represents the first truly synthetic commercial plastic in that it is produced from synthetic chemicals. Leo Baekeland commercialized this thermosetting plastic in 1910 under the trade name of Bakelite. This plastic was very hard and strong and found many useful applications. One major problem with it, however, was its dark color. So in 1928, H. John and F. Pollack invented another synthetic thermoset called urea-formaldehyde resin, which possessed a lighter color and could be produced in many light shades.

In the 1930s, a flurry of new commercial plastics was introduced onto the market. Another thermoset called melamine-formaldehyde resin appeared. Also, in an effort to find a substitute for silk, Dupont set out through planned research to develop a silk-like synthetic polymer. This effort led to the invention of nylon by Wallace Carothers. Likewise, Rohm and Haas Company worked out practical economical methods of producing polymethyl methacrylate. Also, Waldo Semon of B.F. Goodrich Co. commercialized polyvinyl chloride by introducing new effective methods of plastication. Another important innovation in the area of plastics was developed by M. W. Perrin and John Swallow of Imperial Chemical Industries, in their work with polyethylene. They discovered an effective way of polymerizing low density polyethylene through what is now called the high pressure process.

The development of PVC and PE in the nineteen-thirties greatly aided the Allies in World War II. During the war, polymer research accelerated, resulting in the development of unsaturated polyester, polystyrene, polytetrafluoroethylene, and epoxy. After the war, in 1954, stereo-specific catalysts were developed enabling high density polyethylene and polypropylene to be commercialized. With the development of these technologies and the availability of relatively cheap petrochemical feedstocks, plastic production took off and launched the world into the "Plastic Age".

The decade of the 1960s not only saw the rapid growth of commodity thermoplastics such as PVC, PE, PP, and PS, but also the emergence of the "super" plastics such as polycarbonate, polyacetal, phenoxies, and polyphenylene oxide. These specialty plastics are commonly called engi-

Chronology

Approximate Date of Commercialization	Plastic	Pioneers
1869	Cellulose nitrate	Alexander Parkes J.W. Hyatt I.S. Hyatt
1908	Cellulose acetate	G.W. Miles Cross and Bevan
1910	Phenol-formaldehyde resins	Leo Baekeland
1928	Urea-formaldehyde resins	H. John F. Pollack
1931	Nylon	Wallace Carothers (Dupont)
1931	Polymethyl methacrylate	Rowland Hill Otto Rohm (Rohm and Haas)
1932	Polyvinylidene chloride	(Dow)
1935	Melamine-formaldehyde resin	
1936	Polyvinyl chloride	W. Semon (B.F. Goodrich)
1938	Cellulose acetate butyrate	
1939	Low density polyethylene	M.W. Perrin John Swallow (ICI)
1942	Unsaturated polyesters	Whinfield and Dickson
1945	Polystyrene	Hermann Staudinger
1945	Cellulose acetate propionate	(Celanese)
1948	Polytetrafluoroethylene	(Dupont)
1948	Epoxy	Pierre Castan S.O. Greenlee
1950	Acrylonitrile-butadiene-styrene	(Uniroyal)
1954	High density polyethylene	Karl Ziegler
1955	Polypropylene	G. Natta
1959	Polycarbonate	(G.E.)
1960	Polyacetal	(Dupont)
1962	Phenoxies	
1965	Polyphenylene oxide	(G.E.)
1976	Linear low density PE	(Union Carbide)
1980s	Emergence of new polymer "alloys"	
1980s	New electrically conducting polymers	

neering plastics and have gained acceptance as economical substitutes for metal alloys.

MANUFACTURING PROCESS

As described earlier, plastic resin manufacturers are usually large chemical and oil companies. Large companies dominate this aspect of manufacture because of their technical expertise and the very large investments required to achieve the necessary economies of scale in order to effectively reduce per pound cost. Very large reactor vessels are used to polymerize large quantities of polymer. Various polymer additives (to be discussed) are added after polymerization by the polymer producer, a custom mixer, or a fabricator. For thermoplastics, these additives may be hot melt mixed in a large internal mixer called a Banbury, mill mixed, or pre-powder blended in a powder mixer. Many hot melt mixing techniques are borrowed from the rubber industry. (See the chapter on the rubber industry for more details.) For many thermoset resins, additives are incorporated by using conventional powder blending or liquid blending methods.

Unlike polymer producers which are somewhat large in size, a typical plastic fabricator is generally quite small. It is estimated that there are approximately 6,000 plastic fabricators in the United States alone. These fabricating companies operate extrusion and/or molding operations on a variety of scales. The following discussion briefly describes these operations. (A detailed discussion is beyond the scope of this book.)

Compression molding

This is one of the simplest and oldest methods used with thermosetting resins (resins that permanently crosslink and harden on heating). It consists of simply placing a pre-weighed and perhaps preformed quantity of powdered or granulated resin in one half of the mold cavity. On closing under pressure and at a controlled elevated temperature, the resin flows and sets. On release the molded part is ejected from the heated cavity. One disadvantage of this method is that the molding resin must be accurately preweighed.

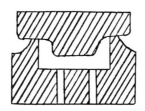

Transfer molding

This operation is similar to compression molding except that the resin is premelted in a separate reservoir and then forced into the hotter molding cavity. Again this method is used with thermosetting resins and is commonly used to mold parts which may have metal inserts. Compression molding is not as suited to molding parts with metal inserts in that their position may be changed. Transfer molding is ideal for this application. One disadvantage to this method is a normally higher scrap rate. Also it requires a thermosetting resin that can remain stable for sufficient time in the reservoir but will rapidly gel at the hotter mold temperature.

Extrusion

This method is commonly used with thermoplastics (resins that can be melted over and over again without setting). It is a continuous process which is used to manufacture products such as pipe, sheeting, film, or even to apply insulation to wire. An extruder consists of a hopper which feeds resin granules into a heated cylinder containing a rotating screw. As the screw rotates, it carries the resin forward in the chamber. The plastic resin is heated and compressed causing it to melt into a viscous fluid. As the melted plastic is driven to the head of the extruder, greater pressure develops and the plastic is forced out through a die opening which is in the shape of the product being extruded. This extrudate is generally cooled by immersion in water or forced air circulation.

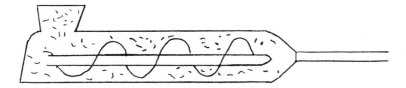

Injection Molding

This method is used to mold plastic articles from thermoplastic resins. The plastic resins are melted in an extruder, (discussed above) and automatically injected under pressure through an opening into a closed unheated mold. Upon cooling, the mold is automatically opened and the rigid article is ejected. Then the mold closes and the cycle is repeated. Some molds may contain several cavities connected by channels which will allow several articles to be molded together on the same cycle. Some injection molding machines can produce thousands of molded plastic articles each hour.

Blow molding

This operation is used to manufacture hollow products, usually bottles, jars, and other containers. Again only thermoplastic resins are used. To begin with, a soft, hot plastic tube is extruded. This tube is called a parison and is trapped between two halves of a mold which closes on it. One end of the parison is sealed automatically while compressed air is forced into the other end. The soft parison expands and fills the cavity. On contact with the mold walls, the plastic becomes rigid. The mold opens and the article is ejected.

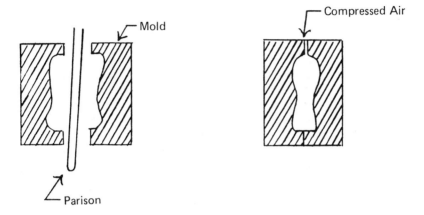

Thermoforming

This process involves the forcing of a heated thermoplastic sheet to conform to the shape of a mold. The sheet may be forced into the mold by applying another matching mold over it, applying a vacuum between the sheet and mold, or using air pressure above the sheet. Articles such as large signs and trays are made by this method.

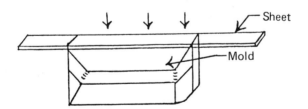

Casting

This operation is similar to molding operations previously described, except that no pressure is applied. Instead, unreacted thermosetting liquid resins are poured into a mold. Through a catalyst or moderate heat, the resin crosslinks into a solid mass and the mold is opened. Because no pressure is required, very large parts can be made by this method. Sometimes the mold is made of silicone rubber which can be peeled back after the resin has hardened.

Lamination

This is the process of constructing a product by laying, one on top of the other, sheets of woven fabrics, mats, or paper which have been presoaked in a thermosetting resin. These laminates may be set under heat and pressure or, in some cases (depending on the resin used), no pressure and little heat may be required. Laminates are discussed in more detail in the chapter on the Adhesives Industry.

Calendering

This method has been borrowed from the rubber industry. A calender consists of three or four large, heated metal rolls which melt and squeeze a plastic into a film or sheet or applies it as a coating to a fabric or paper substrate. The thickness of the film or sheet is controlled by the distance between the rolls which are preadjusted. After calendering, the sheet or film is usually passed over a cooling roll, cut to a specified width, and wound by a take-off roll. The initial investment cost of installing a calender is quite high compared to some of the other operations described; however, once installed, a calender's output can justify this investment. Examples of calendered products are plastic wall covering, vinyl upholstery fabrics, and linoleum.

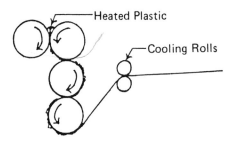

Heated Plastic

Cooling Rolls

Rotational casting

This method consists of placing either a thermoplastic powder or a plastisol (a liquid mixture of PVC powder and plasticizer) in a mold that is closed and rotated biaxially to allow the plastic to coat the heated mold's interior. The mold's heated surface melts and fuses the thermoplastic powder (or gels the plastisol), thus producing a hollow molded article which is then removed from the mold cavity. Rotational casting is relatively inexpensive and can be used to mold very large objects such as large trash cans. Also hollow toys are commonly molded this way.

TESTING

Standardized physical tests are commonly used in the plastics industry to assure that processing and product requirements are met. The following is a brief discussion concerning some of these tests.

First of all, several tests are commonly used for quality control, to assure the uniformity of a plastic resin. The melt index test is commonly used to measure the melt viscosity of a given thermoplastic resin. The test involves measuring the rate in which melted resin can be forced through a given size orifice when melted at a predetermined temperature in a cylinder under a given predetermined force created by a piston with a standard weight applied.

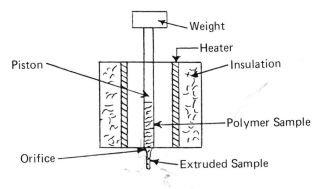

The higher the melt index value (in grams/10 minutes), the lower the melt viscosity. In general MI values may roughly relate to how well a resin may process in the factory under certain conditions. However, MI is by far not the only polymer property that determines processability. In a crude sense, the lower the MI value of a given polymer, the higher its molecular weight.

Another test commonly used which relates to melt properties of polymers is called the vicat softening point test. This test tells at what temperature a flat polymer specimen will soften enough to allow a standard needle to penetrate it in a heated oil bath. Vicat values relate to the heat softening properties of the plastic.

With thermosetting resins there are several tests that relate to their processing and molding properties. "Molding index" tests are performed on thermosetting resins to control flow and setting properties. Procedures such as the spiral mold test measure these properties in a spiral channel under standard laboratory conditions. Also tests to directly measure gel time (time required for a thermoset to begin to harden) are conducted in the factory. As will be discussed, it is important that these properties be controlled when molding thermosets. In transfer molding, for example, if the resin gels too quickly it will not fill the cavity. On the other hand, if the resin takes too long to gel, a longer molding time will be required and production output will be reduced.

In addition to the processing properties, the physical characteristics that will be imparted to the final product are also important. One of the most important of these is *tensile strength*. Quite simply this property denotes the force in pounds per square inch required to pull apart a given plastic specimen in the shape of a dumbbell when being pulled apart at a given rate usually by an instrument such as an Instron.

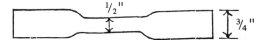

Another important property obtained from stress-strain testing is *tensile modulus* (stiffness) which is the ratio of stress (usually in PSI) to strain (usually in./in.) of the linear portion of the stress-strain curve below the elastic limit (as shown below).

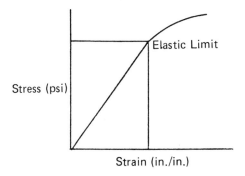

While tensile strength measures the strength of a given plastic while being pulled apart, *compressive strength* measures how crush resistant a plastic is under compressive force. Test samples in the shape of a prism or a cylinder are tested under continuously increasing pressure until failure occurs. *Flexural strength* is the ability of a plastic to resist bending forces without failure. Here a plastic speciman in the shape of a beam supported at each end is bent by an increasing load at the center of the beam until failure occurs. These properties of tensile strength, compressive strength, and flexural strength are independent properties. It is quite possible for a given plastic to be stronger in one property but weaker in another.

Another important physical property of a plastic is its impact strength, i.e., its ability to withstand a sudden shock without breaking. A brittle plastic such as an unmodified polystyrene may have good tensile strength but poor impact strength. The various physical tests that have been developed to measure impact strength in the laboratory may not closely relate to impact strength of the product under actual service conditions. One such test that is commonly used is called the Izod test.

This test simply measures the amount of energy required to break a notched specimen.

Another important property that is commonly referred to in plastic testing is hardness. For rigid, hard plastics, this property is determined by the *Rockwell Hardness Test*, which measures the resistance that a plastic mass displays to deformation by a steel ball. For "soft" plastics, a Shore Durometer is used, which measures the resistance to deformation by an indentor.

A detailed discussion of these tests and others is beyond the scope of this book; however, a listing of ASTM tests that are commonly performed on plastics is provided below. These tests fall under the jurisdiction of ASTM Committee D20.

Test	ASTM Designation
Tensile test	D 638
Compressive strength	D 695
Flexural strength	D 790
Izod impact	D 256
Charpy impact	D 256
Tensile impact	D 1822
Rockwell hardness	D 785
Flow rate (melt index)	D 1238
Vicat softening point	D 1525
Deflection temperature	D 648
Stiffness in flexure	D 747
Durometer hardness	D 1706
Deformation under load	D 621
Shear strength	D 732
Brittleness temperature	D 746
Flammability	D 635
Flow properties	D 569
Water absorption	D 570
Environmental stress cracking	D 1693
Weathering (accelerated)	E 42
Outdoor weathering	D 1435
Heating weight loss	D 706
Permanent heat effects	D 794
Specific gravity	D 792
Transmittance and color	D 791
Haze of transparent plastics	D 1003
Dielectric strength	D 149
Dielectric constant	D 150
Electrical resistance	D 257
Arc resistance	D 495

BASIC PRINCIPLES

There are two basic classes of plastics, i.e., thermosets and thermoplastics. Products made of a thermosetting resin are permanently shaped in a mold when a catalyst and/or heat is applied to promote the formation of a three dimensional crosslinked network. Once the resin is "set" in the mold, it can not be remelted. Thermosets can be made quite hard and tough through this intermolecular crosslinking process. A thermoset may consist of a tri-functional reactant, such as phenol, and a difunctional reactant, such as formaldehyde, which react to form a macromolecular network. One way to process and mold such a resin is to restrict the heat history in the manufacture of the resin in order to prevent the resin from prematurely reaching gel (the point where a three dimensional network start to form). Before the gel point is reached, the resin can be kept fluid enough to be molded. After gel, fluidity is irreversibly lost and the resin sets into a macromolecular network which is insoluble to solvents. An example of such an under-reacted resin which requires further heat to set is the *resole* phenol-formaldehyde resin (to be discussed).

A more common alternative method to hardening thermosets during molding is by the controlled pre-addition of a catalyst which promotes crosslinking. An example of this is the careful addition of catalysts to unsaturated polyesters (also to be discussed). This same principle can be applied to novolac phenol-formaldehyde resins (manufactured with insufficient formaldehyde to allow gelation) in which hexamethylenetetramine is added by pre-powder blending. Upon heating in the mold the "hexa" serves as a methylene donor to take the resin past the gel point and establish three dimensional crosslinks.

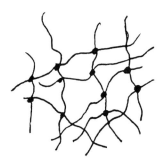

Three dimensional
crosslinked network
of thermoset resin

Thermoplastics are quite different from thermosets in that they can be melted and remelted without alteration. "Cured" thermosets derive their strength and hardness from crosslinking molecular chains to one another to prevent chain slippage when deformational forces are applied.

Thermoplastics also have restrictions of macromolecular chain mobility which prevents slippage; however, these restrictions are achieved through crystallization, intermolecular hydrogen bonding, or intermolecular polar attraction of functional groups, not through covalent crosslinking. Obviously these restrictions on chain slippage are eliminated when the thermoplastic is heated above its glass transition temperature and its crystalline melt temperature (if the polymer is crystalline). Above these transition temperatures, the molecular chains slide past one another and the thermoplastic becomes fluid enough to allow extrusion and molding. Upon cooling, these restrictions on chain movement are reformed.

Unlike, say, rubber, polymers which make good strong thermoplastics will generally have a high glass transition temperature (Tg) and/or a high crystalline melt temperature (Tm). This means that the polymer must be heated to sufficiently high enough temperature in order to provide the molecular kinetic energy necessary to overcome those intermolecular restrictions (referred to earlier) that prevent chain movement. By contrast, various elastomeric polymers (to be discussed in another chapter) have a very low glass transition and usually no melt transition (most elastomers are amorphous). Generally the molecular structure of an elastomer possesses fewer functional groups along the chain to promote polar attraction or hydrogen bonding with other chains. Also most elastomer chains do not tend to pack together and crystallize (unless they are under tension). Therefore, unlike plastic polymer chains, elastomer chains are far more mobile at ambient service temperatures which permits their elastic behavior. On the other hand, thermoplastics are stronger because their molecular chain movements are tightly restricted.

Commercially speaking, there are two categories of plastics commonly referenced, i.e., rigid and flexible. Very rigid plastic may have moduli values in the range of 500,000 psi or higher. More importantly, very rigid plastics characteristically have small ultimate elongation values at break (perhaps 10 percent or less). Many times these very rigid polymers may be amorphous in character. The high rigidity may be due to high density crosslinking or the molecular chains possessing bulky side groups which raise the glass transition temperature. On the same note, many of these very rigid plastics can also be quite brittle. Flexible plastics, on the other hand, have relatively high elongation values at break (some up to 500% or greater). Some examples of such flexible plastics can be seen with polyethylene and polyvinyl chloride compounded with plasticizers. Although some of these flexible plastics may elongate to the same degree that some elastomers do, they are quite different. The most important difference is that these flexible plastics do not possess the elastic recovery of elastomers. In other words, when these plastics are stretched and released, they do not return to their original shape.

Given below are examples of stress-strain curves for flexible and rigid plastics.

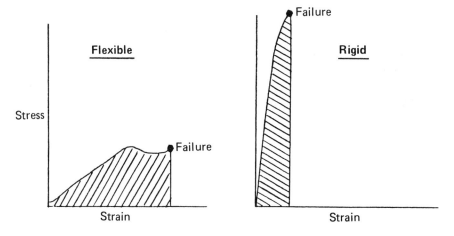

Of course, the differences in slope of these curves denote differences in modulus values. The area under the curves is a crude indication of the energy absorbing capacity of the plastic prior to failure. The greater this capacity, the better the impact resistance and toughness of the plastic. For example, if you have a high molecular weight polymer which provides significant chain entanglements between crystalline regions, then you may have both a hard, rigid plastic as well as a tough plastic with high impact resistance. This plastic's stress-strain curve will have a greater area under the curve and may appear as shown below.

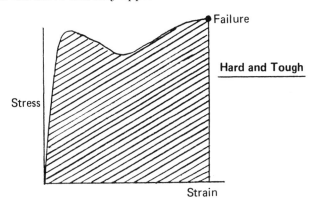

COMPOUNDING

Various additives or compounding ingredients are commonly incorporated into a plastic resin before it is molded or extruded into a final product. These ingredients are added by the resin manufacturer, by an independent custom blender, or by the fabricator. These additive ingredients are necessary to impart specific properties to the final product. These specific properties may be improved stabilization and resistance to oxidative attack, better impact resistance, higher strength, increased or decreased hardness or elongation, reduced pound-volume cost, easier extruding and molding, increased flame resistance or imparting of a specific color. Also special additives may be used in thermosetting resins to harden and set them.

Most plastics used to make a product will contain additives of some sort even if only one or two. For example, a resin manufacturer or fabricator may only add an antioxidant and perhaps a colorant to a plastic resin before extruding. In other cases, however, much more complicated formulations may be needed to meet specific product requirements. Given below are two examples of more complex formulations, one based on a thermoset, the other on a thermoplastic.

Example of a Thermosetting Formulation

Phenolic resin	40%
Reinforcing fiber	30%
Filler	20%
Colorant	2%
Lubricant	1%
Plasticizer	3%
Hardener	4%
Total	100%

Example of a Thermoplastic Formulation

PVC resin	100 phr
DOP plasticizer	30 phr
Secondary plasticizer	5 phr
Filler	10 phr
Heat stabilizer	3 phr
Colorant	1 phr
Total	149 phr

These are examples of formulations that might be used. Obviously, hundreds of thousands of combinations are possible to achieve any combination of specific properties. As may be noted, the first formulation is based on percent; however, the second formulation is given in parts per hundred resin or phr. Many plastic formulations may be based on phr units because it is possible to change the level of one additive without having to recalculate percentages for the rest. It should be noted that sometimes the resin base may be a combination of two or more different polymers. Formulations used in the plastics industry are many times referred to as plastic *compounds.*

In the following sections, we will be discussing the various plastic compounding ingredients used in the industry today. These compounding components are given below.

Thermoplastic resins
Thermosetting resins
Fillers and reinforcements
Coupling agents
Flame retardants
Plasticizers
Lubricants
Heat stabilizers
U.V. stabilizers
Antioxidants
Catalysts and hardeners
Impact modifiers
Colorants

The task of a professional plastic formulator (also called a compounder) is to understand how to utilize each of these ingredients effectively to achieve the best combination of desired properties to meet the end product requirements at the lowest possible cost. To be highly proficient in this skill may frequently require more art than science.

FUNCTIONAL USES

The base polymer is the single most important ingredient contributing to the properties of the plastic compound or composite. Shown below is the annual U.S. production output in millions of pounds for the more common thermoplastic resins.

The first four polymers (PE, PVC, PS, and PP) have such a large volume that they are considered *commodity plastics* with relatively low prices that are more or less determined by free market forces of supply and demand.

A second group might be considered *quasi-commodity* plastics. These consist of plastics such as ABS, MMA, PET, and cellulosics with moderate production volumes of between say 300 million and 2 billion pounds. These plastics may command a somewhat higher price because of the special properties that each polymer possesses. For example, ABS is known for its high impact strength while MMA is known for its clarity and optical properties.

Thermoplastic	1984 Annual Pound Production (millions)
Polyethylene (PE)	15,003
Polyvinyl chloride (PVC)	8,292
Straight and rubber modified polystyrene (PS)	3,808
Polypropylene (PP)	5,216
Acrylonitrile-butadiene-styrene (ABS)	1,220
Cellulosic plastic	*468
Polymethyl methacrylate (MMA)	503
Polyethylene terephthalate (PET)	1,023
Nylon	316
Polytetrafluoroethylene (TFE)	*20
Engineering plastics (polycarbonate, polyacetal, polyphenylene sulfide, polyphenylene oxide, polyimide, polysulfone)	744

*1978 Production because 1984 figures not reported by ITC.

Source: International Trade Commission.

The group called *engineering plastics* represents a third class of thermoplastics. These plastics have extraordinary properties concerning dimensional stability, creep resistance, strength, and ductility, which allow their use as substitutes for metal alloys in a variety of engineering applications. This group includes nylon, polyacetal, and polycarbonate. These polymers are priced higher than "quasi" commodities; but on a pound-volume basis they will save a product manufacturer considerable money when substituted for traditional metal alloy parts.

The last thermoplastic group consist of the *specialty* plastics. Plastics such as polysulfone, tetrafluoroethylene (TFE), and polyphenylene oxide (PPO) are included in this group. These plastics possess special properties such as chemical inertness, heat resistance, or low friction coefficients which enable them to command higher prices. These plastics are commonly used in special engineering applications; however, the poundage produced annually for each plastic is relatively low.

There are still high production volumes for thermosetting resins which are given below. However, it must be kept in mind that for such polymers as PF, UF, and alkyd resins, only about 5 to 15% of the production volume cited is used directly in plastic molding applications (the balance being used in making laminates, adhesives, and coatings). The volume of U.S. production output for thermosets is given below.

Thermosets	1984 Annual Pound Production (millions)
Phenol-formaldehyde resin	1,656
Urea-formaldehyde resin	1,349
Polyester resins (unsaturated)	1,372
Alkyd resins	783
Melamine-formaldehyde resin	229

Source: International Trade Commission.

Over the years, thermosets have been replaced by thermoplastics in many product markets. One of the obvious reasons for this shift has been the greater versatility that thermoplastics possess in lending themselves to a much greater variety of processing methods ranging from high speed extrusions and injection molding to blow molding. Also, thermoplastics are easier to process and do not require hardeners or curing steps. Thermoplastic scrap can even be reprocessed and molded. Still, thermosets have maintained a good share of the plastics market because of their special properties which include dimensional stability, creep resistance, strength, hardness, and high allowable service temperatures (which are higher than their molding temperatures). For reasons to be given, unsaturated polyester resins have become by far the largest volume thermoset used for molding applications.

From the information that will follow, one may note an incredible number of different commercial polymers that are available on the market as molding resins. Why so many? The answer is quite simply that a large number of different polymers are needed to meet the wide variety of process and product requirements demanded by plastic fabricators today. Some of these special requirements are given below.

Melt Viscosity—This property determines how a plastic polymer will process. For example, TFE has too high a melt viscosity to be processed conventionally. Nylon can have a low melt viscosity. A high melt viscosity is required for blow molding.

Injection Molding Temperature Requirement—This is the minimum processing temperature to reduce melt viscosity to a satisfactory level for molding. For example, some engineering plastics require high temperatures while the requirement for PE is lower.

Stiffness and Flexibility—Plastics such as unplasticized PVC (homopolymer) or thermosetting polyester can be very stiff and rigid. Plastics such as LDPE, EVA, and plasticized PVC can be very flexible.

Strength—This property is very important, especially in structural applications. Engineering plastics possess very high strength while olefin polymers (PE and PP) are not as strong.

Hardness—This property can be important. Thermosets generally have very hard surfaces.

Scratch Resistance—Polystyrene has high hardness but poor scratch resistance. Thermosets such as polyester and UF have good scratch resistance.

Gloss—A high glossy appearance can be important in the appearance of a product. Cellulose acetates, polysulfones, and many other plastics have high gloss.

Clarity—A limited number of plastic polymers have good clarity as indicated by percent light transmission. Examples are PS, MMA, polycarbonate, polysulfone, and unloaded polyester thermosets.

Resistance to Yellowing—Unmodified polystyrene has good clarity but will turn yellow on outdoor exposure to light. Polymethyl methacrylate, on the other hand, resists yellowing.

Impact Strength—A brittle plastic such as unmodified polystyrene has poor impact strength. A plastic such as ABS has good impact strength and toughness while still retaining high hardness.

Ductility—This is the ability of a plastic to be drawn and can relate to how well the polymer can be drilled, nailed, or sawed with machine tools without cracking. Engineering plastics generally have good ductility.

Chemical Resistance—Some plastics are oxidized or hydrolyzed by acids, alkali or salts. Nylon and polyester can be affected in this way. Plastics such as TFE are basically inert.

Oil and Fat Resistance—This is an important consideration for plastics coming into contact with food.

Stain Resistance—This is important in kitchen applications where there is contact with food.

Creep Resistance—This is the ability of a plastic polymer to resist deformation under long-term stress. Engineering plastics have good creep resistance.

Flame Resistance—Some polymer plastics such as TFE will not burn when placed in a flame. Others, such as polysulfone, will self extinguish when removed from a flame. Still other plastics such as unmodified polystyrene will continue to burn after a flame is removed.

Heat Resistance—Specialty plastic polymers such as polyphenylene oxide have very good resistance to deterioration at elevated temperatures while many commodity plastics do not. Some thermosets have good heat resistance.

Weathering Resistance—Basically this is how well a plastic holds up in an outdoor environment, i.e., exposure to the sun, rain, snow, etc. Polymethyl methacrylate is considered to have better weathering resistance than many other plastics.

Ultraviolet Resistance—Sunlight contains ultraviolet radiation which can cause degradation in some polymers. U.V. stabilizers are commonly used in polyvinyl chloride, polyethylene, polypropylene, polystyrene, ABS, and polyesters.

Stress Cracking Resistance—Some polymers under constant stress tend to form cracks that propagate. Polyethylene, for example, can display stress cracking problems.

Electrical Arc Resistance—This is the time needed for a given electrical current to render the surface of a plastic electrically conductive because of carbonization by the arc flame. This failure is critical because the arcing results in a carbonized conductive "track" through the polymer which deteriorates the insulating quality of the polymer in future use. Plastics such as urea-formaldehyde have relatively good arc resistance.

Dielectric Constant—This relates to electrical capacitance. If the polymer is to serve as a dielectric in capacitors, then a high value may be desirable. Dielectric constant values are also dependent on the electrical frequency in cycles per second (Hertzes).

Dielectric Strength—This property tells at what voltage a given thickness of plastic will "break down" as indicated by a sudden surge of current. High dielectric strength relates to useful insulating properties.

Electrical Resistivity—This property gives the resistance in Ohms per given thickness of a polymer. Very high values partially indicate good insulation properties.

Water Absorptivity—Some polymers such as cellulose acetate tend to absorb moisture which can reduce physical properties.

Gas Permeability—This property is particularly important in dealing with plastic film. Polymers such as PE

are good moisture barriers but do allow some permeability of gases such as oxygen and carbon dioxide. Cellulose acetate will allow permeation of
moisture.

Mold Shrinkage—Some molding resins shrink more after
molding than others. Thermosets which lose water of
condensation can display shrinkage problems.

Solvent Joining—Plastic parts such as PVC pipe are commonly joined together using "solvent welding". Some
plastics, however, such as polyethylene, do not lend
themselves very well to solvent joining. Many of
these plastics may have to be thermally fused together.

Cost—A good plastics compounder will select the lowest
cost base polymer that will provide the necessary
properties to meet the product requirements.

THERMOPLASTIC RESINS

Polyethylene (PE)

$$\cdots \!\!+\!\! CH_2 \!-\! CH_2 \!\!+\!\!_n$$

This commodity plastic has the largest annual production poundage in
the world today. The reasons for its popularity are its low cost, (especially
on a pound-volume basis due to its low specific gravity) and its unique
physical properties.

One unusual property of PE is its extremely low glass transition
temperature. As discussed earlier, most useful plastics will have relatively high glass transitions (Tg). However, PE has a Tg value of only
$-115°C$ which is less than most commercial rubbers used today. The
reason that PE is not rubber-like at room temperature is the very high
degree of crystallinity normally contained in the polyethylenes used commercially. This high crystalline content (up to 95%) is responsible for PE
having a first order melt transition (Tm) as high as 137°C. These crystalline regions of PE provide strength and stiffness to what otherwise
would be a rubber-like polymer. The high crystallinity also explains the
translucence of polyethylene. The many crystallite domains present disrupt light transmission through the polymer matrix and prevent clarity.
Below are shown regions of crystallinity, called crystallites, in which
ordered polyethylene molecular chains are "packed" together. Between
these crystallite regions are the amorphous unordered regions.

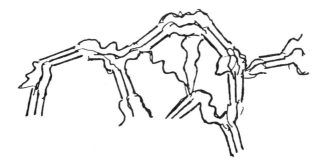

While the crystallinity of PE prevents it from being a rubber, this property can not be made extensive enough to impart high hardness to this plastic. Thus PE is not considered an "engineering plastic"; however, it does display other valuable properties which can explain its wide use.

First of all, PE is uniquely flexible and does not require the addition of a plasticizer to achieve this property. This explains its use in making squeeze bottles, films, and sheets. Secondly, PE has good resistance from attack by most acids, alkali, and salts (with the exception of strong oxidizing agents). This is why it is used to make household detergent containers. Thirdly, PE has good resistance to water and its vapor which partially explains its good electrical properties and use in making electric cables. Lastly, PE is very easy to process and is compatible with most thermoplastic processing methods.

Of course, polyethylene is not the complete answer to our society's plastic needs and there are several disadvantages in using it. As mentioned earlier, PE is not very hard and is relatively low in tensile strength when compared to some other plastics. Unprotected unmodified polyethylene has poor resistance to outdoor U.V. exposure and relatively poor heat resistance. Another problem characteristic of PE is environmental stress cracking (however, increasing average molecular weight tends to reduce this problem). Also PE does not lend itself well to "solvent welding", but different PE parts can be fused together without difficulty through melt joining.

In choosing a polyethylene resin, three important properties should be considered. First, the average molecular weight as indicated by melt index (MI) is important. The higher the MW (lower MI) the better the stress crack resistance and impact strength. On the other hand, higher MW can also mean less ease in processing. Secondly, the molecular weight distribution can be important. For good low temperature properties and high impact strength a relatively narrow MWD is needed. For easy extrusion and molding, a broad MWD is desirable. Thirdly, polymer density is important in that this property relates to percent crystallinity and the degree of molecular packing. With conventional polymerization processes, higher density also generally means a lower tendency for molecular chains to branch (higher molecular chain linearity). Thus, higher density polyethylenes have greater stiffness and hardness. The density of a PE is determined by the type of polymerization conditions under which the resin is made. The *high* pressure process produces *low* density polyethylene (LDPE) while the *low* pressure process provides *high* density polyethylenes (HDPE). An important exception to this rule is linear low density PE, to be discussed later.

Flexible LDPE usually contains only between 40 to 70 percent crystallinity. Its molecular chains are more branched and less packed. The density of LDPE is usually between 0.910 and 0.939. LDPE is polymerized in an autoclave at pressures between 14,000 and 45,000 psi and temperatures between 80° and 300°C usually with a peroxide initiator such as benzoyl peroxide. Because of the flexibility of LDPE, over 50% of it goes into making film and sheeting for such applications as shrink-wrap, packaging linings, covers, and bags. LDPE is also used in blow molding applications to make such containers as squeeze bottles. Usually a higher molecular weight (lower MI) is desired in blow molding to provide satisfactory melt strength. Also, LDPE is used in wire and cable insulation as well as in paper coating.

HDPE is different from LDPE in that its molecular chains are more closely packed giving a higher percent crystallinity, usually between 70 and 95 percent, and a higher density, usually between 0.941 and 0.965. HDPE is less flexible than LDPE and has a higher melting point which is well above the boiling point of water (100°C). Thus articles made of HDPE can often be sterilized.

HDPE was first commercially made in Germany by the Ziegler process in 1954. The Ziegler process polymerizes ethylene under low pressure (80 psi) using an aluminum and titanium catalyst in an inert atmosphere. Although the Ziegler process is now used throughout the world, another process, developed by Phillips Petroleum Co., became very popular in the United States. The Phillips process uses chromium oxide on silica and alumina as a catalyst to polymerize ethylene at 100°C under 100 to 500 psi pressure. Very high crystallinity, up to 95%, is achieved by this process. Although HDPE is less flexible than LDPE and has poorer impact resistance, it does have better resistance to oxidative aging and chemical attack as well as better resistance to environmental stress cracking. Thus HDPE is used in making a wide variety of commercial products such as fuel containers, outdoor chairs, toys, luggage, pipes, and conduits.

A far less common form of polyethylene is *ultra high molecular weight polyethylene* (UHMWPE). This polymer is significantly harder and stiffer than other forms of PE. Also it has high impact strength at low temperature. Since UHMWPE has a melt index close to zero, it can not be processed by conventional extrusion or blow molding. UHMWPE has been found to be very resistant to chemical attack thus explaining its specialized use in making chemical pumps and valves. Also UHMWPE is considered self lubricating and resistant to stress cracking.

Lower crystallinity in polyethylene relates to increased flexibility. Therefore, if the tendency to form crystallite regions is reduced, greater flexibility should result. This in fact has been achieved by the addition of a second monomer to be copolymerized with ethylene, namely vinyl acetate. This second monomer used in producing *ethylene vinyl acetate* (EVA) copolymer disrupts the tendency of the chains to crystallize. This results in a plastic with greater flexibility than the PE homopolymer. EVA is used in many applications as a "rubber substitute". EVA has good U.V. resistance; however, this copolymer does not display the resistance to water attack that PE homopolymer possesses. EVA is also much clearer than PE because of its reduced crystallinity (fewer crystallite regions to scatter light rays). Ethylene vinyl acetate is not the only copolymer of ethylene used commercially. Also copolymers using ethyl acrylate or methyl acrylate with ethylene are available in limited quantities.

Polyethylene can be cross-linked into a three dimensional matrix by using peroxide or radiation techniques described in the next chapter for curing rubber. Usually a peroxide is carefully added during processing or the polymer is exposed to radiation in order to generate free radicals which ultimately result in carbon-carbon crosslinks between chains. The purpose of crosslinking is to improve dimensional stability of the polymer under a variety of service conditions including high temperatures.

Lastly, a new polymer called *linear low density polyethylene*, (LLDPE), has emerged. LLDPE is a copolymer of ethylene with one of the following monomers: butene, hexene, or octene. These LLDPE polymers are produced by the more cost effective low-pressure process; therefore, LLDPE has replaced about 27% of the total low density PE market and about 40% of the PE film market. LLDPE films have better physical properties than film made from conventional high pressure processes. LLDPE shows improved tear resistance, puncture resistance, and tensile strength.

Polyvinyl Chloride (PVC)

$$\left(CH_2-\underset{\underset{Cl}{|}}{CH} \right)_n$$

This polymer represents the second largest volume commodity plastic produced in the world with good expectations of a continued high growth rate. One of the reasons for this is the plastic's high versatility. PVC can be used as a strong, rigid plastic, or it can be compounded with a variety of plasticizers to produce flexible plastics. Even though it has a relatively high specific gravity for a commodity plastic (1.4), its price per volume is still more than competitive against other commodity thermoplastics on the market. Part of this plastic's composition is based on a nonpetroleum feedstock—chlorine—which may give PVC a unique cost advantage in the future.

PVC is the major member of a group called "vinyl" plastics. This name is applied because these plastics are based on a monomer which contains the vinyl radical ($CH_2 = CH$-). Unlike polyethylene, PVC has a much higher Tg value at 82°C. This makes PVC a hard, rigid plastic at room temperature if plasticizer has not been compounded with it. Because the chlorine atoms along the macromolecular chains are non-regular, an atactic structure prevails with only 10–15% crystallinity.

PVC is produced from the polymerization of vinyl chloride monomer (VCM) which in turn is obtained from the dehydrochlorination of 1,2-dichloroethane or the oxychlorination of ethylene. The polymerization process is carried out by the bulk, solution, emulsion, or suspension process; however, well over 80% of all PVC produced in the U.S. is polymerized by the suspension process because of the ease of scale up, temperature control and particle size control. This process uses suspension agents and monomer soluble free radical initiators in an aqueous medium. The emulsion process (which is similar to the suspension process, but achieves smaller VCM droplets by using stronger emulsifiers), represents about 12% of the PVC production. This process produces resin of smaller particle size. Manufacturers of PVC will commonly sell the straight PVC resin, uncompounded with other ingredients (or they may add these ingredients themselves). The addition of these components can be achieved through melt mixing (in a Banbury or continuous mixer), powder blending (in a powder blender), or plastisol preparation (to be discussed).

There are many advantages that PVC has over other plastics. First of all, PVC has good resistance to fats and oils as well as acids and alkalis. When properly compounded, it can be made quite resistant to outdoor weathering and oxygen and ozone attack. PVC also displays good electrical insulation properties (historically one of its first uses), and is considered flame resistant. In addition, PVC displays good water resistance and inherent resistance to fungus growth. Also PVC parts can be solvent "welded". Lastly, the most important advantage that PVC has over other plastics is its unique quality that enables it to be easily compounded with a wide variety of plasticizers in order to obtain plastic compounds varying from rigid forms to extremely soft and flexible products. No other plastic shows this versatility.

With all the advantages just discussed in using PVC, there are also some important problems in its use that must be overcome. First of all, uncompounded PVC has poor heat stability resulting in dehydrochlorination, chain scission, crosslinking, and oxidation. This could be a problem during processing of PVC in that the uncompounded polymer's softening point is not very far below its decomposition temperature. Any thermal decomposition during processing evolves HCl, a toxic gas. Over the years this problem has been eliminated through the use of a wide variety of heat stabilizers, lubricants, and plasticizers (to be discussed later). Another problem with uncompounded PVC is its instability under sunlight (ultraviolet exposure) which causes dehydrochlorination. This effect results in discoloration from the formation of a polyunsaturated polymer which changes the plastic's appearance from clear to yellow or brown. As with the heat stability, this U.V. stability problem is also easily overcome through the addition of compounding additives called U.V. stabilizers (to be discussed). If color is not important, a small amount of carbon black or inorganic colorant such as titanium dioxide can also be added to shield the polymer from U.V. absorption.

The markets for PVC are about evenly divided between two broad categories, i.e., rigid and flexible. When PVC is compounded with little or no plasticizers, a strong, rigid plastic results. Between 30 and 40% of all PVC production goes into pipes and fittings used in building construction. Another large and growing area of use for rigid PVC is in making home siding and building panels. The remainder of rigid PVC is used in various applications such as gutters, downspouts, weather stripping, automobile parts, credit cards, and blow molding of bottles.

The other half of the PVC market consists of flexible plastic applications. PVC is usually made flexible through compounding 30 to 80% of a liquid plasticizer into the formulation. (Plasticizers will be discussed in detail later). These plasticizers are chemically compatible with PVC. When mixed, these plasticizers occupy space and physically separate the PVC molecular chains from one another, thus reducing their intermolecular attractive forces and allowing more chain movement and flexibility. Thus, with the addition of a plasticizer to PVC, the glass transition temperature of the compound is lowered well below room temperature. Also, plasticizer addition to PVC reduces the temperature needed to process the plastic compound (as well as reduce its melt viscosity). On the other hand, plasticizer addition results in a lowering of the polymer's tensile strength and may destroy flame retardancy characteristics.

The uses for flexible PVC are varied. One large market for flexible PVC is in extrusion of insulation for wire and cable. Also these flexible compounds are applied to fabric by calendering to make vinyl upholstery and wall covering. Sheeting and film are also large applications for flexible PVC. Other applications include swimming pool liners, inflatable toys, garden hose, shower curtains, table cloths, etc.

General purpose PVC resins are produced through the suspension process as a white powder. These resin powders consist of particles with high internal porosity. Thus relatively large quantities of liquid plasticizers can be absorbed by these resins during a blending operation. Afterwards these blends are extruded into products.

Most flexible PVC products are made from *general purpose* suspension resins which usually have been blended with plasticizers. General purpose suspension resins have an average particle size of about 130 microns. On the other hand, another class of PVC, called *dispersion* resin, can be used to make a liquid dispersion of PVC resin in plasticizer. This plasticizer dispersion of PVC resin and other compounding ingredients is called a *plastisol*. The PVC dispersion resins used to make the plastisol have an average particle size of only one micron. When these viscous plastisols are heated, the suspended PVC particles begin to swell and absorb the surrounding liquid plasticizers. When the temperature is increased to over 300°F, fusion of the particles occurs and the particles coalesce into a homogeneous mass. This transition stage is called fusion.

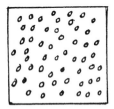

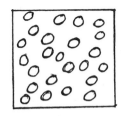

Dispersion PVC
Resin particles
in plasticizer

Upon heating,
PVC resin
particles swell,
absorbing plasticizer.

Fusion
Plasticizer
completely
absorbed in
resin particles;
particles have
coalesced

Plastisols are commonly used in rotational molding to cast such items as toy parts. Also plastisols are commonly used in dipping operations in which such items as dishwasher racks, fabric gloves, and tool handles are coated with plastisols, then heated at elevated temperatures to allow the coating to fuse. The plasticizer concentration as well as the size and distribution of the resin particles, and other factors determine the viscosity of the plastisol before fusion. In some special applications, very low plastisol viscosities are needed where good fabric penetration is required. If too much plasticizer is used to try to achieve this viscosity reduction, the fused hardness will be too low. To solve this problem some plastisol compounds have organic solvents added to them to reduce viscosity while not reducing fusion hardness (the solvent is lost to evaporation upon heating). These formulations using solvents are called *organosols*. Of course, when solvents are used, there are additional safety concerns regarding toxicity, flammability, and handling that must be addressed. On the other hand, if a particularly viscous plastisol is needed, many times a thickening agent such as bentonite clay or silica gel is added. These formulations are called *plastigels*.

Vinyl chloride is commonly copolymerized with any one of several monomers such as vinyl acetate, acrylonitrile, ethylene, propylene, diethyl maleate, and vinylidene chloride. The inclusion of the bulky functional groups, associated with some of these monomers, on to the backbone of the copolymer, results in a reduction of the attractive forces between the molecular chains. Thus, the groups of these vinyl copolymers impart greater chain mobility and polymer flexibility. This is similar to the effect that plasticizers have on PVC homopolymers described earlier. Therefore, vinyl copolymers are sometimes described as possessing "internal plasticization" without the use of plasticizers. On the other hand, if plasticizer is to be used, less is required with a vinyl copolymer than with the homopolymer. This has obvious advantages in that there is little or no plasticizer lost from the compound through volatilization or solvent extraction. These vinyl copolymers do not require as high a process temperature as the unplasticized PVC homopolymers need for satisfactory processing. Also, the copolymer generally displays better mold flow properties. On the other hand, solvent resistance may not be as good.

By far, the largest production of vinyl copolymer is as vinyl chloride-vinyl acetate copolymer. A large use for this copolymer is in making floor tiles where good abrasion resistance is important. Also, the acetate copolymer is commonly used in molding high quality phonographic records because of its excellent flow properties. Other applications are in the production of luggage, and coatings.

Another closely related polymer to polyvinyl chloride is *polyvinylidene chloride* (PVDC). Unlike PVC, PVDC has two chlorine atoms attached to the same carbon.

$$\left(\!\!-CH_2-\underset{\underset{Cl}{|}}{\overset{\overset{Cl}{|}}{C}}-\!\!\right)_n$$

PVDC

Because of this symmetrical difference in structure, PVDC has a higher crystalline content than PVC. PVDC has high strength, good resistance to water absorption, and excellent resistance to chemical attack. Its two largest uses are in making monofilament and in making film. The film used is usually based on a copolymer of 85% vinylidene chloride and 15% vinyl chloride sold under the trade name of Saran. This film provides a superior barrier to water vapor and oxygen, thus enabling its use in food applications. Also PVDC is used to make valves, car cover seats, upholstery, and brushes.

Lastly, a vinyl product called chlorinated polyvinyl chloride (CPVC) is rapidly growing in popularity. CPVC is produced by the chlorination of PVC in order to improve the heat resistance of PVC. As a result, CPVC has gained wide acceptance in the making of hot water piping in building construction.

Polystyrene (PS)

$$\left(\!\!-CH-CH_2-\!\!\right)_n$$

This polymer represented the third largest volume commodity plastic produced in the U.S. in 1981. Today it is fourth. It emerged as an inexpensive plastic after World War II because of the high availability of styrene resulting from the war's synthetic rubber program. Today, however, polystyrene's future growth is projected to be slower than the other commodity plastics because of the higher price of benzene (essential feedstock for styrene production), the maturity of its market outlets, and the potential loss of market from penetration of other less expensive commodity plastics.

Unlike the commodity plastics discussed earlier, PS contains no crystallinity and is therefore very clear (no crystallite regions to scatter light). The bulkiness of the benzene ring appendages bonded to the macromolecular backbone prevents ordering of structure, thus causing PS to be amorphous in nature. PS is a useful plastic because at room temperature it is well below its glass transition temperature (Tg) of approx. 100°C. This quality makes it a hard, yet somewhat brittle, plastic. Above 100°C, PS softens readily, but without a sharp melt transition which is characteristic of highly crystalline polymers. PS lends itself quite well to all forms of thermoplastic processing. In addition to the amorphous clarity of the plastic, it also possesses a certain "brilliance" in appearance because of high refractive index of 1.592 resulting from the density of the benzene ring structure present. In addition, the benzene ring structures also increase the specific gravity of PS to 1.05, which gives it a cost per volume disadvantage when compared to the olefinic commodity plastics (polyethylene and polypropylene).

The production of the styrene monomer is dependent on both benzene and ethylene feedstock availability. The styrene is commonly polymerized by the suspension process or by the bulk polymerization process.

There are advantages in using PS as opposed to some other commodity plastics. First of all, PS is easily processed and lends itself to most any method including extrusion, injection molding, blow molding, etc. Also, because of its amorphous nature and high refractive index, PS displays very good clarity and optical properties. PS plastic has a very good resistance to water absorption and some good electrical properties. It also gives good resistance to acid, alkali, and salt attack. On the other hand, it is readily attacked by ketone solvents which enable PS parts to be easily cemented together. Other advantages are high surface hardness and gloss, although its surface can be easily scratched.

Of course, PS also has several properties which may be considered undesirable. First, unmodified PS is brittle, has poor impact resistance, and cracks easily. One method of helping to prevent stress cracking of PS parts is to anneal the part in order to relax stress. Another disadvantage is that unprotected clear PS will turn yellow on exposure to sunlight. This can be corrected to some extent by compounding with U.V. stabilizers, for clear PS, or opaque filler loadings, for non-clear applications. Overall, PS displays poor weathering resistance and is certainly not the best choice when selecting a plastic for outdoor applications. Although PS processes easily, it also softens at a relatively low temperature just above the boiling point of water. Polystyrene's softening temperature can be increased by loading with a filler; however, its high temperature properties are not considered good. Also, PS is not flame resistant and burns with a sooty smoke. Compounding flame retardants with PS is generally only partially effective. Also, in some applications, PS can build up electrical static readily and may require an antistatic agent. Another problem with PS is that it quickly deteriorates upon exposure to hydrocarbon solvents, such as aromatics and esters.

PS is used commercially in a wide variety of applications, including the making of containers, lids, bottles, and housewares. PS is used to make wall tiles, door knobs, and fluorescent light fixtures. Some optical lenses are made of this clear plastic. Many novelties and toys are made of PS, especially model kits because of the ease of gluing. Also because of the ease of molding and hard quality of PS, many projector slide trays are made from this plastic.

As discussed, one of the major disadvantages of straight PS plastic is its poor impact resistance. One effective way around this problem has been the development of rubber modified polystyrene. Rubber modified PS is now considered a high volume commodity plastic used in many product areas. PS can be rubber modified for improved impact resistance by blending rubber directly into PS through mill or Banbury mixing. Also the rubber can be added to PS by dissolving it into the styrene monomer before polymerization. The elastomers normally used to modify PS are 1,4-polybutadiene (PB), styrene-butadiene rubber (SBR), or ethylene-propylene copolymer (EPM). Usually rubber particles in modified PS will range between 0.5 and 5 microns. Rubber modification of PS results in increasing the polymer's toughness, impact strength, and elongation. However, tensile and modulus values are reduced. Also, these rubber impact modifiers usually destroy polystyrene's transparency. Rubber modified PS plastics are commonly used in making TV and appliance housing, refrigerator interiors (trays, shells, liners, etc.), and containers.

Another important use for PS is as expandable beads. These beads are made by adding pentane (the blowing agent) during the suspension polymerization of PS. This addition results in the volatile pentane being surrounded by polystyrene, thus producing beads filled with "gas". These expandable beads in turn are sold to a fabricator to be molded into a product. The fabricator first preheats the beads to expand them, allows them to soften (called maturing), and lastly molds the expanded beads at approx. 100°C in order to allow the expanded beads to fuse together. It should be noted that this sequence is not the only one that a fabricator can use to mold a product. Under special conditions, densities as low as 1 lb. per cubic foot are attainable. Obviously, these expanded polystyrene plastics can make excellent thermal insulators and are commonly used in making hot drinking cups, hot and cold storage containers and ice chests. Other applications include water flotation products and novelties. The largest single market for its use is packaging.

Copolymers and terpolymers of styrene are also commonly used. Acrylonitrile-butadiene-styrene terpolymer (ABS) and styrene-acrylonitrile copolymer (SAN) are very widely used today and will be discussed individually later. Another less commonly known styrene copolymer is with methyl methacrylate. This copolymer gives better light stability than PS homopolymer and is commonly used in making light fixtures.

Polypropylene

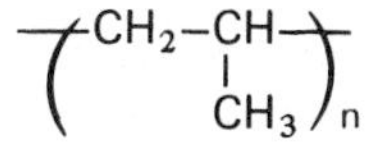

$$\left(\!\!\begin{array}{c} CH_2-CH \\ | \\ CH_3 \end{array}\!\!\right)_n$$

This polymer is perhaps the most rapidly growing commodity plastic in the industry today. It is penetrating into other markets traditionally held by other polymers such as polystyrene. Although on a per weight basis, PP is slightly more expensive than PE, on a cost per unit volume basis PP is very competitive because of its low specific gravity. PP has a low density of only 0.90 g/cm^3 because of the methyl groups attached to the backbone of the polymer as shown in the structure above.

The polypropylenes are commercially successful because of the advent of special stereospecific catalysts which polymerize propylene units in an ordered manner. These catalysts were first developed by Ziegler in 1954 for polymerizing other polymers such as polyethylene. They were later successfully applied to the polymerization of polypropylene in Italy by Professor G. Natta. Previously attempts had been made to polymerize a useful polypropylene product through a free radical mechanism; however only an amorphous, atactic rubbery polymer resulted with no commercial value.

The "secret" to the commercial polypropylene marketed today is in its tacticity or regularity in structure. Commercial PP plastic used today is over 90% isotactic, meaning that the methyl groups along the polymer backbone repeat in an orderly fashion on the same side, as shown below.

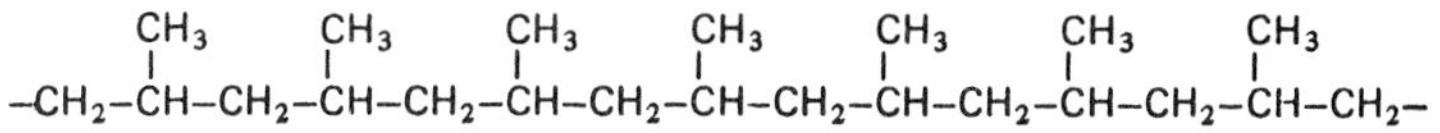

$$-CH_2-\underset{\underset{CH_3}{|}}{CH}-CH_2-\underset{\underset{CH_3}{|}}{CH}-CH_2-\underset{\underset{CH_3}{|}}{CH}-CH_2-\underset{\underset{CH_3}{|}}{CH}-CH_2-\underset{\underset{CH_3}{|}}{CH}-CH_2-\underset{\underset{CH_3}{|}}{CH}-CH_2-\underset{\underset{CH_3}{|}}{CH}-CH_2-$$

In fact, atactic polypropylene content (where methyl groups appear randomly on either side) is usually removed or reduced when possible through the use of solvent extraction. The regularity in the polymer structure results in commercial PP being greater than 60% crystalline. This is the reason PP has high enough strength to be used as a fiber as well as possessing greater stiffness than HDPE.

The commercial PP used today is a tough polymer with very good impact resistance as well as good electrical properties and excellent chemical resistance to attack by acids, bases, and salts. PP by its physical nature has several important advantages over HDPE. First of all, PP can be used at operating temperatures that are 40°C higher than what HDPE can withstand. This is an important advantage. Also PP does not have the problem of stress cracking as found when using PE. PP is stiffer and harder than HDPE. It also has a high gloss appearance. PP also may have an advantage in resistance to hot water and detergents which explains its frequent use in washing machines.

Polypropylene also has several disadvantages which should be pointed out as well. For one thing, unprotected PP has rather poor weather aging properties which can result in discoloration and crazing. Of course, antioxidants can be used to prevent this. PP is more susceptible to oxidative attack than PE because the polypropylene structure contains tertiary carbons which are more susceptible to oxidative attack resulting in removal of the t-carbon's hydrogen and the formation of a radical which propagates the oxidative degradation process. In a similar manner, unprotected PP (containing no U.V. stabilizers), is also more susceptible to degradation caused by U.V. exposure than HDPE. Another problem with PP is that it can become quite brittle at low temperatures (PP has a glass transition temperature of -20°C compared to -115°C for PE). In order to prevent this sort of problem with PP, natural rubber or synthetic elastomers such as 1,4-cis-polyisoprene or butyl rubber are sometimes blended with the polypropylene to improve its low temperature properties.

Polypropylene has found uses in a very wide variety of applications. In the auto industry, it is used for fabricating door panels, trims, fan blades, mats, battery boxes, and pedals. In appliances and housewares, PP is used in making radio and TV cabinets, dishwasher components, washer agitators, and piping. Also PP is commonly used to make containers and is used in wire and cable applications. Many drinking straws are made of PP. Many pieces of luggage and toys are made of PP. PP is also blow molded into articles such as bottles. Approximately 10% of PP production goes into the making of film for use in such applications as food wrap. Another 30% of PP output goes into the manufacture of monofilament and fibers. This rapidly growing market has resulted in the replacement by PP of much of the jute fiber used in the manufacture of carpet backing and shipping sacks. Also indoor/outdoor carpet, "grass" turf, rotproof rope and nets, and brushes are now made from PP.

Although very large quantities of polypropylene are used as homopolymer, a growing market has developed for the propylene-ethylene copolymer as well. The addition of ethylene during polymerization results in a polymer with reduced stiffness and greater impact resistance. The higher the ethylene content, the greater these effects. These copolymers are available as either random or block copolymer forms. Large markets have emerged for some of these copolymers.

Acrylonitrile-Butadiene-Styrene (ABS)

$$\left[\left(-CH-CH_2- \right)_x \left(-CH_2-CH- \atop \quad\quad CN \right)_y \left(-CH_2-CH=CH-CH_2- \right)_z \right]_n$$

This plastic reached production levels in excess of one billion lbs. in the year 1978, and today approximately 1.2 billion lbs. are produced annually. Although this volume is not considered by many to be high enough to be a true "commodity" plastic, it nevertheless represents much higher volume than engineering or specialty plastics possess. The basic reason for the wide spread use of ABS is its rather unique combination of properties, i.e., being both tough and hard. Usually a plastic that is tough and resistant to impact may not also be hard and rigid. ABS obtains this combination of properties by using acrylonitrile for high strength and chemical resistance; butadiene for toughness, cold temperature properties, and impact strength; and styrene for hardness, gloss, and improved processability. Basically ABS consists of rubbery, discontinuous domains (or particles) of polybutadiene grafted to the continuous phase of styrene-acrylonitrile. The grafting was pioneered by Borg-Warner Corp. for better moldability and gloss. Before this development, the polymer units were combined through blending of straight dry polymers or latices before coagulation.

ABS is a very tough plastic which can take a lot of abuse without cracking. Unlike other more flexible plastics of equivalent impact strength, ABS is also hard and rigid with high gloss and excellent scratch resistance. In addition, the surface of selected grades of ABS is known for its resistance to chemicals, solvents, and moisture. Another advantage of ABS is its ability to be compounded for different hardness and flexibility. ABS is also considered to have good ductility as well as good cold temperature properties.

Of course, the use of ABS also has several disadvantages that should be noted as well. For one thing, ABS has limited heat resistance and is generally not used in service at temperatures greater than 220°F. Also ABS is not very resistant to sunlight exposure and will discolor. Except for a few very special grades, most ABS plastics are opaque. Compared to other plastics, ABS has lower tensile strength (a necessary trade-off to achieve high impact strength). Lastly, ABS must be compounded with a flame retardant to be rendered self extinguishing.

Approximately one-third of ABS consumed in the U.S. goes into the production of pipes and fittings—the largest single market. Another one-sixth of production goes into appliance and automobile parts. For example, ABS is used in household appliances such as vacuum cleaner housings, refrigerator door liners, and radio and TV cases. Automobile uses include instrument panels, decorative light trims, steering wheels, etc. Another 10% of the ABS produced goes into business machine and telephone housings. Other applications include power tool housings, luggage, furniture, camper tops, golf carts, trays, boats, etc.

As discussed previously, styrene-acrylonitrile copolymer (SAN) is commonly used as a blending stock in the production of ABS; however, SAN is also used as a plastic in its own right. As mentioned earlier, the acrylonitrile imparts strength and solvent resistance to the plastic. In fact, SAN generally has better solvent resistance than ABS although it has poorer impact resistance. Also, SAN is resistant to stress cracking and scratching and has good resistance to household chemicals. SAN is used in making telephone parts, piano keys, packaging, and a variety of other applications.

Lastly, it should be pointed out that ABS is commonly "alloyed" with PVC for commercial applications to attain greater impact resistance and self extinguishing properties from fire exposure. Also alloys of ABS and polycarbonate are used commercially. In addition special properties have been obtained with a modified ABS using a fourth monomer—alpha-methylstyrene.

Poly(Methyl Methacrylate) (MMA)

$$\left[\begin{array}{c} CH_3 \\ | \\ -CH_2-C- \\ | \\ C=O \\ | \\ O-CH_3 \end{array} \right]_n$$

This plastic has become widely used in many applications. As discussed with ABS, MMA's production volume is considerably greater than specialty plastics (to be discussed), but not sufficient to be considered a true commodity plastic. This lower volume is primarily due to the high cost of MMA. The reason for the commercial success of MMA is its unique combination of brilliant clarity and excellent weatherability resistance. This combination of properties has led to its wide use in making bright-looking commercial signs which have replaced most neon signs.

MMA derives some of its "gem-like" appearance from its relatively high refractive index of 1.49. It also possesses a relatively high percent light transmission and a specific gravity of 1.18. MMA is today made by bulk, suspension, emulsion, or solution polymerization processes. The bulk process is common in producing MMA using free radical initiation. Often MMA sheeting is made by completing the polymerization of a partially polymerized MMA "syrup" in a mold. Also many MMA molding powders are manufactured by the suspension polymerization process.

As discussed, MMA is widely used because of its clarity and weathering resistance. MMA has excellent clarity and optical properties as well as high gloss. Also its colorability potential is excellent. Likewise its outdoor durability is outstanding. This one feature gives MMA a great advantage over polystyrene. MMA will not turn yellow from outdoor exposure to ultraviolet light. MMA has some resistance to scratching, but it is well to have a protective coating on it for best resistance. MMA is resistant to food oils as well as non-oxidizing acids. MMA being resistant to moisture absorption, also possesses good electrical properties. MMA is considered moderately tough and its sheets can be fabricated in a machine shop with conventional wood-working or metal-cutting machinery.

On the other hand, MMA has some disadvantages that should be noted. It is important to note that MMA does not have extremely good strength on impact, although it is more resistant than unmodified polystyrene. MMA does not have good resistance to attack by ester, ketone, aromatic, or halogenated solvents. Also, unmodified MMA is not flame resistant although it is slow burning. Lastly, MMA costs more than commodity plastics such as polystyrene.

Today MMA homopolymers and its copolymers are used in a wide variety of commercial applications. About one-fourth of MMA is sold directly to fabricators as cast sheets to be thermoformed into a variety of end uses. Commercial signs represent the largest single use area for MMA. This plastic has almost completely displaced the neon sign in advertising. Many consider the acrylic sign to be more attractive and require less maintenance. Glazing is another large area of application. Windshields for boats and snowmobiles are commonly made of MMA for applications where high abrasion or scratch resistance is *not* needed. Also the automobile industry uses MMA in tail light prisms and lenses as well as horn buttons. Because of the superior optical properties of MMA, it is commonly used in making eye lenses and binoculars. Also MMA is used in fiber optics for "piping" light to "hard to reach" areas in medical operations. A variety of other applications include watch faces, clock and radio faces, dentures, piano keys, telephone dials, bathroom knobs, etc.

As mentioned, MMA is commonly used as a copolymer with methyl acrylate, ethyl acrylate, styrene, alpha-methylstyrene, or acrylonitrile. The MMA copolymer based on styrene achieves a better heat resistant and cost balance than the homopolymer. Also PVC is sometimes "alloyed" with MMA to provide a balance in tear strength, toughness, and flame retardance. Sometimes MMA is blended with elastomers to improve impact resistance.

Lastly, the related acrylic plastic called poly(ethyl methacrylate) has a lower softening point than MMA and is now used commercially at a more greatly reduced volume.

Polyethylene Terephthalate (PET)

This polymer has been used for many years in drawing fibers for textile applications; however in recent years, the use of PET as a plastic has grown, especially in the blow molding of bottles. Amorphous PET is not very strong or useful as a molding plastic. Amorphous PET resin can be crystallized at an elevated temperature range. However, another important way that PET can be crystallized is by orientation techniques. PET tensile strength can be increased perhaps four-fold by crystallization.

PET has a density of 1.45 if crystalline, and 1.32 if amorphous. Its glass transition temperature is 70°C and its melting range is between 255° and 270°C. PET crystallizes at approximately 190°C.

PET is produced from the polycondensation of terephthalic acid with ethylene glycol. In some cases there may be a postpolymerization of the resin in the solid phase. One of the most important advantages to using PET as a plastic is that on orientation during processing it becomes one of the strongest thermoplastics available because of crystallization. Also PET displays good high temperature service properties.

Besides PET's high volume use in the production of fibers, this polymer is also used in drawing biaxial film and more recently in the blow molding of bottles. PET is ideal in producing carbonated drink bottles because of its high strength which allows the bottle to withstand over 100PSI in gas pressure and its good permeability resistance to carbon dioxide diffusion. As a result, PET is now used to make soft drink bottles as well as a wide variety of other bottles. PET is also used in packaging. One unique application is in making food packages that can be boiled. Also other application for PET film include photographic film and magnetic recording tape.

Another terephthalate used in plastic applications is polybutylene terephthalate (PBT).

PBT is made from the polycondensation of dimethyl terephthalate and 1,4-butanediol. PBT possesses better ductility and processes easier than PET. Therefore, PBT is commonly used in injection molding. PBT has a relatively low melt viscosity for easy processing at 460–500°F. This plastic is known to have good electrical and chemical resistance properties and displays low friction. Also PBT is highly crystalline with a rapid crystallization rate for fast injection molding cycles. PBT is used in molding car parts such as distributor caps and rotors, gears, and other parts. Also this polymer is used to make skillet handles, kitchen equipment housings, and other household applications.

Cellulose Acetate

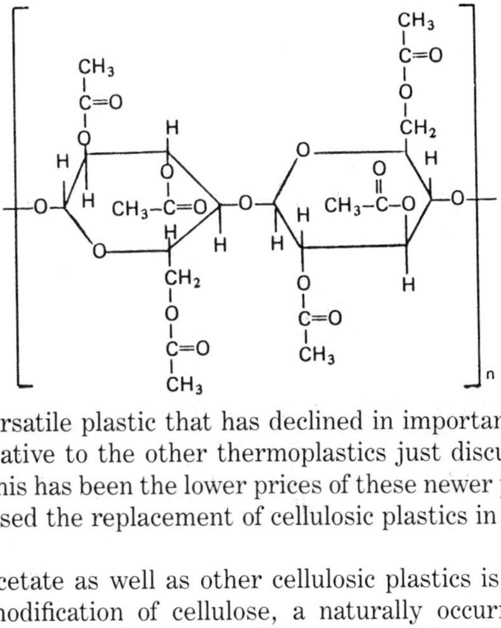

This is a versatile plastic that has declined in importance in the last thirty years relative to the other thermoplastics just discussed. Part of the reason for this has been the lower prices of these newer plastics which have slowly caused the replacement of cellulosic plastics in many market areas.

Cellulose acetate as well as other cellulosic plastics is derived from the chemical modification of cellulose, a naturally occurring polymer. Cellulose is a polysaccharide based on the glucose unit with a 1-4 linkage as shown below.

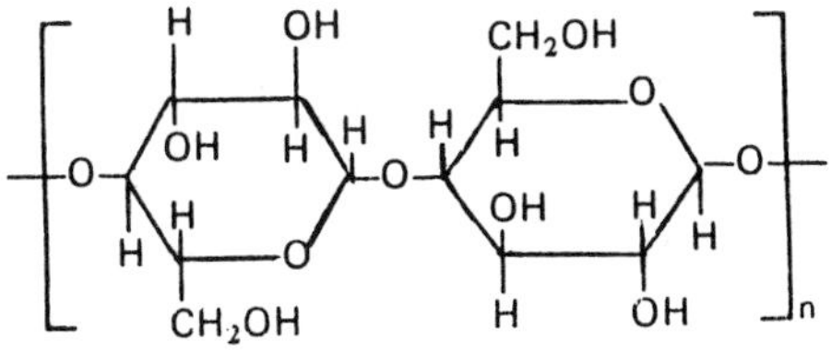

Unlike other polymers discussed earlier, cellulose itself is not thermoplastic. This is because the strong hydrogen bonding with the hydroxyl groups between chains is stronger than the bonds in the backbone of the molecular chains themselves. Thus the theoretical melting point of cellulose is higher than its decomposition temperature. Therefore in order to convert cellulose into a useful thermoplastic for molding, some of the hydroxyl groups must be replaced with bulky ester or ether functional groups in order to reduce these intermolecular forces.

The first cellulose plastic ever made (as well as the first man-made plastic) was produced in the 1870's as cellulose nitrate. Cellulose nitrate (or celluloid) is not used today in significant volume as a plastic and can not be used in injection molding because of its explosive flammability. Instead, cellulose acetate, which does not represent as severe a fire hazard, was commercialized at the turn of the century.

Cellulose acetate is manufactured by treating cotton linters (short fibers) or alpha cellulose (from wood pulp) with acetic acid or acetic anhydride to yield cellulose triacetate. Cellulose triacetate has too high a softening point to be used easily as a thermoplastic. Therefore, the triacetate is partially hydrolyzed back to secondary cellulose acetate where some of the acetate groups are replaced with hydroxyl groups. In molding operations, it is this secondary form that is commonly referred to when discussing the plastic use of cellulose acetate.

Cellulose acetate is known most for its toughness and high impact strength. The plastic can also be quite hard. Cellulose acetate is used with select plasticizers in compounding for specific properties. It can be provided with good clarity and has excellent colorability. The plastic possesses a high gloss appearance and is somewhat scratch and scuff resistant. It is somewhat resistant to oil and grease. It also lends itself to vacuum forming in molding applications.

One of the chief disadvantages of cellulose acetate is its cost. Many of the newer thermoplastics are considerably less expensive. This disadvantage could change in the future, however, as the cost of petrochemical feedstocks continues to increase in relation to cellulose, a renewable resource. Another disadvantage that cellulose acetate has is its affinity for moisture. Molding resins must be dried before extrusion. If molded products of cellulose acetate are exposed to very high moisture levels for extended periods, mechanical strength and electrical properties of the product will be worsened. (In fact, one advantage of cellulose acetate film in some applications is its high moisture permeability, which is desirable in certain food applications.) Also slow processing and extrusion rates are another disadvantage of cellulose acetate. This plastic many times requires the addition of plasticizers. Generally, the softer the plastic compound, the faster the extrusion rate and the better the impact strength. Lastly, cellulose acetate plastics are not resistant to attack of ketones or alcoholic solvents and these plastics are flammable.

A very large portion of the cellulose acetate produced in this country goes into the manufacture of acetate yarns for textile use. Still a significant portion is used as molding resin to produce a wide variety of products. Applications which need toughness, clarity, or good colorability, high gloss, and hardness commonly use cellulose acetate. Examples of its use are in molding eyeglass and sunglass frames, goggles, pens, combs, toothbrush handles, tool handles, toys, knobs, etc. Also, to some extent photographic film and magnetic recording tapes are made of this plastic. Cellulose acetate film is used for "windows" in packaging.

Another cellulose plastic is *cellulose acetate butyrate* (CAB). This plastic is prepared by treating cotton linters with acetic and butyric anhydrides under special conditions. CAB is similar to cellulose acetate for many of the properties just discussed. CAB, however, is also known for better resistance to moisture attack and oil than cellulose acetate. Also CAB has better dimensional stability than cellulose acetate because of its better compatibility with higher boiling plasticizers. One disadvantage of CAB is that it has been reported to produce a rancid-like odor. CAB is commonly used in molding tool handles. Also this plastic is used to make car steering wheels, armrests, car light lenses, packaging, etc.

Cellulose propionate (also called cellulose acetate propionate) is used commercially because of its better melt flow and ability to fill the mold completely. Cellulose propionate requires less plasticizer and is easier to extrude than cellulose acetate. Cellulose propionate plastics give shorter molding cycles. They are used to mold toothbrush handles, pens, toys, tool handles, etc.

Lastly, *ethyl cellulose* plastic has the best impact strength over the widest temperature range when compared to other cellulosics. Ethyl cellulose plastics have superior low temperature properties compared to cellulose acetate. On the other hand, ethyl cellulose plastics are not as clear as other cellulosic plastics. Also ethyl cellulose has poorer water resistance than the two cellulosic plastics discussed. Another ethyl cellulose disadvantage is the limitation on processing temperature before polymeric degradation occurs. Because of its superior low temperature properties, ethyl cellulose is commonly used in refrigerators. Also wheels and luggage are sometimes made of ethyl cellulose.

Nylon 6,6

$$\left(-\underset{\underset{H}{|}}{N} - CH_2CH_2CH_2CH_2CH_2CH_2 - \underset{\underset{H}{|}}{N} - \underset{\underset{O}{\|}}{C} - CH_2CH_2CH_2CH_2 \underset{\underset{O}{\|}}{C} - \right)_n$$

This is a polyamide polymer formed by the controlled polycondensation reaction of hexamethylenediamine and adipic acid. The "6,6" designation indicates that this nylon was formed from a diamine monomer containing six carbons and a dicarboxylic acid containing six carbons. Although nylon 6,6 is a commonly used polyamide, other nylons, such as nylon 6, nylon 11, and nylon 6,10 are also used in plastic molding.

As is well known, nylon is primarily used in the making of textile fibers. Less than one-tenth of the nylon produced in the U.S. is used as a molding resin for plastics. However, nylon has one of the largest volumes of use when compared to the so-called engineering plastics. Nylon is considered an engineering plastic and is used widely in engineering applications because of its unique combination of properties such as toughness, low friction coefficient, high strength, and ease of processing and molding.

Nylon 6,6 as well as other nylon polymers are highly crystalline, unlike other plastic polymers previously discussed. Because of this high crystallinity, nylon 6,6 has a relatively sharp melt point at 265°C with a relatively low melt viscosity above this temperature. The high crystallinity accounts for nylon's high strength. Likewise, with the destruction of nylon's crystallinity above its melt point, it displays a very low melt viscosity which enables its use in molding intricate parts. Another feature of nylon is its low friction coefficient. This polymer is very slippery to the touch. Therefore, mechanical parts such as gears and rollers made of nylon are considered "self lubricating" in that they do not require lubricants. Not only is nylon 6,6 strong, it also is very tough with extraordinary impact strength. Nylon also possesses high ductility and is abrasion resistant. Nylon is considered somewhat heat resistant and retains its physical properties at elevated temperatures; however, above 175°F it may become brittle if not compounded with a good heat stabilizer. Nylon also has good fatigue resistance. In addition, nylon is highly colorable and resistant to oil and grease, and mildew growth.

On the other hand, nylon 6,6 and nylons in general have several disadvantages that should be noted. First of all, nylon resins are somewhat hygroscopic and must be kept dry when being used in an extrusion operation. Exposure of nylon resin to moisture can reduce some of the polymer's physical properties. Nylon 6,6 is not a good barrier to water vapor. Nylon also displays poor U.V. resistance and needs a U.V. stabilizer or some sort of pigmentation to protect it. Nylon is generally not recommended for long term outdoor service. Nylon also displays poor hot water resistance and degrades if held for an extended period of time in the molten state. In addition, nylons can be attacked by acids and bases. Another problem with nylons can be dimensional distortion occurring on cooling after molding because of slow crystallization. (To avoid this problem, nucleating agents are added to some molding nylon resins to provide rapid crystallization.)

Because of the unique properties that nylons possess, they are used to make many engineering parts including various gears, cams, rods, tubes, rollers, bearings, valves, brushes, and pipes. Parts may be made of nylon that are used to make, say, an electric razor. Kitchen door catches may be made of nylon as well as sliding door rollers.

Although the discussion this far has centered on nylon 6,6, other nylon polymers such as nylon 6, nylon 6,10, nylon 11, and nylon 6,12 are used as molding resins. Nylon 6 is the second most commonly used nylon molding resin. It is made from the polymerization of the caprolactam monomer.

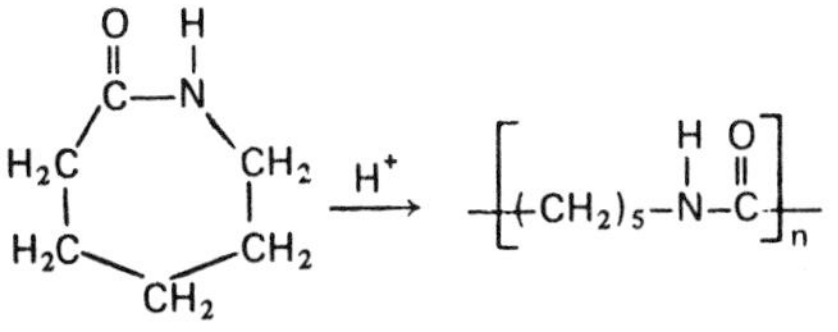

ε-Caprolactam   Nylon 6

Unlike other polycondensation reactions, this polymerization reaction does not evolve water as a byproduct. Of course, nylon 6 is widely used conventionally as a thermoplastic molding resin. Nylon 6,6 has greater stiffness and better creep resistance than nylon 6. On the other hand, nylon 6 has better impact resistance and may process better than nylon 6,6.

As previously mentioned, nylon 6,6 is not considered very resistant to water attack. Nylon 6,10 on the other hand, has improved resistance to water absorption and better dimensional stability. Nylon 6,10 is produced by the polycondensation of hexamethylenediamine and sebacic acid. By using sebacic acid (a dicarboxylic acid based on 10 carbon atoms instead of 6), nylon 6,10 has a lower amide content and thus improved water resistance. On the other hand, nylon 6,10 is less stiff and heat resistant than nylon 6,6.

Other specialty nylons are also occasionally used as molding resins. Examples are nylon 6,12, nylon 11, and nylon 12. Some of these nylons are used in Europe. Also nylon copolymers such as nylon 6,6/6,10 are sometimes used to lower the melt points and improve processing while maintaining good water resistance.

Polycarbonate

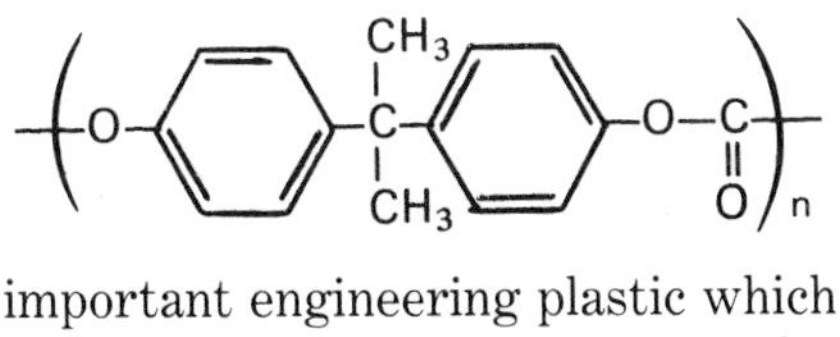

This is another important engineering plastic which has gained wide acceptance because of its good dimensional stability, hardness, ductility, heat resistance, and clarity. Polycarbonate was introduced commercially on the U.S. and European markets in the early 1960s and has expanded greatly in use. This plastic is made from the polycondensation of phosgene and bisphenol A.

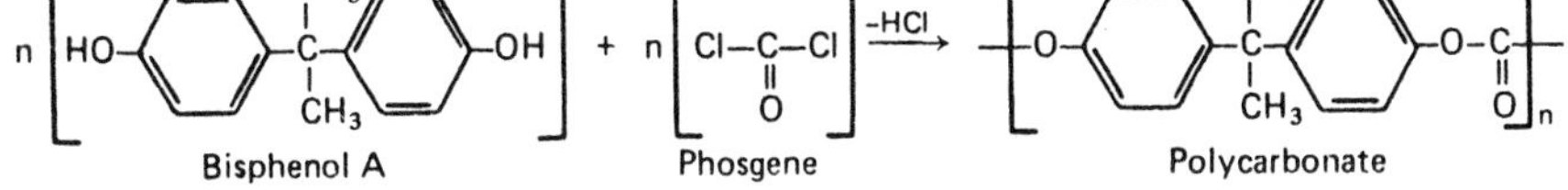

Bisphenol A   Phosgene   Polycarbonate

Polycarbonate can also be produced from the transesterification of diphenyl carbonate and bisphenol A. The superior physical properties such as strength, stiffness and dimensional stability are attributable to the macromolecular structure. Unlike previous plastics discussed, polycarbonates have benzene rings contained within the polymer's backbone and bulky methyl substitutions on a carbon between rings. This type of structure is stiff, prevents chain slippage, and imparts favorable properties for strength and dimensional stability even at elevated temperatures. Thus polycarbonate is a very useful material for engineering applications, many times being used to replace metals.

Polycarbonate, while somewhat more expensive than commodity plastics, offers many advantages. As mentioned earlier, this plastic possesses excellent dimensional stability with a very high resistance to creep. In addition, polycarbonate is very ductile in that it can be easily sawed, nailed, or drilled without cracking the polymer or clogging up metal work machinery. This polymer has good impact strength. In addition, it is hard with high stiffness and strength. Because of the benzene rings in the backbone, polycarbonate has a high softening point and excellent heat resistance. This polymer can be used to make parts that have to be sterilized with steam. Polycarbonate is stable at temperatures ranging from -215°F to 260°F. It is basically self extinguishing; however, on forced incineration, the key gaseous products are essentially carbon dioxide and water vapor which are safe to the environment. Another very important property of polycarbonate is that some grades have extremely good transparency. Also polycarbonate has good weathering resistance but can turn yellow from exposure to U.V. radiation in sunlight. Therefore U.V. absorbers may be required. Polycarbonate has resistance to water attack and hydrolysis as well. In addition, it is resistant to attack by mineral oils but can be solvent "welded" by chlorinated solvents. Polycarbonate is recognized for its excellent electrical properties. Lastly, this plastic lends itself well to most of the methods of molding thermoplastic resins.

There are also some potential problems with polycarbonate. First of all, this plastic can be hydrolyzed by alkali solution and attacked by strong acids. Also polycarbonate molding resins must be dried before use because polymer degradation from hydrolysis can occur due to the hot processing temperatures needed. Also moisture contamination in the resin can take away from optimal impact strength resistance. Another disadvantage is that certain detergents can attack polycarbonate. Lastly, polycarbonate is not resistant to attack by chlorinated solvents.

Because of polycarbonate's unique advantages, it has gained a wide variety of applications where other plastics might not be as suitable. In some cases, polycarbonate has even replaced metals. Some common examples of polycarbonate uses are car light lenses, window panels, streetlight globes, hot service lenses, police car light covers, sunglasses, unbreakable bottles, kitchen ware, coffee brewers, helmets, electrical insulation, aircraft parts, medical equipment, golf club parts, beer mugs, etc.

Polycarbonate is commonly compounded with other ingredients such as reinforcement fibers (which will be discussed later). Also, polycarbonate is compounded with thermal and U.V. stabilizers. Sometimes polycarbonate is blended with small levels of polyolefin to give better low temperature impact strength.

Polyacetal (Polyoxymethylene)

$$CH_3-\overset{\displaystyle O}{\overset{\displaystyle \|}{C}}-O-\!\!\left(CH_2-O\right)_{\!n}\!\!-CH_2-O-\overset{\displaystyle O}{\overset{\displaystyle \|}{C}}-CH_3$$

This is another important engineering plastic. It was introduced by Dupont in the 1960s under the trade name Delrin. Because of its unique chemical structure and very high crystallinity, polyacetal has some extraordinary properties in strength, creep resistance, toughness, and fatigue resistance which enables it to be used in engineering applications.

Polyacetal is a white, translucent to opaque polymer derived from the polymerization of formaldehyde under special conditions. Every chemist who has worked extensively with formalin (an aqueous solution of approximately 40% formaldehyde), knows that in cold storage a white precipitate may form. This precipitate is a highly branched low molecular weight polymer called paraformaldehyde, which has no useful properties as a polymer. However, a very useful formaldehyde polymer—polyacetal—can be obtained through the cationic polymerization of formaldehyde in dry hexane using a catalyst such as triphenylphosphine. This high molecular weight polymer must also be stabilized to prevent a "zipper" effect (catalytic polymer degradation). This is done by Dupont through esterification of the polymer end groups with acetic anhydride.

Because of the oxygen linkages contained in the polymeric backbone, polyacetal is highly ordered and crystalline. Because of polyacetal's unique polymeric structure and high crystallinity, it is perhaps one of the strongest thermoplastics available commercially. Also, the creep resistance of polyacetal is one of the highest for a thermoplastic which makes it ideal for engineering use. Other important engineering properties are high hardness and stiffness, good impact strength (even down to -40°F), excellent high temperature service (melting point is 176°C), high wear resistance, and very good fatigue endurance. In fact, the fatigue properties of polyacetal have allowed it to be used in spring applications. Polyacetal has sufficient ductility properties to allow it to be drilled, sawed, or riveted without cracking. Another important property of this plastic is its very low absorption of moisture which enables it to have better dimensional stability than some other engineering plastics which absorb moisture. Another important advantage to polyacetal is its resistance to solvent attack, including halogenated solvents that normally attack other plastics. Polyacetal is resistant to oil and grease attack as well. Polyacetal's high crystallinity means the plastic requires shorter molding cycles. Lastly, polyacetal has excellent electrical properties.

Of course, no molding plastic is perfect and polyacetal has its weaknesses as well. First of all, the excellent solvent resistance property referred to earlier is also the reason that polyacetal does not lend itself well to "solvent" welding. Secondly, even though polyacetal is very resistant to organic solvents, it possesses poor resistance to attack by strong acids, alkali, or oxidizing agents. Also, polyacetal is degraded by exposure to ultraviolet radiation. The polymer may chalk from outdoor exposure and pigment loadings may not help. On exposure to fire, polyacetal is not considered self extinguishing. Also, if the molding resin is held or "hung-up" for long periods in hot processing equipment, the polymer will give off formaldehyde which could cause health problems in the workplace. Therefore, polymer must be purged from heated equipment between runs. Because any thermal degradation of polyacetal releases toxic gas, the operator must have adequate ventilation when processing the polymer.

Because of polyacetal's unique combination of properties, this plastic finds a very wide variety of small volume uses. The excellent water resistance enables its use in shower heads, faucet cartridges, sprinklers, and dishes. Its excellent resistance to solvents and fuels allows its use in fuel systems, pump assemblies, pipe fittings, valve fittings, and carburetors. Its excellent mechanical properties permit its use in making small gears, cams, tool handles, and zippers. This polymer is commonly used instead of nylon to mold small gears because of its superior dimensional stability. (Nylon is commonly used for larger gears because of its impact resistance.) Lastly, polyacetal has a wide variety of miscellaneous uses, such as door handles, furniture castors, and bottles.

The polyacetal discussed above is a homopolymer. Copolymers of formaldehyde and a small quantity of either ethylene oxide or 1,3-dioxolane are commercially available under the trade name of "Celcon" by Celanese Co. These polymers are stabilized to prevent "zipping" degradation from occurring by the addition of the second monomer which randomly introduces C—C bonds into the backbone.

1,3-Dioxolane   Ethylene Oxide

These copolymers have about the same properties as those discussed earlier for the homopolymer. Likewise, there must be adequate ventilation in the workplace to protect the worker from potential exposure to toxic formaldehyde.

Polyphenylene Oxides (PPO)

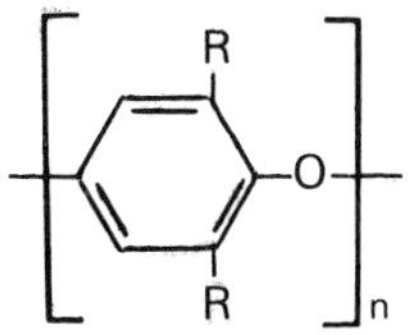

This group of polymers was introduced commercially by General Electric Co. in 1964. Structurally it is an aromatic polyether. These plastics are known for their extraordinary dimensional stability, creep resistance at elevated temperatures, and their very low water absorption which makes them ideal for autoclave applications. Polyphenylene oxides have physical properties which allow their use in engineering applications.

Polyphenylene oxides are opaque polymers typically with a specific gravity of 1.06, the lightest of the so-called engineering plastics. Their softening points are very high and they can withstand very high service temperatures because of the strong bonding forces associated with the aromatic structures in the backbone of the polymer structure. PPO plastics are manufactured by the oxidation coupling of 2,6-disubstituted phenols as shown below.

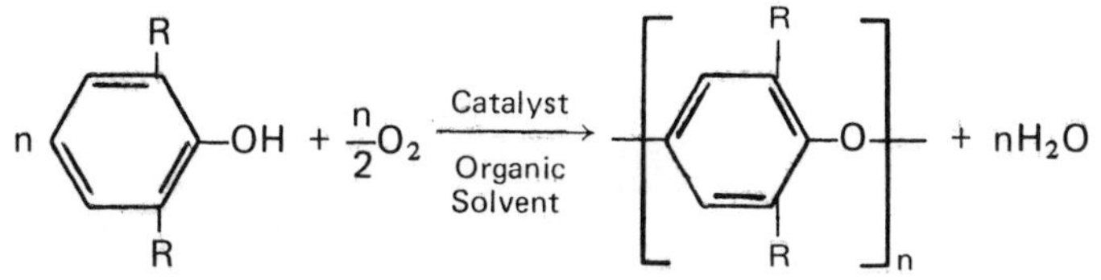

There are several advantages in using PPO plastics which we shall discuss. First, because of the unique molecular structure, PPO polymers can be useful at service temperatures ranging from -275°F to 375°F. Secondly, PPO polymers have very low moisture absorption and are very resistant to repeated exposure to steam (water can not hydrolyze the polymer.) Because of this combination of properties, PPO polymers are very "autoclavable". Also PPO plastics are resistant to hot detergents. In addition, these plastics possess those properties considered important for use in engineering applications. PPO plastics are extremely tough with very high impact strength. They are also stiff, hard, and strong with very good creep resistance even at high temperatures. These polymers will burn but are considered self extinguishing. PPOs also have good electrical insulation properties. Lastly, PPO plastic's thermal expansion is so small that they have replaced thermosets in some applications.

Besides high cost, there are some potential problems in working with PPO plastics that should be mentioned. First, they have poor resistance to strong oxidizing acids as well as chlorinated, aromatic, or ketone solvents. Also, because of the high process temperatures required (over 600°F), these plastics are not as easy to process as some other thermoplastics. The temperature range required in processing is critical. Molding operations require heated molds. Also, only a limited selection of organic colorants can be used that can withstand these high processing temperatures. Lastly, PPO plastics reportedly do not have the best electrical arc resistance even though they have some other good electrical properties.

PPO plastics are used in very special applications. Because of their resistance to steam, PPOs have in some cases replaced stainless steel in making surgical tools and equipment which must be repeatedly sterilized with steam. Also because of the superior resistance to water or aqueous chemicals, PPO polymers are used in making parts for washing machines, pipes and fittings, pumps, valves, and shower heads. Because of impact resistance and electrical properties, these polymers are used to make TV cabinets, electrical parts, computer and calculator housings, and business machine housings. In addition, PPO polymers are used in the auto industry to make dashboards, grilles, and wheel covers.

Many of the important PPO resins marketed today are actually "alloys" of PPO with polystyrene for improved processability. In fact, this PPO/PS blend was one of the first plastic "alloys" to be commercialized.

Polysulfone

This plastic is another specialty engineering plastic introduced by Union Carbide Corp. in 1965. It is a member of the aromatic polyether family. It is a tough, transparent plastic that has extraordinary heat resistance for a thermoplastic. It is completely amorphous (which explains its clarity) and has a high glass transition temperature of 375°F (because of its structure). The sulfone group contained in the backbone is responsible for the polymer's stiffness and contributes to its rigidity. On the other hand, the flexible ether linkages in the backbone·help impart impact resistance resulting in a tough polymer as well. Also the diaryl sulfone groups in the polymer chain add to the plastic's inherant resistance to oxidative attack at elevated temperature. This quality gives polysulfone an important advantage over some other plastics in long term heat aging resistance. Polysulfone is obtained commercially from the substitution reaction between the sodium salt of bisphenol A and 4,4-dichlorodiphenyl-sulfone.

As just mentioned, polysulfone has all the important properties required of an "engineering plastic". It possesses extremely low creep even at elevated temperatures. It has good strength, rigidity, and impact resistance. As discussed, this polymer has an inherent resistance to oxidative attack, a very high glass transition temperature, and a high resistance to hydrolysis from water exposure. These properties explain its exceptional performance at high temperatures (up to 300°F) in both air and water. As with PPO plastics, polysulfone can also withstand repeated steam autoclaving without degradation. In addition, polysulfone is resistant to aqueous chemical attack by strong acids, alkalis, salts, and detergents. Also polysulfone is oil resistant at elevated temperatures. Since it is amorphous, it is obtainable in transparent forms and is very easily colorable. Also the amorphous quality of this polymer provides for very low mold shrinkage on cooling. Because of polysulfone's resistance to thermal degradation, its scrap can be continuously recycled without loss in quality, thus providing an economic savings. Although polysulfone will burn, it is considered self extinguishing. It possesses good electrical properties. Overall, polysulfone can be processed as a thermoplastic, yet possesses some of the thermal stability advantages previously attained by using thermosets.

An important disadvantage to using polysulfone is cost. This molding resin is more expensive than some other engineering plastics on the market; but it also possesses special properties which may justify the price. Also this plastic requires a very high processing and molding temperature between 600° and 740°F. Mold temperatures for injection molding must be preheated to 200–300°F. These high temperature requirements can cause special problems. In addition, polysulfone molding resin should be dried before use. Lastly, polysulfone has poor resistance to attack by ketones, aromatic or halogenated solvents.

Because of polysulfone's superior heat, oxidation, and hydrolysis resistance, it is used in heat resistant aerospace parts, microwave kitchenware, coffee makers, kitchen dishwasher and dryer parts, hot water pipes, sterilizable medical tools, and auto parts for under-the-hood service. The plastic's chemical resistance enables it to be used in making food processing equipment parts and pipes, chemical pumps, valves, etc. Its electrical properties explain its use in electronic circuit boards, coils, wire insulators, switches, circuit breakers, TV parts, and meter housings.

There are two other commercial sulfone plastics of relatively low production volumes which are shown below.

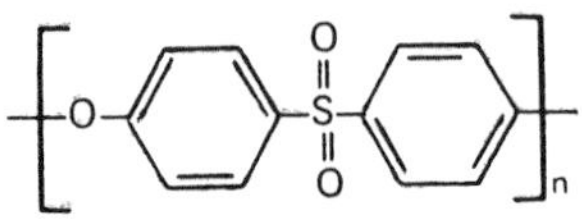

Polyether Sulfone 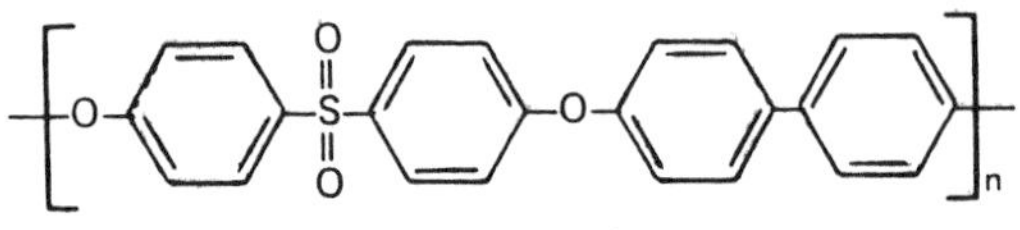   Polyphenylsulfone

These two polymers have somewhat similar properties as those discussed for polysulfone, but vary in the degree of high temperature service, environmental stress crack resistance, impact strength resistance, electrical properties etc.

Polytetrafluoroethylene (TFE)

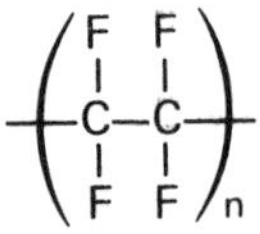

This specialty plastic is commonly referred to by its Dupont trade name of Teflon. Dupont Co. commercialized Teflon in the 1950s. This polymer possesses truly unique properties—extremely low coefficient of friction, an extraordinarily wide temperature range of serviceability, and a very high degree of chemical inertness. These special properties are mainly due to the molecular structure of TFE. The structural symmetry gives TFE a high degree of crystallinity which does not melt until a temperature of approximately 330°C is reached. Also, the F—C bonds are very strong; they do not break easily even at elevated temperatures. These bonds help explain the polymer's excellent heat resistance and chemical intertness. Also the low coefficient of friction is reportedly due to the weak intermolecular forces among fluoride groups along the chains. The close packing resulting from the fluoride-carbon bonds also explains the high specific gravity of 2.15 for TFE.

TFE is a white flexible "waxy" textured polymer that is obtained from the catalytic free radical polymerization of tetrafluoroethylene. This polymer has gained wide use in special areas because of its unique properties. As mentioned, TFE has an extremely low static and dynamic coefficient of friction. It possesses a "no-stick" quality in which most adhesives cannot adhere to its surface. TFE is chemically inert to almost everything except certain "exotic" fluoride solvents. Acids, alkalis, and organic solvents do not affect TFE. It is not affected by moisture or U.V. radiation, and possesses superior resistance to weathering. Thirdly, this polymer remains flexible at temperatures as low as -400°F and is useful in service up to 520°F. It is nonflammable and will not support combustion. Lastly, TFE has excellent electrical properties.

Of course, there are problems in using TFE. TFE has a very high-melt viscosity—so high, in fact, that it can not be processed by conventional extruders or injection molding equipment. Instead, TFE powders must be shaped by pressure into a preform. These preforms are then "sintered" in an oven or salt bath, where the particles coalesce into a product. Care should be used not to heat TFE too high or poisonous gases could be given off in the workplace. Adequate ventilation in the work area is always required to protect the worker from potential exposure to toxic gases. In addition, TFE is relatively high in cost and is opaque. It cannot be made transparent. Also, because of its non-stick quality, conventional adhesives cannot be used to join TFE parts. Lastly, TFE does not possess equivalent physical properties compared to other engineering plastics. It is not very strong or creep resistant and does not possess the best wear resistance relative to other engineering plastics.

The public generally thinks of TFE (or Teflon) as being used only for lining "no-stick" frying pans. However, because of the polymer's chemical inertness and heat resistance, it finds use in a wide range of "hidden" applications not normally seen by the public. Examples of these applications are O-rings, seals, rollers, gaskets, tubes, stopcocks, ski linings, tank linings, valve and pump parts, food equipment lining, oven wire insulation, heat cables, and aerospace parts.

Despite the unique properties that promote TFE's widespread use, its difficulty in processing has restricted its use. In order to circumvent this, several other fluorinated polymers have been developed in the last twenty years. The following fluorinated plastics possess a degree of chemical inertness and heat resistance similar to TFE; however, they lend themselves to direct extrusion and injection molding as well. If any of these fluorinated plastics are heated, however, to their thermal decomposition temperature, they too will emit a poisonous gas.

$$\left(CF_2-\underset{\underset{Cl}{|}}{CF} \right)_n \qquad \left(CF_2-CF_2-CF_2-\underset{\underset{CF_3}{|}}{CF} \right)_n \qquad \left(CH_2-\underset{\underset{F}{|}}{CH} \right)_n$$

$$\text{CTFE} \qquad\qquad\qquad \text{FEP} \qquad\qquad\qquad \text{PVF}$$

$$\left(CH_2-CF_2 \right)_n \qquad \left(CH_2-CH_2-CF_2-CF_2 \right)_n \qquad \left(CH_2-CH_2-CF_2-\underset{\underset{Cl}{|}}{CF} \right)_n$$

$$\text{PVF}_2 \qquad\qquad\qquad \text{ETFE} \qquad\qquad\qquad \text{ECTFE}$$

Polychlorotrifluoroethylene (CTFE) has a structure similar to TFE except that it has a chlorine atom on every other carbon in the backbone. The substituted chlorine reduces some of the crystallinity of the polymer (Tm = 220°C) and reduces the melt viscosity just enough to permit it to be extruded or injection molded on commercial equipment. Still this polymer is not considered easy to extrude. CTFE is not quite as chemically inert as TFE. CTFE is used to mold O rings, gaskets, transparent film, etc.

Fluorinated ethylene propylene copolymer (FEP) is polymerized from tetrafluoroethylene and hexafluoropropylene monomers. The CF_3 pendant groups off the main chains enable greater chain mobility thus reducing the melt point (Tm = 290°C). Thus FEP can be processed by conventional extrusion, injection molding, and blow molding; however it possesses a little less resistance to chemical attack and a lower maximum service temperature than TFE. FEP is commonly used to make laboratory containers and autoclave bottles.

Polyvinyl fluoride (PVF) can also be processed commercially. However, as may be noted from its structure, PVF is not completely fluorinated; thus it possesses significantly poorer heat and chemical resistance compared to TFE. PVF is highly crystalline. This polymer is not used widely as a molding resin but rather in making weather resistant film used in lamination for aircraft construction.

Polyvinylidene fluoride (PVF_2) is polymerized from vinylidene fluoride which in turn is produced from the dehydrochlorination of chlorodifluoroethylene. PVF_2 is crystalline and possesses superior creep and wear resistance as well as better strength than TFE. With a Tm equal to 170°C, PVF_2 does not have the inherent heat resistance of TFE; however PVF_2 has good chemical resistance and is used in making pipes, tubes, wire coating, and high temperature valves and seals.

Copolymers of ethylene and tetrafluoroethylene (ETFE) are used as a molding resin with similar properties described for fluorinated polymers. Ethylene-chlorotrifluoroethylene (ECTFE) has better resistance to wear and creep than TFE and can be processed on conventional processing equipment.

THERMOSETTING RESINS

Phenol-Formaldehyde Resin

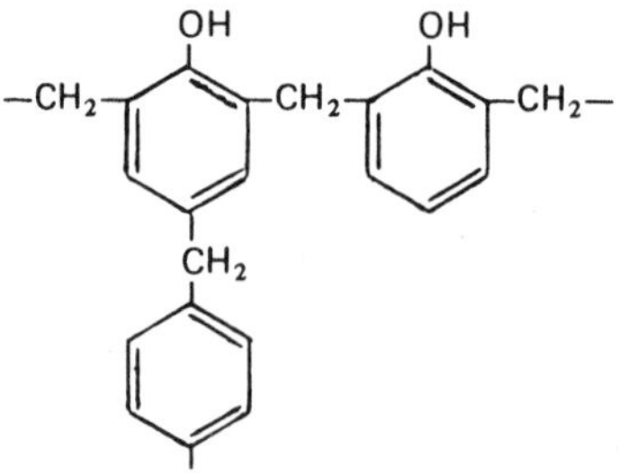

This thermosetting polymer was the first commercial plastic produced from completely synthetic feedstocks. Leo Baekeland first marketed a PF resin under the tradename of Bakelite in 1910. PF resins today have the largest commercial volume of all the thermosetting polymers used. However the percentage of PF resins that are used in molding applications is relatively small compared to its high volume use in plywood and particle board products as well as its many adhesive applications.

PF resins, as the name implies, are formed from the polycondensation reaction between formaldehyde (usually in the form of formalin, an aqueous solution) and phenol. Phenol is considered trifunctional in that it will react with formaldehyde at the ortho and para positions, thus achieving a three dimensional matrix with the difunctional formaldehyde when carried beyond the gel point. After gelling (or curing), the polymer can no longer be melted or processed further. Therefore, all processing of these PF resins must be done before they reach the gel point.

Gel is delayed in one of two ways. Resole resins are formed from the proper stoichiometric ratio of formaldehyde to phenol necessary to reach the gel point. This polymerization is carried out using an alkaline catalyst such as ammonia or caustic soda. However, the polymerization of this resin is stopped short of gel by cooling. When the resole PF resin is reheated in a mold, it resumes polymerization three-dimensionally to progress to a completely cured state.

On the other hand, novolac PF resins delay cure in a different manner. Novolac PF resins are polymerized from phenol and formaldehyde using an acid catalyst such as sulfuric acid. The ratio of phenol and formaldehyde feedstocks, however, is deliberately deficient of sufficient formaldehyde to allow the resin to reach gel. The resin is cooled and ground to a powder. This resin powder is then carefully blended with hexamethylenetetramine powder (called "hexa") which will supply enough formaldehyde when the resin is reheated in the mold to cure it. ("Hexa" is derived from the reaction of formaldehyde and ammonia; upon heating, hexa decomposes providing formaldehyde for polymerization and ammonia to serve as a catalyst.)

The resole PF resins are commonly referred to as "one-step" resins while the novolac PF resins are called "two-step" resins in which the "hexa" is added in the second step. Commercially available one-step PF resins are not widely used directly in molding applications because their cure times are harder to control than two-step curing resins. These one step resins are occasionally used in molding when ammonia and formaldehyde gas can not be tolerated (these gases are given off from the decomposition of "hexa"). One step PF resins today are commonly dissolved in hydrocarbon solvents to impregnate cloth or paper sheets which are in turn cured into laminates.

Commercially available two-step PF resins are commonly used as molding resins. Many times these two-step molding resins will be precompounded with a filler such as wood flour and wax or stearate lubricant to permit easier mold release after cure. Adequate ventilation is needed in cure areas.

There are many advantages to using PF plastics. First they are relatively inexpensive compared to other thermosets commercially available. PF plastics provide very good thermal stability at elevated temperatures. This plastic can normally be used at temperatures of 320°F or higher. PF plastics are also known for their resistance to water and chemical attack. Also PF plastics are very resistant to solvents and oils and will not stress crack from solvent exposure. PF plastics are hard and display very good abrasion resistance. Lastly, this polymer possesses very good electrical properties and is considered self-extinguishing.

Of course, there are several important disadvantages associated with PF plastics. First of all, PF plastics without filler loadings are very brittle. Therefore PF resins are almost always compounded with a filler such as wood flour. Filler loadings not only improve impact resistance but also reduce cost and may improve heat resistance. Another disadvantage of PF plastics is their characteristic dark color, which prevents their use in light-color applications. These resins tend to turn darker in color from heat and oxidation. Also PF plastics can be attacked by strong oxidizing acids and alkali at elevated temperatures. Lastly, PF plastics tend to be sticky in processing and stick to the mold. This is why a wax lubricant is normally compounded with this plastic.

PF plastics are still used in a wide variety of applications because of their special properties. As mentioned earlier, a much larger portion of PF polymers is used in non-molding applications such as in making plywood, grinding wheels, particle board, sand molds for casting metals, and brake bonding adhesives. In molding applications this thermoset is used to make ashtrays, kitchen cookware, cooking handles, and parts for use in aerospace. Because of its resistance to water and chemical attack, it is also used to mold washing machine agitators. PF plastics are used in car distributors and transmissions because of their solvent resistance. Superior abrasion resistance supports their use in tool handles, knobs, and gun stocks. Lastly, the plastic's desirable electrical properties enable its use in making electrical switches, electrical coils, and printed circuits.

Commercial PF resins are not always based solely on phenol and formaldehyde feedstocks. Some polymers have substituted xylenol or cresol for part of the phenol. These substitutes may not be trifunctional. Also furfural is occasionally substituted for formaldehyde. This substitution reportedly improves mold flow properties.

Urea-Formaldehyde Resin (UF)

$$\left(-CH_2-N-\overset{\overset{\displaystyle CH_2}{|}}{\underset{\underset{\displaystyle CH_2}{|}}{N}}-\overset{\overset{\displaystyle O}{\|}}{C}-N-CH_2-\right)_n$$

This thermoset was successfully commercialized in the late 1920s. Today it is second only to phenol-formaldehyde resins in total number of pounds produced annually. Its volume will exceed one billion pounds per year. However, just as with PF resins, over three-fourths of all UF resins go into adhesive and bonding applications. Only a small portion is used in molding. UF resins are produced from the polycondensation of urea and formaldehyde in a manner similar to PF production. However, unlike PF resins, UF resins can be made water-white in color which lends itself to pastel coloring.

UF resins are produced by reacting formalin with urea under carefully controlled alkaline conditions to achieve linear polymerization (urea will behave difunctionally under these alkaline conditions). Upon spray drying, the resin is blended, usually with filler, and an acidic catalyst is added. Upon reheating, the urea units will behave tetrafunctionally and gelatin results from three dimensional crosslinking. The resin cures into a hard plastic.

There are several advantages to using UF plastics that should be mentioned. First of all UF plastic has a water-white color which enables light colored or translucent molded products to be made. This is an important advantage over PF plastic which is much darker. Secondly, UF plastics have better electrical properties than PF plastics in that UF has better arc resistance. UF chars, but doesn't carbonize; thus it is more track resistant. In addition, UF plastics have all the advantages of a thermoset. UF composites have good surface durability, stain resistance, high hardness, good impact resistance, high strength, high surface gloss, and no cold flow.

By contrast, there are several disadvantages in using UF resins. UF polymers have less water resistance than PF polymers. When UF plastic is exposed repeatedly to humidity, swelling can result as well as crazing and cracking. Also, UF plastic is not as heat resistant as PF plastic. UF polymers can be attacked by strong acids and bases. UF plastic normally must be used with a filler, such as alpha cellulose, as a composite to achieve the best physical properties.

By far the bulk of UF polymer produced in the U.S. is used in non-molding applications which include adhesives, paper treatment, textile treatment, and decorative laminates. Molding applications include such items as buttons, knobs, handles, lamp shades, piano keys, toilet seats, and cosmetic bottles. Because of the extraordinary electrical properties of UF, it is used to make electrical sockets and plugs, circuit breakers, insulators, and switch wall plates.

Another very similar thermoset is *melamine-formaldehyde* resin (MF). Unlike urea, melamine theoretically can be considered hexafunctional in that it has six substitutable hydrogens.

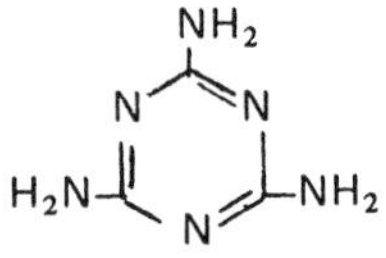

Melamine

Although MF resins have many properties very similar to UF resins, there are some special properties possessed by MF resins which justify its higher cost. MF polymers are water white in color and can be made resistant to yellowing. MF polymers can provide one of the hardest surfaces for a plastic. Also, MF plastics are generally more water resistant than UF. MF possesses superior heat aging properties than UF. Because of these properties, MF plastic is used extensively in making break resistant dinnerware (commonly used by institutions because they can be repeatedly washed in a dishwasher without breakage). For the same reasons, MF is the preferred polymer for making the table and counter tops found in many homes.

Unsaturated Polyester Resins

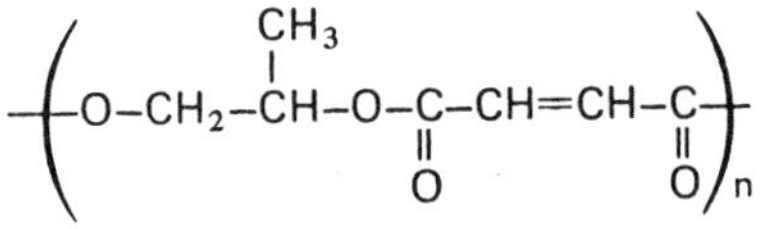

These polymers represent the largest volume thermoset used in molding and casting. Although the total production tonnage of PF or UF resins each exceed unsaturated polyester production, only a relatively small portion of PF and UF resin output is actually used in molding or casting applications. By contrast, much of the polyester thermosets produced go into molding and casting and wet lay up techniques, with only a small portion going into adhesives and coating applications. In total, over one billion pounds of these polyester thermosetting resins are made annually. The chief reasons for the popularity of these resins are their ability to be cured or set at room temperature in a relatively short time period without external pressure, their low viscosities and their low cost. This enables very large articles to be molded without the need for large presses and at low cost.

As is well known, unsaturated polyesters are commonly used to make composites with glass fibers. This is why the public commonly refers to these unsaturated polyester composites as "fiber glass". Unsaturated polyester was developed during World War II and has grown in popularity ever since. These resins are made from an unsaturated linear polyester prepolymer. The prepolymer is usually made from the polycondensation of propylene glycol stoichiometrically with a blend of maleic anhydride and phthalic anhydride. The phthalic anhydride is present to reduce the unsaturation of the prepolymer to the desired level.

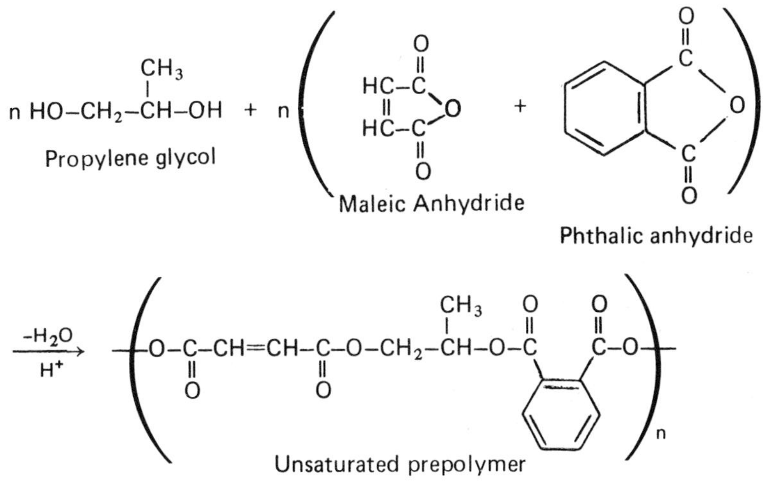

$$\underset{\text{Propylene glycol}}{n\ HO-CH_2-\underset{\underset{CH_3}{|}}{CH}-OH} + n\left(\underset{\text{Maleic Anhydride}}{\begin{array}{c} O \\ \| \\ HC-C \\ \| \quad \rangle O \\ HC-C \\ \| \\ O \end{array}} + \underset{\text{Phthalic anhydride}}{}\right)$$

$$\underset{H^+}{\overset{-H_2O}{\longrightarrow}}\ \left(\underset{\text{Unsaturated prepolymer}}{-O-\underset{\underset{O}{\|}}{C}-CH=CH-\underset{\underset{O}{\|}}{C}-O-CH_2-\underset{\underset{CH_3}{|}}{CH}-O-\underset{\underset{O}{\|}}{C}\underset{\underset{O}{\|}}{C}-O-}\right)_n$$

The unsaturated prepolymer is blended with styrene monomer which reduces the prepolymer's viscosity for better mold flow and, more importantly, provides a crosslinking agent. The blend of unsaturated prepolymer and styrene is stabilized for a reasonable shelf life by adding an inhibitor such as hydroquinone. The prepolymer/styrene blend is cured in a casting through the separate additions of accelerator (also called promoter, such as metallic soap) and an initiator (also erroneously called a "catalyst," such as an organic peroxide). These additions initiate exothermic curing through the formation of free radicals which cause vinyl polymerization to occur with styrene forming crosslinks between unsaturated sites on different polymer chains.

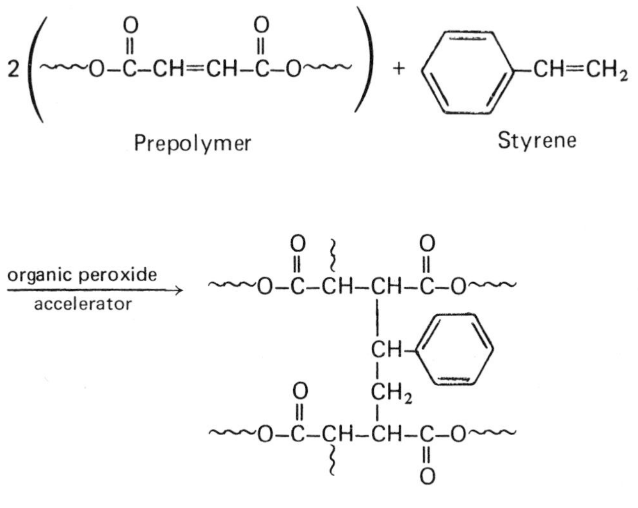

$$2\left(\underset{\text{Prepolymer}}{\sim\sim\sim O-\underset{\underset{O}{\|}}{C}-CH=CH-\underset{\underset{O}{\|}}{C}-O\sim\sim\sim}\right) + \underset{\text{Styrene}}{-CH=CH_2}$$

$$\underset{\text{accelerator}}{\overset{\text{organic peroxide}}{\longrightarrow}}$$

This crosslinking reaction differs from thermosetting reactions noted earlier for PF and UF resins in that no water is evolved as a byproduct. This means that the unsaturated polyester can be cured without the need for pressure to prevent porosity. For this reason, these resins can be cured into very large objects and intricate shapes in simple molds or forms requiring no heat or external pressure.

There are many other advantages to using unsaturated polyesters besides ease of molding. First these resins are relatively inexpensive. They can be colorless and cast into transparent objects as well as light colored ones. They can be made to have good weathering resistance. Unsaturated polyesters generally have high surface hardness and can be compounded to have high impact strength and tensile. They also possess good electrical properties and have fair heat resistance. These resins are also known for some resistance to chemicals, salt water, and fungus growth.

As with every plastic, however, there are certain disadvantages that should be noted. First, unreinforced polyester (containing no fibers or fillers) does not possess high strength. Also unsaturated polyester will shrink on curing to some degree. After cure, some polyester products might have a tacky surface. Also these polyesters are generally not self extinguishing, but can be made so through compounding modifications. Probably the most important disadvantages to the polyester thermosets are some aspects of the curing process itself. For one thing, special ventilation must be installed to minimize worker exposure to styrene vapors during cure. Styrene is also very flammable and can be accidentally ignited from peroxide initiators and promoters if extreme safety precautions are not exercised. Also the handling of peroxide initiators themselves in the factory can be dangerous. Some peroxides can ignite or explode if shocked or handled improperly. Also precautions must be taken in handling peroxides to avoid any worker contact in that exposure can cause severe throat irritation, lung damage, eye damage, skin burns, etc. Also extreme care must be maintained in the factory to prevent the organometallic or amine promoters (or any material) from in any way coming into contact with the peroxide initiators in that an explosion and/ or fire may result. The manner in which these dangerous components (the promoter and peroxide initiator) are added to the uncured resin can be very hazardous and should only be done in accordance with accepted industrial procedures. These materials should *never* be added directly to one another or consecutively to a resin unless one ingredient (the promoter) is thoroughly mixed in the resin before adding the other. If these precautions are not followed, an explosion can result. Resin manufacturers many times make available "prepromoted" polyester resins in which promoter is premixed. Prepromoted resins require only the careful addition of peroxide catalyst to start polymerization. Also, if proper procedures are not followed, uncontrolled exotherms from the curing process and the risk of fire can result. There are many other potential safety hazards that should be considered before working with these materials.

As mentioned before, unsaturated polyester with fiberglass is commonly used to inexpensively mold large objects such as boat hulls, sport car bodies, and modular bathrooms and bathtubs. In fact, approximately one-fifth of these polyester resins are used in making boats and one-third is used in making laminates for construction. Also these polyester thermosets are commonly used to make simulated wood for furniture and home paneling. These polymers, when properly molded to simulate wood grain with the proper colorants, can fool most observers into thinking they really are wood. These polyesters are also commonly used to make scratch resistant table tops as well. Lastly, unsaturated polyesters are used to make electrical appliances as well as tanks, fishing poles, pole vaults, trays, roofing, and luggage.

As previously discussed, the typical high production volume unsaturated polyester is based on maleic anhydride, phthalic anhydride, propylene glycol, and styrene. On the other hand, there are many unsaturated polyester resins on the market that are made from different monomers. Besides maleic anhydride, other dicarboxylic acids are commonly used to impart special properties.

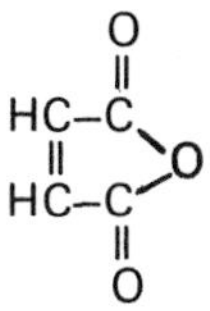

Maleic Anhydride

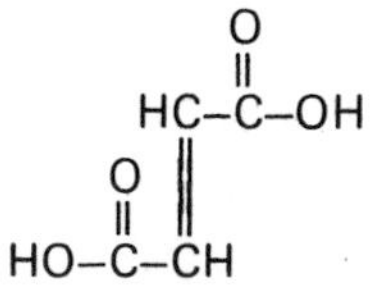

Fumaric Acid

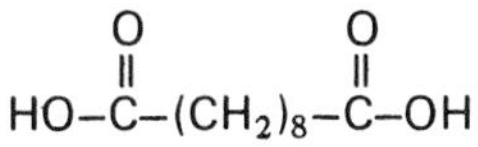

Sebacic acid

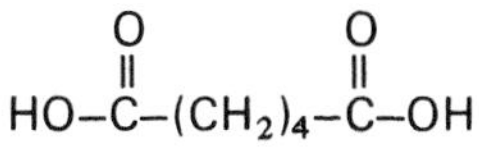

Adipic acid

Adipic or sebacic acids (both saturated dicarboxylic acids) are sometimes partially substituted for maleic anhydride in order to reduce unsaturated sites and increase flexibility and resilience. Conversely, fumaric acid, with its asymmetrical structure, is used to improve the polymer's hardness. When flame retardance is needed, an appropriate chlorinated organic acid is substituted.

Although propylene glycol is commonly used because of its relatively low cost, it is by no means the only glycol used. The cured polyester resin can be made tougher and more flexible by substituting tripropylene glycol, triethylene glycol, diethylene glycol, or dipropylene glycol.

As discussed, styrene is the cheapest crosslinking agent. However, other crosslinkers are also commonly used. Vinyl toluene is used occasionally for better heat resistance. Methyl methacrylate is used for better stability to U.V. Other crosslinkers are also available for improved heat resistance.

FILLERS AND REINFORCEMENTS

Introduction

As discussed earlier, many thermosets would be nearly useless if they were not used with extending fillers or reinforcing fibers. The combination of thermosets with fibrous or particulate fillers is known as a *composite*. These composites have far greater strength and resistance to impact than cured thermosets by themselves which would normally be too brittle to be useful. Also in the last twenty years, fibers and particulate fillers have been used more extensively in thermoplastics such as polyethylene, polypropylene, polyvinyl chloride, nylon, etc., to improve their properties as well as reduce their cost. Therefore, filler technology is playing an important role in today's plastic industry and their future growth in thermoplastics looks particularly promising.

Reinforcing fibers improve thermoplastic properties in the following ways.

> They increase the tensile strength of the polymer
>
> They increase its stiffness (modulus)
>
> Impact resistance and toughness are improved
>
> The polymer's creep is reduced
>
> Heat resistance is improved
>
> Resistance to stress cracking and crack propagation is improved
>
> Abrasion resistance is enhanced
>
> Compression strength is improved
>
> Mold shrinkage is improved
>
> Fatigue resistance is increased
>
> Thermal expansion is reduced
>
> Melt viscosity and hot strength are increased

Some disadvantages of using fibrous fillers are that they hurt transparency of a polymer and may make the process viscosity of the polymer too high, particularly if the fibers are very long. The more elongated the fibers, the higher the process viscosity but also the greater the reinforcement to the composite.

Fibrous fillers work by having a given stress transferred through the polymer matrix medium to the dispersed fibers themselves which have greater strength stability properties than the polymer medium itself. The fibers represent the stronger discontinuous phase dispersed in the weaker continuous polymeric medium. From this dispersion, the fibers interface with the polymer, thus reducing polymer chain mobility (improved creep resistance). These fibers are most effective if they are orientated from processing (ordered in one direction) to provide greater strength in one direction (anisotropic strength). How well the fiber filler is able to assume some of the stress force on the composite is dependent on the following factors.

(1) The concentration of fibers in the polymer matrix is important. There must be a high enough concentration of fibers present to assume some of the deformational forces.

(2) Directional orientation of fibers is very important. By orientating fibers in the direction of the stress, maximum transfer of stress to the fibers through shear at the interface is achieved.

(3) A sufficiently high length to diameter ratio (L/D ratio) is needed for the filler fibers used in order to allow sufficient opportunity for the polymer medium to adhere or "grab" onto the fiber surface. If the L/D ratio is too small, the fiber will not have sufficient surface to be "gripped" by the polymer matrix and poorer composite properties will result.

(4) Compatibility of the fiber to the polymer at the interface is very important. Fibers that adhere well to the polymer medium will provide good reinforcement. On the other hand, if the fibers have a surface that gives a weak interfacial bond, poor physical properties will result. Untreated glass fibers may not adhere well to many polymers. To overcome this problem, the glass fibers are commonly treated with organo-silanes to enhance adhesion.

(5) For the fibers to be effective in improving properties, the strength and modulus of the fibers themselves should be considerably greater than the polymer medium.

(6) The strength of the polymer medium is also important. The selection of an inherently weak base polymer will reduce the physical properties attainable by compounding with a given fiber filler.

If the L/D ratio (called aspect ratio) of a filler is smaller and the filler particles approach more of a spheroidal shape (as opposed to long "needle" shapes) then the filler is considered an *extender* filler rather than a reinforcing filler. Extender fillers impart the property improvements noted earlier for fiber fillers, but not as effectively or to the same extent. In fact, many extender fillers may actually cause a loss in tensile strength rather than an increase. Some extender fillers can hurt impact strength also. Many extender fillers are used primarily to reduce the plastic compound's cost. These fillers are commonly used in polyolefins and PVC. Also many extender fillers help keep processing equipment clean by providing a mild "scrubbing" action. On the other hand, some extender fillers are very abrasive and can cause excessive equipment wear. One large advantage of extender fillers over fibrous ones is that extender fillers do not increase processing viscosities excessively at higher loadings as do fibers. Thus polymers can be loaded higher with extender fillers and still be processible.

It is not always easy to neatly characterize a filler as 'reinforcing' or 'extending'. Many fillers may fall between these two categories depending on their particle shapes, surface compatibilities, etc. Some fillers consist of very long filaments, some of short fibers, and others of needle-like particles. Still others may be prism shaped, spheroidal, or perfectly spherical. These differences in shape all affect the physical properties imparted to the composites.

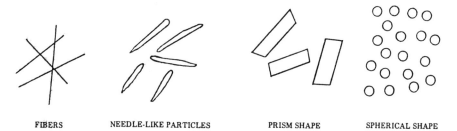

FIBERS	**NEEDLE-LIKE PARTICLES**	**PRISM SHAPE**	**SPHERICAL SHAPE**

Disregarding shape, generally the smaller the particle, the better the reinforcement.

Lastly, over and above the property effects already discussed from using both classes of fillers, there are also specific advantageous properties that are uniquely associated with particular fillers. These special properties may make a specific filler the one of choice. Some of these filler specific properties are given below.

   Improved Xray absorption (shielding)

   Increased flame retardancy

   Improved chemical resistance

Increased electrical conductance

Improved magnetic properties

Increased thermal conductance

Reduced compound cost

Increased compound density

Reduced compound density

Increased lubricity

Improved processibility

Improved moisture absorption

Improved weathering resistance

Improved color control

Improved surface appearance

Increased cure rate for thermosets

The following is a brief description of fillers commonly used in plastics. *Warning*: it is important to note that many of these fillers described may be hazardous to one's health if appropriate safety precautions are not followed. Inhalation of dust as well as ingestion of particles can be extremely dangerous. Specific information on the toxicity and health hazards of the fillers to be discussed is beyond the scope of this book. Therefore, read the appropriate toxicity and safety literature as well as government safety regulations and guides beforehand if you should ever anticipate working with these materials.

Glass Fibers

This is the most commonly used reinforcement filler in the plastic industry. It is used in both thermosets and thermoplastics. Approximately 80% of its consumption goes into making unsaturated polyester resin composites (used to make boat hulls, car bodies, etc.). Its fibrous structure imparts high tensile strength and high impact strength at relatively low cost. To improve compatibility, glass fibers are usually pretreated with a silane or chrome complex as a coupling agent for better interfacial adhesion. There are three types of glass fibers that are used as a filler in plastics—Types E, S, and C. Type E is the cheapest and by far the most common. It is composed of approximately 55% silica, 20% calcium and magnesium oxides, 14% alumina, 10% boron oxide, and 1% sodium and potassium oxides. Type C is similar but with a lower lime content to give it better acid resistance. Type S is a special high tensile fiber glass used commonly in making composites that are used in aerospace applications.

Carbon/Graphite Fibers

These fibers are relatively expensive but impart unusually high modulus and strength to a composite. Originally carbon fibers were obtained from the pyrolysis of stretched rayon (orientated). Later it was found that even better fiber tensile strength could be achieved by pyrolyzing stretched polyacrylonitrile. The formation of graphite under tension results in highly orientated crystals. If the carbon content is less than 97%, then these fibers are called *carbon fibers*. If the carbon content is, say, 99%, the fibers are called *graphite fibers*. These fibers are commonly used in making composites for aerospace applications where high strength at light weight is required.

Glass Solid Microspheres (Ballotini)

This filler consists of spherical particles of glass between 4 to 40 microns in size depending on grade. Because the particles are spheres and not elongated, the resulting compound properties are isotropic (same in all directions). One big advantage to using glass microspheres is that processing viscosity is not increased nearly as much as would result from an equivalent loading of a filler with highly elongated particles. Therefore microspheres give better processing and injection molding properties. Usually glass microspheres are pretreated with a coupling agent to enhance reinforcement. Glass microspheres are normally made from type A glass (soda lime glass) instead of Type E glass (commonly used in making glass fibers) because type A with a higher sodium and potassium oxides content provides the necessary melt characteristics for making the spheres. Glass microspheres are commonly used with many different plastics including ABS, SAN, PPO, HDPE, PP, and nylon.

Glass Microballoons (Hollow Microspheres)

This filler is used to lower the density of a plastic compound and also reduce compound cost per unit volume. The filler is not considered a true reinforcing filler. Usually particle sizes do not fall between 20 microns in size. Microballoons are commonly used in polyester and epoxy thermosets as well as other plastics.

Ground Micas

This filler is used in both thermosets and thermoplastics to impart better heat resistance, air permeability, crack resistance, and lubrication. Ground mica particles are flat and plate-like in shape with high aspect ratios (length/diameter ratios). Special delaminating processes can provide micas with even higher aspect ratios (called HAR micas) for higher reinforcement. Also micas can be pretreated with silane coupling agents for greater reinforcement. Micas are generally less expensive than glass fibers but not as effective in improving tensile strength or impact resistance. However, micas do provide lower process viscosities than glass fibers. The dust from handling mica is toxic and special safety precautions are necessary as with many other fillers.

Talc

This mineral filler is also commonly used in plastics to increase stiffness. Talc is a hydrated magnesium silicate which is very soft (only one on the Mohs hardness scale) and relatively non-abrasive. Its particles can be needle-like, platelets, or irregular spheroidal shapes. Particle sizes usually range from 1 to 30 microns. Talcs are used in polyvinyl chloride, polypropylene, etc. This filler may possibly impart better heat resistance to polypropylene. Also talcs may give better stiffness and creep resistance than calcium carbonates. Worker exposure to dust can be hazardous if impurities such as crystalline silica or asbestos are present.

Kaolin Clay

Another inexpensive mineral filler which is used to increase hardness and reduce creep in many plastics, Kaolin particles are platelet in shape with particle sizes ranging from 0.2 to 10 microns. These particles are about 2.5 on the Mohs scale of hardness. Kaolins are not as white as calcium carbonates. Clay is very inexpensive and is used in both thermosets and thermoplastics. For improved electrical properties, calcined kaolin clay is available in which the water of hydration (normally present in clay) has ben driven off by heating the clay at very high temperatures. This calcination process also may make the clay harder and more abrasive than the non-calcined clays. For worker protection, avoid using clays which contain unsafe levels of silica impurities.

Calcium Carbonate

This filler is obtained by grinding limestone or from a carefully controlled chemical precipitation process which provides a smaller particle size down to 0.1 micron. Calcium carbonate loadings do not stiffen the plastic compound as much as other fillers. Also calcium carbonate can give better impact resistance than some other extender fillers available. It can be used as a secondary acid acceptor in PVC. Calcium carbonate can impart a good white color but does not provide a high degree of opacity or hiding power. Many calcium carbonate fillers are pretreated with stearic acid to improve their dispersibility. Particle shapes are roughly spheroidal with a Mohs hardness of approximately 3. This filler does not contain water of crystallization which makes predrying easier. Calcium carbonates are inexpensive and are used to extend PVC and polyolefin plastics. These fillers have very poor acid resistance.

Wollastonite

This filler is naturally occurring calcium metasilicate with very long needle-like or acicular shaped particles. Particles L/D ratios of 15:1 are common. This mineral filler has one of the highest average particle aspect ratios than can be found in nature. Coupling agents are commonly used to improve interfacial adhesion between the filler particle surface and the polymer. This filler has a thixotropic effect on uncured liquid thermosets. It is commonly used in polyvinyl chloride and polypropylene.

Other Inorganic Fillers

Calcium sulfate. This is another inexpensive mineral filler. It has a Mohs hardness of 2.5. It is less abrasive than some other mineral fillers and is used in PVC, PP, nylon, etc. It is less reactive than calcium carbonate.

Colloidal silica. This is a synthetic silica of very small sized particles in loose agglomerates. It has a thixotropic effect that increases melt viscosities of polymers.

Magnesium oxide. This filler is obtained from calcination of basic magnesium carbonate. It is used in polyester, polypropylene, etc.

Carbon black. This material is generally made from the thermodecomposition of petroleum feedstocks. Although it is the main reinforcing agent in rubber, it does not reinforce plastic. It is used as an economical way to protect a plastic against ultraviolet degradation. Carbon black provides very good opacity. Carbon black particles' hardness is less than one on the Mohs scale.

Ground graphite. This filler is used mainly as a lubricant.

Powdered coal. This material is used mainly in thermosets for its chemical resistance.

Hydrated alumina. This non-reinforcing filler is used mainly as a flame retardant. It is used in polyester, acrylics, PVC, ABS, etc.

Metallic Oxides

Zinc oxide. This material imparts qualities such as good weather resistance, high U.V. absorption, and good heat stability.

Titanium dioxide. This filler is very commonly used in plastics. It is mainly used for its white color and superior opacity and hiding power due to its extraordinarily high refractive index. These qualities make it worth its higher cost when compared to other white fillers.

Zirconium oxide. This material gives high modulus and hardness to composites. Also it accelerates curing in polyester thermosets. This material is toxic.

Miscellaneous Fillers

Ground quartz. This material is used to make special phenolic composites which are used as ablative insulators for space re-entry vehicles. It is very abrasive with a Mohs hardness of 7. Accidentally breathing quartz dust is extremely dangerous and a health risk to workers.

Silicon carbide. This material is used for abrasion in that it has a very high hardness of nine on the Mohs scale.

Molybdenum disulfide. This filler is used mainly to improve abrasion and wear resistance and reduce friction. Also it is lubricating and improves polymer processing. This filler is toxic.

Barium ferrite. This powder is primarily used to impart magnetic qualities to the loaded plastic. It is used in making magnetic tapes, strips, etc. Take the appropriate safety precautions when using this filler.

Zinc sulfide. This inorganic filler is mainly used as a yellow colorant.

Barium sulfate. This loading is used to increase specific gravity, absorb Xrays, and/or improve the resistance of the plastic to chemical attack.

Zirconium silicate. This filler is used primarily for its electrical insulating properties as well as for its superior heat resistance. This filler also can increase polyester cure rates and is toxic.

Wood flour. This is the principle filler used with phenol-formaldehyde resins to form an effective composite. This filler is *not* simply sawdust in that sawdust contains too many impurities which would interfere with the hardening of the phenolic resin. Instead, the wood chips (from spruce or pine) are "cooked" in strong acid to extract out lignin and other impurities. Then this product is pulverized into fine fibers of between 80 and 150 mesh.

Coarse untreated wood chips. This material is generally not used directly in plastic but only to make particle board.

Alpha cellulose. This is the principle filler used with urea-formaldehyde resin to form composites. Alpha cellulose is very compatible with UF and has a light color to complement the light color of UF resin. Alpha cellulose gives a higher strength to a composite than wood flour in that alpha cellulose fibers are longer.

Peanut shell flour. This organic filler is obtained from grinding peanut shells. It is less fibrous than wood flour and is used with thermosets. Also walnut shell flour is used as a filler.

Starch. This filler is biodegradable. It has particles between 0.1 and .01 micron in size. It has been used in LDPE and PP.

Fir bark. This filler has been used to improve processability for thermosets.

Keratin. This filler is obtained from grinding chicken feathers or cattle hooves and is occasionally used in thermosets.

Ground vulcanized scrap rubber. This filler can be used in plastics as an extender. It may also improve impact strength.

Organic Fibers

Cotton fibers. This naturally occurring filler gives better impact resistance to thermosets than alpha cellulose, or wood flour. Also cotton fibers are used to control viscosity and improve moldability of thermosets. It is commonly used in phenol-formaldehyde resin composites.

Rayon fibers. This is another filler for thermosets that is better for improving impact resistance than alpha cellulose.

Nylon fibers. This fiber is used in both thermosets, such as phenolics, and thermoplastics, such as polypropylene. It greatly improves impact strength and provides good strength for light weight. For these reasons, nylon fiber composites are sometimes used in aircraft applications. Also it is used to improve vibration dampening.

Polyester fibers. This fiber is used in thermosets such as unsaturated polyester.

Polyvinyl alcohol fibers. This filler gives very good reinforcement and compatibility with unsaturated polyester thermosets.

Polyacrylonitrile fibers. This filler is used in thermosets such as polyester and thermoplastics such as polypropylene. Sometimes it is used for its electrical properties.

Aramid fibers. This fiber under the trade name of Kevlar by Dupont is very strong for its weight. It can be used to make lighter composites of equal stiffness than can be made with glass fibers. This fiber is used to improve shatter resistance because each fiber is so hard to break. Aramid is used in polyester and epoxy composites to impart high strength.

Polytetrafluoroethylene fibers. This fiber is used to increase lubricity of the plastic.

Sisal fibers. This natural fiber from Haiti and Africa is considered inferior to synthetic fibers. Sisal is normally used to make ropes and sacks. It has poor water resistance and may occasionally be used in making laminates.

Jute fibers. Another naturally occurring fiber from the Far East which is normally used to make sacks. Although it is inexpensive, it is inferior in quality to synthetic fibers. Sometimes these fibers are used in composites.

Metallic Powders

Lead powder. This metallic filler is used in plastic to absorb Xrays, gamma rays, and neutrons. Sometimes it is used in phenolic composites which are used to make medical equipment where Xray shielding is needed. Lead powder is toxic and extreme caution and safety precautions must be followed in its use by workers.

Aluminum powder. This is another metallic filler which is used in plastics to increase thermal conductance. Caution: this powder is explosive and a fire risk.

Bronze powder. This metallic filler is used to increase both thermal and electrical conductance. This powder is flammable.

Whiskers (Microcrystals)

These are fillers consisting of crystals of very high perfection with very high aspect ratios. These crystals may have an L/D around 40. Diameters can be quite small, 0.1 micron in some cases. Some whiskers can be very long. Single crystals can be almost totally free of any flaws. The high aspect ratio, small diameter, and flawless crystalline structure enable many of these "whiskers" fillers to impart greater stiffness than, say, glass fibers. Also these whiskers can impart high strength to the plastic. Basically, whiskers are expensive as reinforcing agents and most are only used in aerospace and military applications by the government. Whiskers are used in polyphenylene oxide, nylon, polypropylene, ABS, polyvinyl chloride, polysulfone, polymides, etc. Potassium titanate whiskers are the most commonly used whiskers. Others are made of silicon nitride, alpha aluminum oxide (Sapphire), silicon carbide, and magnesium oxide.

COUPLING AGENTS

Many of the inorganic fillers discussed are not highly compatible with the polymer medium in which they are dispersed. In order to improve interfacial adhesion between the filler and polymer, many times chemical additives called coupling agents are added. A coupling agent will generally improve interfacial adhesion by an intermolecular "bridge" between the filler surface and the polymer matrix. In so doing, stresses on a plastic composite or compound are more effectively transmitted through the matrix to the filler particles or fibers thus imparting higher composite strength. Also, through the use of coupling agents, higher filler loadings are possible without losing composite strength.

Generally a coupling agent molecule possesses an inorganic functional group and an organic functional group. The inorganic functional group is attracted to inorganic filler surfaces such as silicates, silicas, aluminas, etc. The organic functionality is attracted to the polymer medium. In some cases the organic functional group may actually participate in the curing reactions involved with thermosets.

Silane Coupling Agents

This class of coupling agents is by far the most commonly used today whether it be used in plastics, rubber, or adhesives. Basically a silane coupling agent has a structure as shown below.

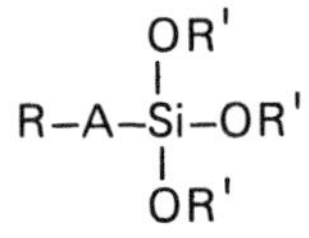

$$
\begin{array}{c}
OR' \\
| \\
R{-}A{-}Si{-}OR' \\
| \\
OR'
\end{array}
$$

R represents an organo functional group which will be attracted to the polymer matrix. These groups are "organophilic" and in the case of thermosets may actually take part in the curing chemistry. These groups may be vinyl, amino, epoxy, chloro, methacryloxy, or other organophilic groups. Different plastics may require a silane with different organo functional groups for best results.

-A- simply represents an aliphatic linkage between the organic group R and the inorganic group.

The OR' represents either a methoxy ($-OCH_3$) or ethoxy ($-OC_2H_5$) group which hydrolyzes on exposure to moisture evolving alcohol (either toxic methanol or ethanol) and leaving the following inorganic reactive group.

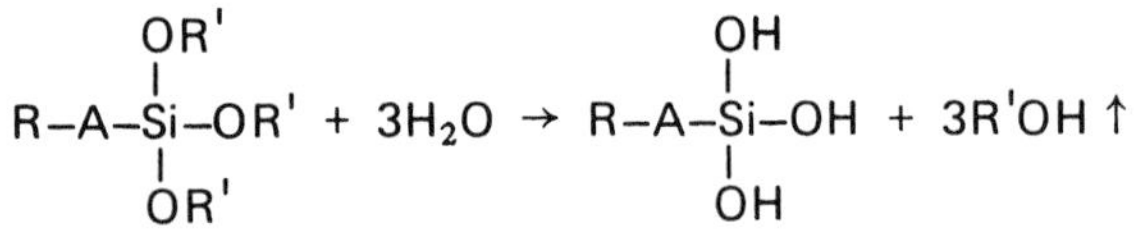

$$
\begin{array}{c}
OR' \\
| \\
R{-}A{-}Si{-}OR' \;+\; 3H_2O \;\rightarrow\; R{-}A{-}Si{-}OH \;+\; 3R'OH \uparrow \\
| \\
OR' \qquad\qquad\qquad OH
\end{array}
$$

The loss of the alcohol unblocks the silane functional groups which in turn are free to react with silanol groups present on siliceous filler surfaces.

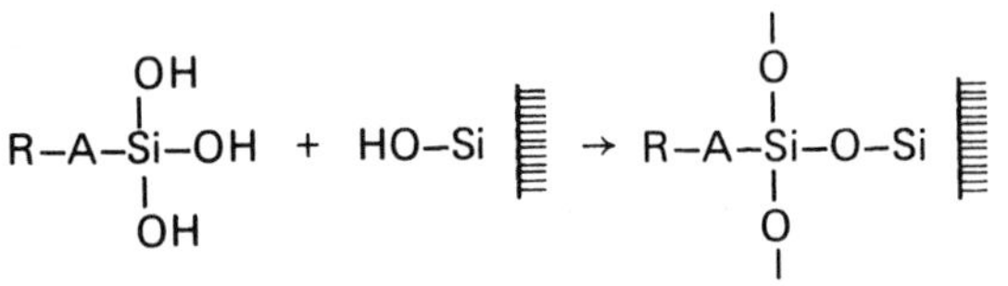

$$
\begin{array}{c}
OH \qquad\qquad\qquad\qquad\qquad O \\
| \qquad\qquad\qquad\qquad\qquad\quad | \\
R{-}A{-}Si{-}OH \;+\; HO{-}Si \;\Rrightarrow\; \rightarrow\; R{-}A{-}Si{-}O{-}Si \;\Rrightarrow \\
| \qquad\qquad\qquad\qquad\qquad\quad | \\
OH \qquad\qquad\qquad\qquad\qquad\quad O
\end{array}
$$

For the most effective bond to the filler surface, it is important that the silane coupling agent molecule not hydrolyze until it is in close proximity to the filler surface. Usually there is sufficient moisture present on a siliceous surface to hydrolyze the silane. Silane coupling agents are effective with mineral silicate fillers, silicas, glass fibers, and alumina. They are not effective with calcium carbonate. It is very common for fillers such as glass fibers or clays to be pretreated. Below are some examples of silane coupling agents used in the plastics industry today. Only a very small quantity of these agents is needed to be effective.

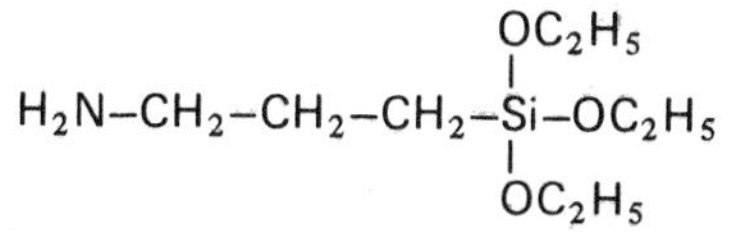

$$OC_2H_5$$
$$H_2N-CH_2-CH_2-CH_2-\underset{\underset{OC_2H_5}{|}}{\overset{\overset{|}{OC_2H_5}}{Si}}-OC_2H_5$$

Gamma-Aminopropyltriethoxysilane

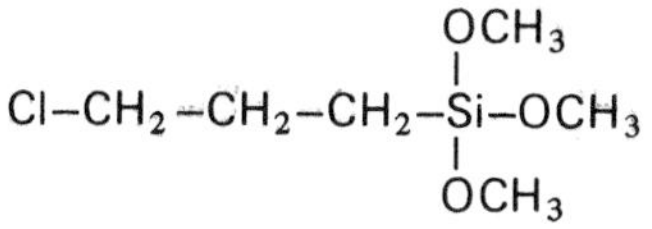

$$OCH_3$$
$$Cl-CH_2-CH_2-CH_2-\underset{\underset{OCH_3}{|}}{\overset{\overset{|}{OCH_3}}{Si}}-OCH_3$$

Gamma-Chloropropyltrimethoxysilane

$$OCH_3$$
$$CH_2{=}CH-\underset{\underset{OCH_3}{|}}{\overset{\overset{|}{OCH_3}}{Si}}-OCH_3$$

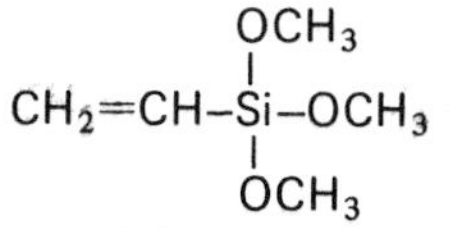

Vinyl Trimethoxysilane

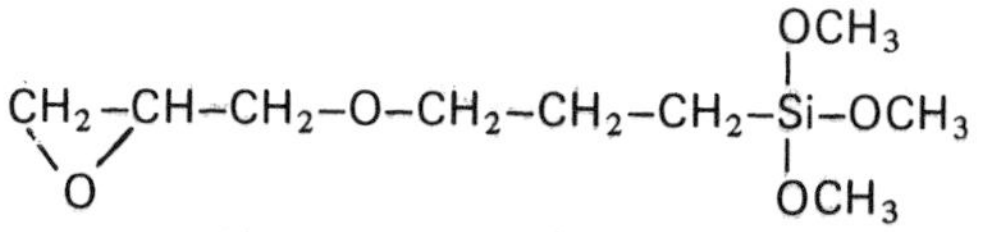

$$OCH_3$$
$$CH_2-CH-CH_2-O-CH_2-CH_2-CH_2-\underset{\underset{OCH_3}{|}}{\overset{\overset{|}{OCH_3}}{Si}}-OCH_3$$

Gamma-Glycidoxypropyl Trimethoxysilane

$$CH_3\ O \qquad\qquad OCH_3$$
$$CH_2{=}\overset{\overset{|}{CH_3}}{C}{-}\overset{\overset{\parallel}{O}}{C}{-}O-CH_2-CH_2-CH_2-\underset{\underset{OCH_3}{|}}{\overset{\overset{|}{OCH_3}}{Si}}-OCH_3$$

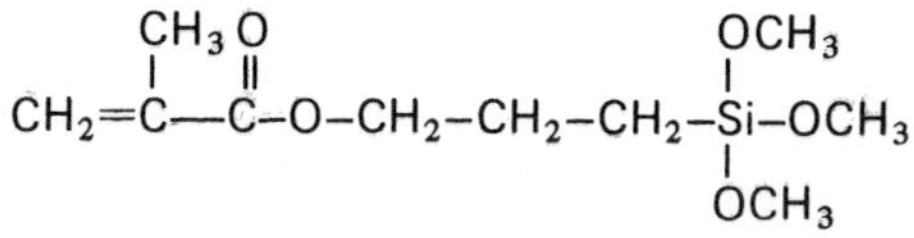

Gamma-Methacryloxypropyltrimethoxysilane

Titanates

These coupling agents adhere readily to an inorganic filler surface forming a monomolecular organic layer around the particle or fiber. Titanates do this in theory by reacting with free protons on the surface of the inorganic particles. In so doing, titanates greatly reduce the compound viscosity and serve as a kind of "super" plasticizer. Thus, through the use of titanates much higher filler or fiber loadings are possible in order to reduce compound costs. Also properties such as impact strength are greatly improved. Titanates not only work with siliceous fillers but are commonly used with calcium carbonates as well. There are a very wide variety of titanates available on the market today. One common general structure is shown below.

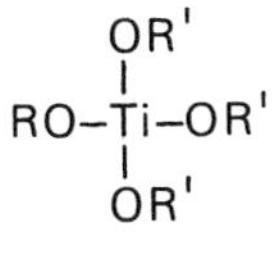

$$OR'$$
$$RO-\underset{\underset{OR'}{|}}{\overset{\overset{|}{OR'}}{Ti}}-OR'$$

Chromium Methacrylate Type Coupling Agents

These coupling agents were originally used with glass fibers before silanes. Their usage has declined.

Other Coupling Agents

There are other types of coupling agents that are used by suppliers to pretreat fibers and fillers for the plastics industry. These agents are proprietary in nature. The chemical structure of these coupling agents is not disclosed.

FLAME RETARDANTS

In the last two decades, flame retardants have experienced a tremendous growth due to new government regulations brought about from concerns over consumer safety. The degree of flammability that a base polymer possesses is determined by how low a temperature at which it decomposes into gaseous products and how combustible those products are. Some polymers such as rigid PVC are non-flammable while others such as polyethylene, polystyrene, ABS, and polyester, to name a few, are quite flammable and require flame retardants.

Flame retardants used in the plastics industry work through one or more of the following mechanisms.

(1) The flame retardant causes a gaseous product to evolve on igniting the plastic which dilutes the combustible gases evolving and which may interfere with the free radical chemistry occurring at the flame front by serving as a "free radical trap". This is the mechanism through which most halogen containing flame retardants work.

(2) The flame retardant may form a glaze or promote the formation of a heavy char on the substrate surface which will thermally insulate the surface and help isolate the flame from the fuel source. Molybdenum compounds, for example, work partially through the formation of a char.

(3) A flame retardant can also work by absorbing heat, i.e., a heat sink. Usually this is achieved through the absorption of heat to vaporize water usually contained in a flame retardant as water of hydration. Alumina trihydrate is a good example of a flame retardant that works through this mechanism.

(4) Lastly, a flame retardant may work by speeding up the rate at which a polymer melts in order to allow it to flow away from the flame. Some plasticizing flame retardants work this way.

The following flame retardants are commonly used in the plastics industry today.

Antimony Oxide

By itself, this chemical has no flame retardancy effect; however, when it is used with a halogen donor, it is one of the most effective flame retardants used today. In fact, the combination of antimony oxide and a halogen donor is much more effective than the use of halogen compounds by themselves. The way antimony oxide works in a given plastic compound is to react with decomposing halogen compounds from ignition to form antimony halides and oxyhalides in the vapor phase. These heavy inert gases form a barrier to cut off additional oxygen to the flame and help promote flame extinction. Also the use of antimony oxide helps promote the formation of a char. Usually halogen compounds decomposing from thermal oxidation will yield hydrogen halides and perhaps some diatomic halogen. Antimony oxides react with chlorine, for example, in the following manner.

$$Sb_2O_3 + 6\,HCl \rightarrow 2SbCl_3 + 3H_2O$$
$$Sb_2O_3 + 2\,HCl \rightarrow 2SbOCl + H_2O$$
$$Sb_2O_3 + Cl_2 \quad \rightarrow SbO_2Cl + SbOCl$$

A similar chemistry is involved with bromine compounds. In fact, bromine is reportedly more effective than chlorine in its flame retardance effect with antimony oxide. From the three equations shown above, it is hard to determine what the optimal molar ratio is for halogen to antimony. However, reportedly three mols of halogen to one of antimony is probably close to optimal (depending on the plastic compound). One problem with antimony oxide is its relatively high cost; however, using antimony oxide will enable a smaller level of halogen to be required. Another problem with antimony oxide is its high toxicity and dusty quality. The new "dustless" forms should be used whenever possible. Today antimony oxide is used widely in many different plastics.

Organic Halogens

These chlorine and bromine containing chemicals are commonly used to impart flame retardancy to plastics. Generally bromine compounds are considered more effective than chlorine compounds; however chlorine compounds are less expensive. Therefore, organic chlorines are more widely used than organic bromines. If antimony oxide or some other synergist is not being used with the organic halogen, then a relatively higher loading is required in order for the organic halogen compound to be effective.

On combustion of the plastic, the organic halide compound decomposes, thus emitting mostly hydrogen halide along with some free halogen. The hydrogen halide acts as a free radical trap in the vapor phase, interfering with the free radical chemistry which propagates the flame. An example of this is shown below.

$$HX + \cdot OH \rightarrow H_2O + X\cdot$$

$$X\cdot + RCH_2\cdot \rightarrow RCH_2X$$

Halogen in the vapor phase also retards the flame by diluting the combustible hydrocarbon gases and serving as a heat sink.

The type of structure an organic halogen compound possesses determines the ease with which halogen is released on combustion. *Linear aliphatic halogen* compounds have relatively weak carbon-halogen bonds and release their halogen at relatively low temperatures. In fact, these compounds can release toxic halogen during high temperature mixing or processing. This limits their use to plastics that can be processed at lower temperatures such as LDPE and flexible PVC. An example of this type of linear aliphatic is chlorinated paraffin.

Another class of organic halogen is called *cycloaliphatic halogen* compounds. These structures possess a stronger halogen-carbon bond than the linear aliphatic structure. While cycloaliphatic structures in general may not be as effective as the linear aliphatic, they do hold their halogen during higher mixing and extrusion temperatures. They can be used in polymers such as polypropylene, nylon, or polystyrene which may require higher processing temperatures. A common trade name that one of these chemicals is sold under is Dechlorane Plus, made by Hooker Chemical Co. The exact structure is not disclosed.

Lastly, *aromatic halogens* generally have the strongest halogen-carbon bonds, especially in the case of chlorine-carbon. Some of these structures are shown below.

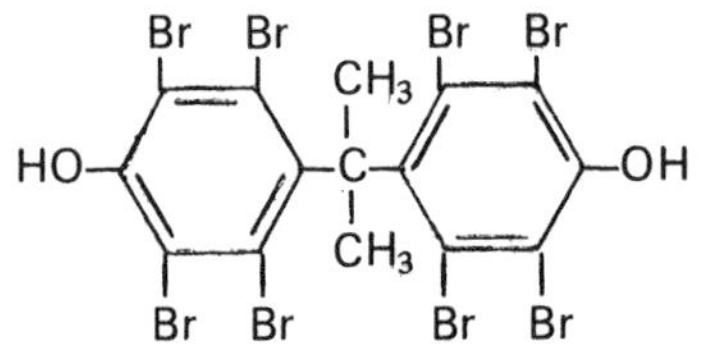

Tetrabromo-
Bisphenol A

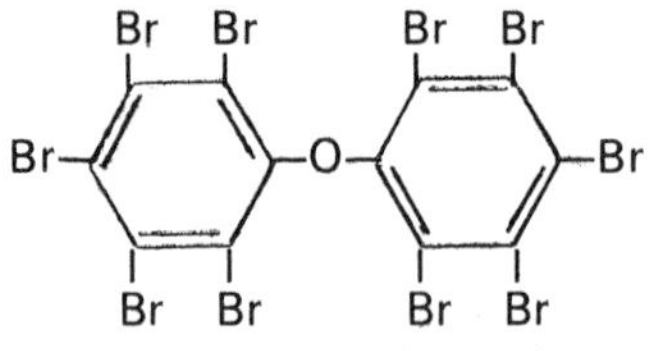

Decabromodiphenyl Oxide (DBDPO)
[also called Decabromobiphenyl ether (DBBPE)]

Molybdenum Compounds

The most common molybdenum compound used in plastics as a flame retardant is molybdenum trioxide (MoO_3); however, other molybdenum compounds such as ammonium dimolybdate [$(NH_4)_2Mo_2O_7$] are also used commercially. Just as with antimony oxide, molybdenum compounds are also used as synergists with halogen donors.

On the other hand, molybdenum compounds may not be as effective as antimony oxide in this synergism to retard the flame. Molybdenum compounds do promote the formation of a char; however they do not volatilize in the same manner as antimony oxide does. Therefore molybdenum compounds are known for their *smoke suppression* characteristics through the formation of char. Molybdenum compounds are not only used in combination with organic halogens, but they are also used occasionally with antimony oxide, borates, phosphates, and other flame retardants. Molybdenum compounds are toxic.

Alumina Trihydrate

This material is one of the less expensive flame retardants used today. Actually alumina trihydrate (ATH) is in reality a crystalline aluminum hydroxide which on being heated above 200°C liberates water, leaving aluminum oxide. The release of water as a vapor absorbs heat from combustion to help extinguish the flame. Also the steam given off dilutes the combustible volatiles which helps retard the combustion. Lastly, ATH helps form an insulating char. However for ATH to work effectively requires ATH loadings in plastic compounds in excess of 30 parts. ATH can not only serve as a flame retardant, but as a smoke suppressant as well. As a result, ATH perhaps has the highest tonnage of consumption as a flame retardant. It is commonly used in a variety of thermosetting and thermoplastic polymers including polyester.

Boron Compounds

Zinc borate hydrate is commonly used in plastic compounding. This chemical is less expensive than antimony oxide and is commonly used as a partial replacement. Zinc borate hydrate is commonly used in PE, PP, and PVC as well as unsaturated polyesters.

Other boron compounds are also used as retardants in plastics. These compounds include boric acid, barium metaborate, borax, and ammonium fluoroborate.

Organic Phosphates

These chemicals are believed to function as flame retardants by promoting the formation of a char or glaze from their thermal decomposition to an acid residue. Also they dilute the combustible hydrocarbon volatiles. In addition, they serve as very effective plasticizers. While rigid PVC is flame resistant, flexible PVC with conventional ester plasticizers is not. Utilizing phosphates as substitutes can improve the flame resistance of flexible PVC. A variety of different phosphate plasticizers are used depending on the plasticizing power needed as well as the volatility characteristics that are required. Also cost is an important factor. For example, tricresyl phosphate (TCP) has greater plasticizing power in many cases while triaryl phosphate (TAP) may be less expensive. Cresyl diphenyl phosphate (CDP) is another common plasticizer used. (The use of TCP has declined because of its cost and its neurotoxicity problems.) Also halogenated organic phosphates such as tris(chloropropyl) phosphate or tris(2,3-dibromopropyl) phosphate are very effective as flame retardants.

Lastly another advantage of phosphate as a flame retardant is that it is clear and can be used to make clear flame resistant plastics.

Inorganic Halides

Generally these salts do not work well in plastics because they do not sublime. Most of these salts are used in paper or wood flame retardance. However, exceptions are ammonium chloride and ammonium bromide which do sublime and do have limited use in plastics in that they thermally decompose into ammonia and hydrogen halide gases. These salts are water soluble, however.

Alloys

One way to promote flame retardancy is to blend an inherently flame resistant polymer, such as PVC, with another polymer that needs protection. An example of this is an alloy of ABS and PVC.

Reactive Flame Retardants

The flame retardants discussed so far have been additives, i.e., flame retardants that are added as compounding ingredients while mixing or processing. On the other hand, reacting flame retardants can be part of the chemistry of the thermoset. They react as a monomer with a thermoset. In the case of unsaturated polyester thermosets, there are acid reactive FRs such as tetrachlorophthalic acid and tetrabromophthalic acid as well as alcoholic reactive FR monomers such as dibromopropanol or phenoltetrabromobisphenol A.

PLASTICIZERS

Introduction

Plasticizers have been used in the plastics industry since it was discovered by J. W. Hyatt in the ninteenth century that camphor added to cellulose nitrate made the polymer more flexible. Today plasticizers are used in cellulosic plastics, PVC, ABS, nylon, polyolefins, and other plastics. Plasticizers are used to achieve any or all of the following property changes.

Improve processing and extrusion characteristics

Reduce melt viscosity

Reduce the minimum required processing temperature

Increase the plastic's flexibility

Increase elongation

Reduce plastic's hardness

Improve low temperature flexibility

Improve impact resistance

For certain plastic applications, the use of an appropriate plasticizer is essential for the desired end use. In PVC applications, plasticizers are used at lower levels to help reduce processing temperatures in order to prevent thermal degradation of the homopolymer. At higher levels, plasticizers not only improve the processing of PVC but also reduce its hardness and increase its flexibility to enable its use as a flexible plastic. In fact the use of plasticizer in PVC for this purpose is so common that it accounts for appproximately 85% of the total plasticizer consumption.

A plasticizer is a non-volatile liquid (or solid) which when mixed with a given polymer will not separate (is chemically compatible) and will partially disrupt the secondary valance forces (Van der Waal forces, hydrogen bond forces, etc.) between polymer chains which restrict chain mobility. By physically occupying space between the chains, separating them, and permitting them to slide past one another under a deformational force, the plastic becomes flexible or "rubber-like". Also the Tg is lowered.

In order for a plasticizer to work effectively in a plastic polymer, it must be compatible. This means that the plasticizer must be chemically similar to the polymer in that molecularly it has a similar cohesive energy density, polarity, hydrogen bonding, etc. The Hildebrand Solubility parameter of the plasticizer and the polymer can be calculated from the cohesive energy density of each material (refer to the discussion on solubility parameters in the chapter on the Adhesive Industry). If the plasticizer and the polymer have solubility parameters that are very similar, then they will most likely be compatible. For example, dioctyl phthalate with a solubility parameter of 8.9 is compatible with PVC with a solubility parameter of 9.7. In other words, the polymer and plasticizer must be sufficiently compatible so that the attractive forces between the polymer and platicizer are greater than either the attractive forces within the polymer itself or the attractive forces within the plasticizer alone. If this is not true, then the plasticizer will be incompatible with the polymer, the plasticizing and solvating effect will not occur and physical phase separation will result in exudation and blooming. While exudating additives in some cases may impart a lubricating effect and possibly aid mold release, they are not effective plasticizers.

In selecting a plasticizer, there are several properties that should be taken into account before making a final decision. First of all, how much solvating power does the plasticizer display in a selected polymer. With PVC, this property determines the minimum temperature required for fluxing or gelating with the plasticizer. Also it determines the flux rate and how effective it is at reducing hardness. Highly polar plasticizers of low molecular weight, such as dibutyl phthalate or butylbenzyl phthalate are faster in fluxing PVC than less polar plasticizers such as dioctyl phthalate.

Secondly, plasticizers have differing effects on a plastic's cold temperature properties. Most commercial plastics, including PVC, have a glass transition value (Tg) that is higher than its ambient service temperature. Thus these plastics are rigid. By mixing these polymers with a compatible plasticizer, the glass transition of the resulting compound is lowered below the plastic's intended service temperature rendering the plastic flexible or even elastic. Compatible straight chain aliphatic type plasticizers are more effective at imparting superior cold temperature properties than might otherwise be attainable from a more polar plasticizer. For example, dioctyl sebacate imparts better cold temperature properties to PVC than dibutyl phthalate.

Another important quality in plasticizer selection is the degree of permanence. Permanence relates to the plasticizer's resistance to removal from a given plastic compound through volatilization, extraction, or migration. Generally higher molecular weight plasticizers of the same class have less volatility and more permanence than lower molecular weight plasticizers. For example, diisononyl phthalate imparts greater permanence in PVC than dibutyl phthalate. Also the more compatible a plasticizer is in a polymer, the better its permanence. High loss of plasticizer through volatilization can cause "fogging" on car windshields, etc. Loss of plasticizer can also occur through solvent extraction. The solvent might be gasoline, oil, or soapy water. Polymeric plasticizers appear to have the best resistance to extraction.

Lastly, miscellaneous plasticizer properties such as heat and light stability, staining, discoloration, flame retardance, electrical resistance, and toxicity are all important considerations. Some of these properties are discussed in other sections.

Approximately 2 billion pounds of plasticizer are produced annually, most of which is used in the plastics industry.

Type Plasticizer	Millions of Pounds 1983
Phthalate esters	1,146
Dibutyl phthalate (DBP)	20.2
Diethyl phthalate (DEP)	15.4
Diisodecyl phthalate (DIDP)	157.9
Dimethyl phthalate (DMP)	7.8
Dioctyl phthalate (DOP)	300.1
Trimellitate esters	40.8
Trioctyl trimellitate	26.2
Adipate esters	116.3
Dioctyl adipate	22.2
Diisodecyl adipate	2.0
Epoxidized esters	117.4
Epoxidized linseed oil	6.7
Epoxidized soya oil	97.3
Oleate esters	10.5
Butyl oleate	1.4
Decyl oleate	0.3
Phosphate esters	26.6
Sebacate esters	5.6
Stearate esters	8.5
N-butyl stearate	5.7

Source: International Trade Commission

Phthalates

These plasticizers represent by far the largest poundage volume used in the industry today. Phthalates are mostly used in PVC; however, they also find application in cellulosic plastics, polyester, and rubber. They are generally formed from a reaction of an alcohol with phthalic anhydride. Some examples of phthalate ester plasticizers are given below.

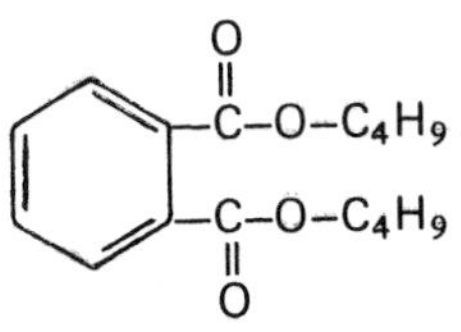

Dibutyl Phthalate (DBP)

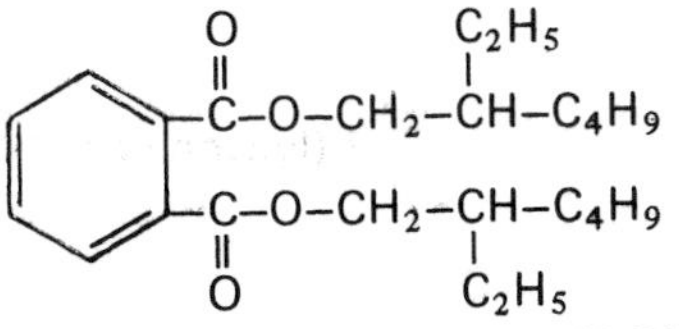

Di(2-Ethylhexyl) Phthalate (DOP)

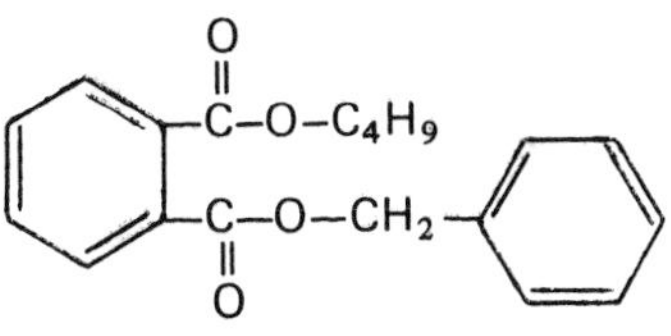

Butyl Benzyl Phthalate (BBP)

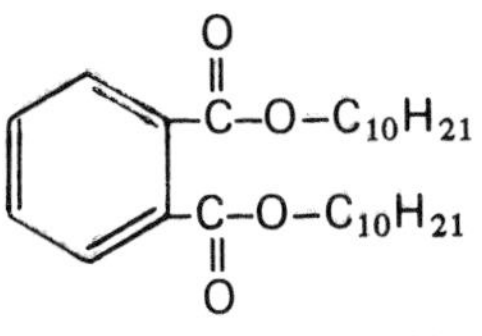

Diisodecyl Phthalate (DIDP)

DBP is highly polar and possesses high solvating power with PVC.

On the other hand, it imparts relatively poor cold temperature properties and is somewhat volatile. Diethyl phthalate (DEP) is similar in properties to DBP.

DOP is the "work horse" plasticizer used in the plastics industry. It is relatively inexpensive and possesses a good combination of solvating power and permanence in PVC. Another plasticizer called diisooctyl phthalate (DIOP) possesses similar properties to DOP but is less commonly used.

DIDP has a higher molecular weight which explains its lower volatility and greater permanence. On the other hand, DIDP has less polarity which means it has less solvating power in PVC. Another plasticizer called diisononyl phthalate (DINP) has similar properties to DIDP.

BBP has high solvating power in PVC and is resistant to staining and migration which accounts for its popularity in PVC flooring.

Another phthalate called di(2-ethylhexyl) isophthalate is commonly used because of its resistance to migration and durability in PVC while di(2-ethylhexyl)terephthalate (DOTP) is used for its permanence properties.

Phosphates

These plasticizers are known for their flame retardancy and good solvating power in PVC. The higher cost of these plasticizers is justified when good flame retardancy is needed. Most other plasticizers hurt flame retardancy properties.

It should be noted also that phosphates are not known for their cold temperature properties which may be less than satisfactory.

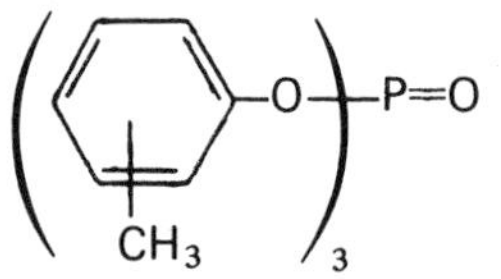

Tricresyl Phosphate (TCP) 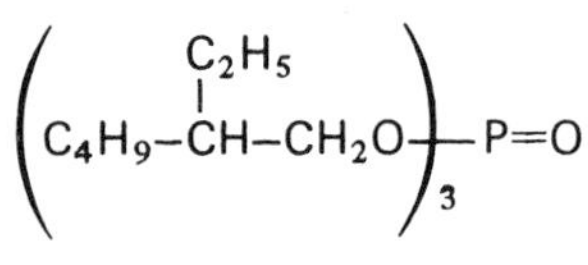   Tri(2-Ethylhexyl) Phosphate (TOP)

TCP was one of the original plasticizers used in PVC. It has good solvating power and flame retardancy properties but poor cold temperature properties. Because of the neurotoxicity problems associated with TCP, its usage has mostly been replaced with triaryl phosphate (TAP) which is made from less expensive synthetic feedstocks.

TOP is used when flame retardancy is required but cold temperature properties must be preserved. TOP has better cold temperature properties than either TCP or TAP.

Adipates

This class of plasticizer is the less expensive of the three types of plasticizer commonly used to improve cold temperature properties, i.e., adipates, azelates, and sebacates. Adipates are produced from the reaction of adipic acid (a dicarboxylic acid) with an alcohol. Adipates do not have the same degree of compatibility with PVC that phthalates and phosphates possess. Two commonly used adipates are shown below.

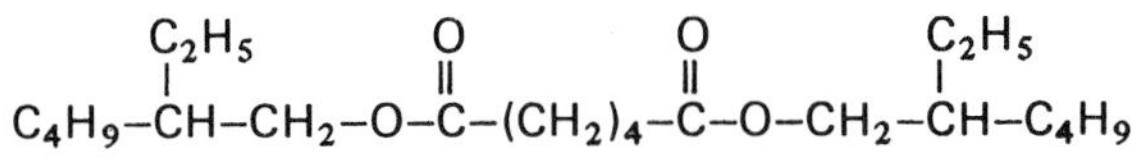

$$C_4H_9-\overset{\overset{\displaystyle C_2H_5}{|}}{C}H-CH_2-O-\overset{\overset{\displaystyle O}{\|}}{C}-(CH_2)_4-\overset{\overset{\displaystyle O}{\|}}{C}-O-CH_2-\overset{\overset{\displaystyle C_2H_5}{|}}{C}H-C_4H_9$$

Di(2-ethylhexyl) Adipate (DOA)

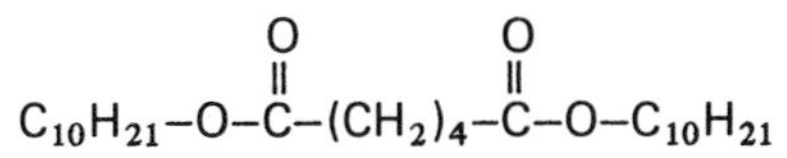

$$C_{10}H_{21}-O-\overset{\overset{\displaystyle O}{\|}}{C}-(CH_2)_4-\overset{\overset{\displaystyle O}{\|}}{C}-O-C_{10}H_{21}$$

Diisodecyl Adipate (DIDA)

Sebacates

These plasticizers are more expensive than adipates, but perhaps impart the best cold temperature properties to a plastic. One problem with sebacates is that they have limited compatibility with PVC which limits the level of loading possible. Because of sebacates' higher molecular weight, they are less volatile than their adipate or azelate analogues. DOS, shown below, is a common sebacate used in plastics.

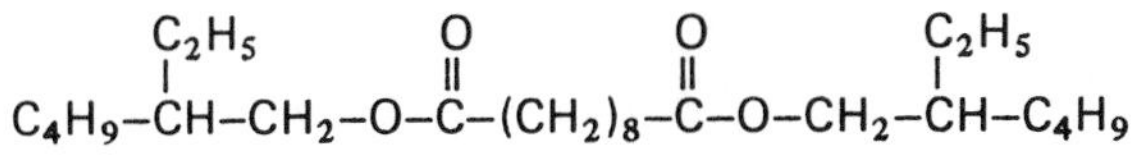

$$C_4H_9-\overset{\overset{\displaystyle C_2H_5}{|}}{C}H-CH_2-O-\overset{\overset{\displaystyle O}{\|}}{C}-(CH_2)_8-\overset{\overset{\displaystyle O}{\|}}{C}-O-CH_2-\overset{\overset{\displaystyle C_2H_5}{|}}{C}H-C_4H_9$$

Di(2-ethylhexyl) Sebacate (DOS)

Azelates

These plasticizers have superior cold temperature properties just as discussed for adipates and sebacates. Azelates also have limited compatibility with PVC especially at higher loading levels. A commonly used azelate is DOZ, shown below.

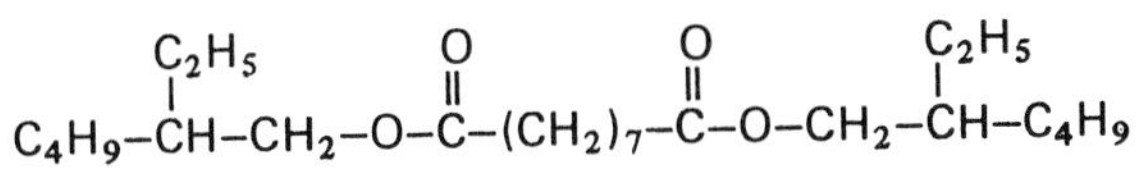

$$C_4H_9-\overset{\overset{\displaystyle C_2H_5}{|}}{C}H-CH_2-O-\overset{\overset{\displaystyle O}{\|}}{C}-(CH_2)_7-\overset{\overset{\displaystyle O}{\|}}{C}-O-CH_2-\overset{\overset{\displaystyle C_2H_5}{|}}{C}H-C_4H_9$$

Di(2-ethylhexyl) Azelate (DOZ)

Trimellitates

These plasticizers are relatively expensive but provide a good combination of solvating power and permanence not normally found with other plasticizers. They are very durable. Trimellitates, however, do not impart extraordinary cold temperature properties. Trioctyl trimellitate is an example of this class of plasticizer used in plastics.

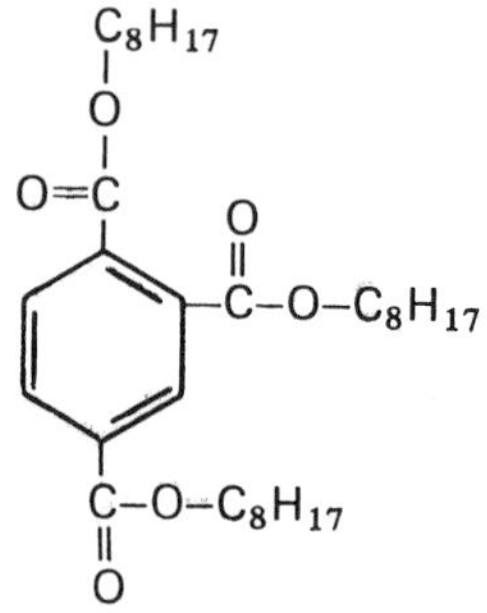

$$CH_3(CH_2)_7CH=CH(CH_2)_7\overset{\overset{\displaystyle O}{\|}}{C}-OC_4H_9$$

Trioctyl Trimellitate

Sulfonamides

These polar plasticizers are used to plasticize acrylics, nylon, cellulosic plastics, etc. Also these plasticizers may be used in thermoset systems such as unsaturated polyesters, melamine-formaldehydes, epoxies, etc. An example of a sulfonamide is given below.

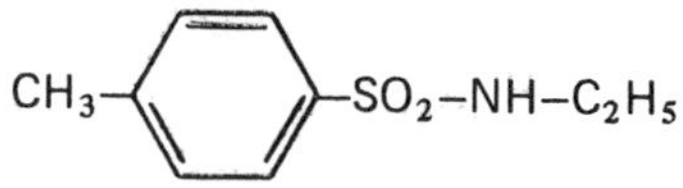

$$CH_3-\langle\!\!\!\bigcirc\!\!\!\rangle-SO_2-NH-C_2H_5$$

N-ethyl-p-toluenesulfonamide

Oleates

These plasticizers are limited in their use because of their low polarity and limited compatibility. At low levels, they have been used in polystyrene to improve processing. Butyl oleate is an example of a commonly used oleate which is shown below.

$$CH_3(CH_2)_7CH=CH(CH_2)_7\overset{\overset{\displaystyle O}{\|}}{C}-OC_4H_9$$

Butyl Oleate

Linear Polyester Plasticizers

These are generally low molecular weight liquid polymers formed from a polycondensation reaction between a dicarboxylic acid (such as adipic, sebacic, or azelaic acids) and a diol (such as ethylene glycol). The exact chemical identity of many of these polyester plasticizers presently on the market is not disclosed by the manufacturers. Their main advantage is their degree of permanence resulting from their inherently low volatility and resistance to extraction. For this reason, these plasticizers are sometimes used in such applications as vinyl upholstery where permanence is very important. On the other hand, these plasticizers can be quite viscous and generally impart poor cold temperature properties.

Secondary Plasticizers

These plasticizers are not compatible by themselves with a given plastic and can only be used along with another primary plasticizer in order to avoid exudation. The primary plasticizer shifts the solubility parameter of the secondary plasticizer just enough to keep it in the plastic matrix. Secondary plasticizers are generally used mainly to extend primary plasticizers and reduce cost. This is important in that many primary plasticizers may actually cost more than say the PVC base resin.

A secondary plasticizer commonly used to bring down the cost is aromatic petroleum oil.

Epoxidized Plasticizers

These plasticizers are known for the heat and light stability they impart to PVC. They establish synergisms with primary heat stabilizers (to be discussed). Two commonly used epoxidized plasticizers are epoxidized linseed oil and epoxidized soybean oil. In order to have good compatibility, these plasticizers should have less hydroxyl content and a relatively high epoxy content.

LUBRICANTS

Lubricants, in some applications, act in a similar manner to plasticizers in that they disrupt the intermolecular forces among the polymer chains, thus reducing melt viscosity. This is the "internal" lubricating effect. On the other hand, lubricants also prevent the polymers from sticking to the metallic surfaces of the processing equipment. Lubricants, unlike plasticizers, may possess a certain degree of chemical incompatibility or insolubility which causes them to be exuded to the surface of the polymer during processing and reducing the surface coefficient of friction. This surface modification (called the "external" lubricating effect) results in easier processing, better extrusion, and increased slip. Other benefits from the use of lubricants may be the formation of a glossy appearance, better mold release, antiblocking (prevention of plastic layers from sticking), and reduction of the clogging of process equipment.

Usually these lubricants are non-polar in nature with large chain segments in their structure. On the other hand, they may contain some polar groups within the molecular structure which may help promote an attraction for a metallic surface to prevent sticking. Many of these materials are relatively insoluble in plastics such as PVC. Some of these materials are soaps. Generally they may be used at low levels of 0.1 to 1 part or more in a plastic compound.

Although examples of several commonly used lubricants are given below, it should be pointed out that some commercial lubricants are not chemically identified by the manufacturers because they are proprietary blends.

Most lubricants are used in rigid PVC to help reduce processing temperatures and prevent thermal degradation during processing. Lubricants are also used to aid processing of ABS, PS, PP, engineering plastics, phenolics, and other plastics.

Testing lubricants in the laboratory reportedly will not predict their effectiveness in the factory. Many times, trial and error methods determine the best lubricants to use for a specific application and at what level. Also, many times best results are obtained by using a blend of two or more lubricants. There is something of an art to using the correct lubricants.

The following are some of the lubricants presently used in the plastics industry.

Calcium stearate (most common)

Zinc stearate

Stearic acid (good release)

Glycerol monostearate (improves flow)

Ethylene bisstearamide (good slip and antiblocking)

Ethylene bisoleamide

Low molecular weight polyethylene

Oxidized polyethylene

Low melt petroleum wax (for thermoplastics)

High melt petroleum wax (for thermosets)

Mineral oil

Vegetable oil

Silicone oil (also good mold release)

Polyfluorocarbon powder

Molybdenum disulfide

Montanic acid (clear bloom, external lubricant)

Polyvinyl alcohol

HEAT STABILIZERS

These chemical components are generally added to halogen containing polymers for protection against thermal degradation at high processing temperatures. Polymers that may need such protection include polyvinyl chloride, polyvinylidene chloride, chlorinated polyvinyl chloride, and chlorinated polyethylene; however, by far the highest volume usage of heat stabilizers is in PVC, particularly rigid homopolymer PVC applications. The reason for this is that rigid PVC requires a processing temperature that is close to its thermal decomposition temperature. In fact, in many cases, without the use of heat stabilizers, routine extrusion of rigid PVC might be impractical.

PVC and other chlorine containing polymers have limited thermal stability because of the relative ease of losing an allylic chlorine which combines with a hydrogen to form hydrogen chloride (HCl). This initial formation of HCl serves to catalytically accelerate the thermal dehydrochlorination of PVC in what is known as the "zipper-like" reaction which ultimately results in a conjugated unsaturated backbone as shown below.

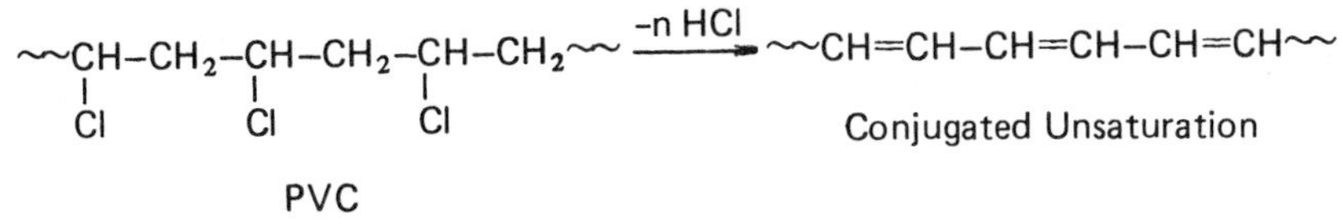

$$\sim\sim CH-CH_2-CH-CH_2-CH-CH_2\sim\sim \xrightarrow{-n\,HCl} \sim\sim CH=CH-CH=CH-CH=CH\sim\sim$$

$$\begin{array}{ccc} | & | & | \\ Cl & Cl & Cl \end{array} \qquad\qquad\qquad \text{Conjugated Unsaturation}$$

PVC

The dehydrochlorination of PVC not only reduces physical properties of the polymer but can cause a color change because of the conjugated unsaturation which absorbs light at different wavelengths. Slight thermal degradation of PVC can show up as a yellowish color but very severe degradation results in a brown or even black appearance. Lastly, the evolution of HCl itself is undesirable in that this gas can combine with moisture to form hydrochloric acid which is highly corrosive to processing equipment. Also HCl gas is highly toxic and must be avoided by workers.

Most heat stabilizers help prevent thermal degradation by serving as an HCl scavenger or acid acceptor to the initial formation of HCl in order to prevent it from serving as a catalyst for further degradation. An example of this is shown below.

$$M^{++}(COOR^-)_2 + 2HCl \rightarrow MCl_2 + 2(H-COOR)$$

Heat stabilizers may help prevent thermal degradation by acting as free radical acceptors and by combining directly with unsaturated bonds.

There are many different heat stabilizers on the market today to meet a wide variety of requirements. Some considerations that should be taken into account before selecting one or more heat stabilizers are given below.

(1) Effectiveness

(2) Cost

(3) Propensity to bloom (exudate)

(4) Stain and discoloration effects

(5) Color

(6) Clarity

(7) Lubrication effects

(8) Odor

(9) Effects on fusion rates

(10) Effects on melt flow

(11) Toxicity

The following are some commonly used classes of heat stabilizers used today.

Barium-Cadmium

This class of heat stabilizer is perhaps the most popular type used today and is used more widely than any other class of stabilizer. By using a combination of barium and cadmium, a synergistic effect results. Supposedly cadmium is very effective at preventing early discoloration, but is not as effective after the initial period. On the other hand, barium is more effective at preventing severe discoloration, but does not prevent early discoloration as well. Apparently, the cadmium is quick to react with the liberated allylic chloride to form cadmium chloride while the barium component reacts with the cadmium chloride to free the cadmium for further stabilizing reactions.

One problem with cadmium is that it is very toxic. Another problem is that cadmium can stain from the formation of cadmium sulfide. One method to help reduce cadmium staining is to use a barium-cadmium-zinc stabilizer. The addition of zinc may reduce the stabilizer's effectiveness some but will help prevent cadmium staining.

Many barium-cadmium stabilizer manufacturers will not reveal the exact chemical composition of their product; however, some of the anions reportedly contained in these barium-cadmium compounds are given below.

Octoates

Phenolates

Benzoates

Naphthenates

Neodecanoates

Laurates

Stearates

Palmitates

Myristates

As can be seen, some of these components are simply soaps. Compounds such as stearates may have limited compatibility while salts such as phenolates display better solubility. Some of these compounds are solids while others are liquids at room temperature and have advantages for use in flexible PVC applications.

Organo-Tin

This class of stabilizer represents a group of compounds which are very effective, clear, relatively expensive, and toxic in nature. These compounds are mostly used in rigid PVC. Some organo tins also impart lubrication and plasticizing effects.

There are basically two types of organo tin stabilizers, i.e., mercaptides and non-mercaptides. The mercaptide tin stabilizers are more effective at heat stabilization than the non-mercaptide types; however, the mercaptides possess objectionable odors and impart poorer light and ultra-violet stability properties than non-mercaptide types. Examples of mercaptide tin stabilizers that are in use today are dibutyltin lauryl mercaptide and dibutyltin mercaptopropionate. Examples of non-mercaptides are di(n-octyl)tin maleate, dibutyltin maleate, and dibutyltin dilaurate.

Calcium-Zinc

This class of non-clear stabilizers in general is toxicologically safer than the barium-cadmium class and is considered non-staining. On the other hand, calcium-zinc stabilizers are generally less effective than barium-cadmium or tin stabilizers. The synergism of calcium and zinc is analogous to the barium and cadmium synergism already discussed in which the zinc cation functions in a similar manner to cadmium and the calcium functions in a similar manner to the barium. Manufacturers of calcium-zinc stabilizers often will not reveal their exact chemical composition; however, usually they consist of carboxylic salts such as stearates, etc. Calcium-zinc stabilizers are commonly used in such applications as flooring where staining resistance is important.

Lead

This class of heat stabilizer has been declining in use over the years because of concerns regarding its high toxicity. Also lead stabilizers are easily stained by sulfur impurities and are not clear. On the other hand, lead stabilizers are relatively inexpensive, very effective, and have good resistance to water absorption. These stabilizers are still used in PVC wire and cable applications. Specific examples of lead stabilizers that are used are basic lead carbonate, lead phthalate, lead phosphate, lead phosphite, and lead stearate. As with other toxic stabilizers, every effort should be undertaken to prevent factory worker exposure to these substances. Non-dusty forms as well as encapsulated forms are gaining wide commercial acceptance in order to help prevent worker exposure.

Antimony Mercaptides

This class of stabilizer is being promoted in some cases as a less expensive alternative to tin; however, this class may not have as good lubricating properties as many of the tin stabilizers. Also there are concerns about the light stability properties of antimony mercaptides. These stabilizers are toxic.

Epoxies

This group has already been discussed under plasticizers. Epoxies are actually secondary heat stabilizers in that they do not work without the presence of a primary metallic stabilizer such as barium-cadmium or tin. Epoxies also function as light stabilizers. Examples of epoxies used today are epoxidized soya (soybean) oil, epoxidized linseed oil, octylepoxy tallate, butylepoxy stearate, and octylepoxy stearate. An epoxy stabilizer functions as an HCl scavenger in a manner shown below.

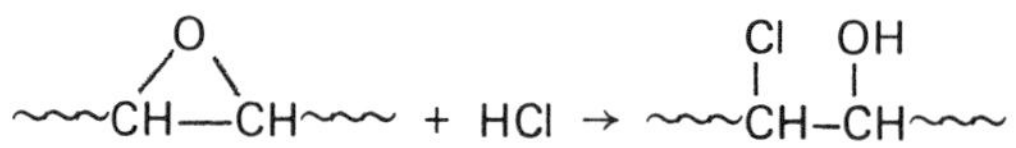

$$\overset{\displaystyle O}{\sim\sim CH-CH\sim\sim} + HCl \rightarrow \overset{\displaystyle Cl\ \ OH}{\sim\sim CH-CH\sim\sim}$$

Epoxy Compound   Chlorohydrin Product

Phosphites

Just as discussed with epoxies, phosphites are also ineffective alone but impart a synergistic effect with other metallic primary stabilizers such as barium-cadmium or calcium-zinc. Phosphites are usually clear but can affect water resistance in that they can be hydrolyzed.

U.V. STABILIZERS

Ultraviolet radiation from outdoor exposure can cause photooxidation in many plastics. This is because photons of U.V. are commonly absorbed at different wavelengths by different polymers. This U.V. absorption is destructive to the polymer because the energy level of photons at the U.V. wavelength range (290–400 nm) is sufficient to break polymer bonds, i.e., the energy absorbed is greater than the bond energy.

U.V. stabilizers are added to different polymers to protect them from such radiation through several mechanisms. The first way that many U.V. stabilizers work is to absorb the U.V. energy and then dissipate it in a harmless form. For example, 2,4-dihydroxybenzophenone "screens out" harmful U.V. photons (hv) in the following manner.

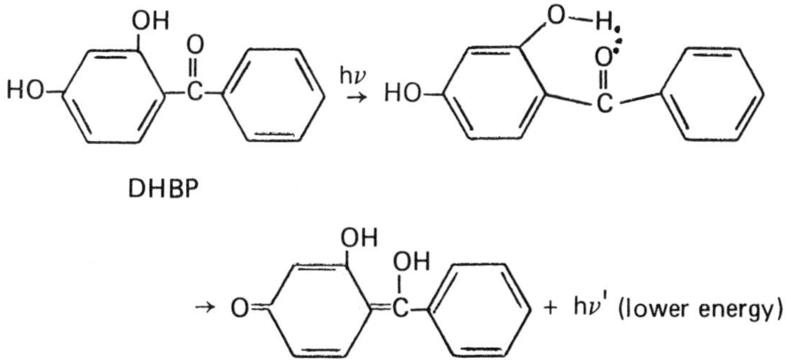

DHBP

Another way that the same stabilizers might protect a polymer from U.V. radiation is by "quenching" a macromolecule that has been excited to a higher energy level. The U.V. stabilizer may draw energy away and dissipate it as heat.

A third mechanism through which a given stabilizer may work is through accepting free radicals from U.V. exposure. Also some stabilizers may serve as peroxide destroyers.

Unprotected polymers exposed to U.V. radiation outdoors may change color (commonly to a yellow shade or darker), lose surface gloss, or even lose physical properties such as impact resistance or tensile strength. Different polymers have different degrees of susceptibility to degradation from U.V. exposure and are hard to stabilize effectively. For example, polystyrene will turn yellow quickly after a relatively short exposure to ultraviolet. On the other hand, polymethyl methacrylate plastics absorb very little U.V. and require little stabilization. Unprotected polyvinyl chloride plastics are sensitive to U.V. exposure and will turn yellow or darker in color; however PVC can be effectively stabilized for years. Polyesters absorb U.V. which causes color change as well as loss of gloss. Polyolefins, on the other hand, absorb little U.V. radiation; however, their catalyst residues and other impurities may.

A wide variety of different U.V. stabilizers is needed for reasons to be explained. First of all, different plastics absorb U.V. radiation at different wavelengths and therefore should contain U.V. stabilizers which will be most effective in protecting these specific plastics. Also the following other properties of the stabilizer should be considered before a selection is made.

(1) Initial color imparted

(2) Color stabilization

(3) Stability during high temperature processing

(4) End use product life expectancy

(5) Volatility

(6) Compatibility (avoid exudation)

(7) Cost (some U.V. stabilizers are extremely costly)

(8) Toxicity

Although the tonnage of U.V. stabilizers manufactured each year may not be nearly as high as other raw materials discussed earlier, the total dollar value is quite high because of the high price of many of these substances.

2-Hydroxybenzophenones

This class of U.V. stabilizers is the most widely used and represents the largest pound volume in use today. These stabilizers work both by absorbing U.V. as well as "quenching" molecules that are excited by U.V. exposure. These stabilizers are used in PVC, PE, PP, cellulosics, PET, acrylics, unsaturated polyesters, and polystyrene. An example of a compound in this class is shown below.

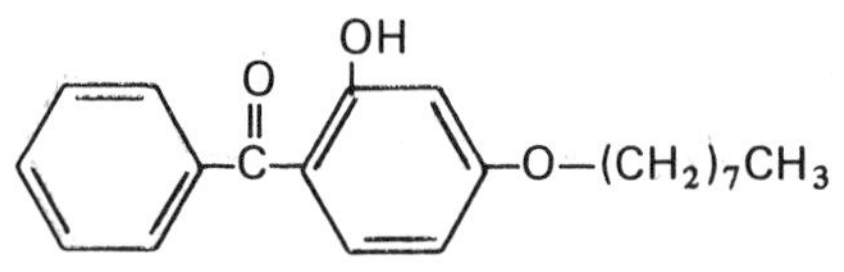

2-Hydroxy-4-(Octyloxy) Benzophenone ("Octabenzone")

2-Hydroxyphenylbenzotriazoles

This class is similar to the hydroxybenzophenones; however, the benzotriazoles are reportedly not as good in polyolefins as phenones are. Triazoles are supposedly more effective than the phenones in polycarbonates. Benzotriazoles are known for their good color stability. A specific example of a stabilizer in this class is alkylated 2-(2-hydroxyphenyl)-2H-benzotriazole (also called Tinuvin P).

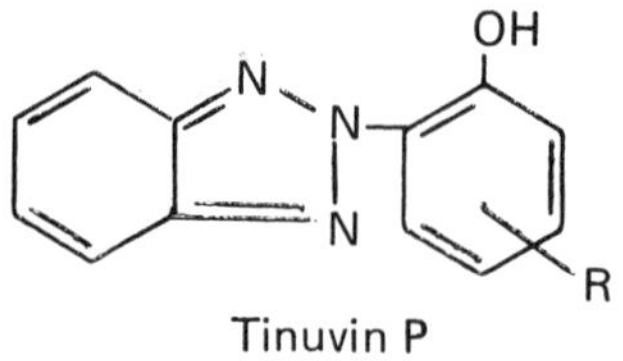

Tinuvin P

Aryl Esters

This class of stabilizer absorbs over a wide range of wavelength and is commonly used in cellulosic plastics. Also these stabilizers are used in PET. An example of a member of this class is shown below.

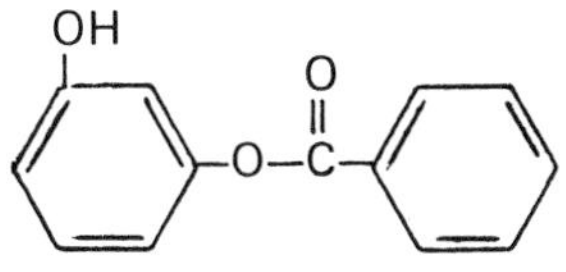

Resorcinol Monobenzoate

Acrylic Esters

This less common class of stabilizers absorbs in the short wavelength region of U.V.; however its absorption range is narrow. Although these stabilizers have good color stability, they are sensitive to moisture. They are commonly used in cellulosic plastics, polyester, and epoxies. An example of a compound in this class is dimethyl 2-(4-methoxybenzylidene) malonate.

Hindered Amines

This class does not protect plastics from degradation by directly absorbing the U.V. radiation, but rather by serving as an effective free radical scavenger. Members of this class may have good clarity and low volatility. These stabilizers are used in polyolefins, ABS, and cellulosic plastics. An example of a specific compound in this class is 4-(2,2,6,6-tetramethylpiperidinyl) sebacate.

Metallic Salts

This class of U.V. stabilizers works by "quenching" the excess energy from the macromolecules that are "excited" by exposure to U.V. radiation. Members of this class are commonly used in PE and PP. Nickel salts represent the largest subgroup that is used for this purpose. Nickel salts, however, do impart color. On the other hand, zinc salts, while not as effective as nickel, do not affect color nearly as much. A good example of a nickel salt is nickel dibutyldithiocarbamate (shown below).

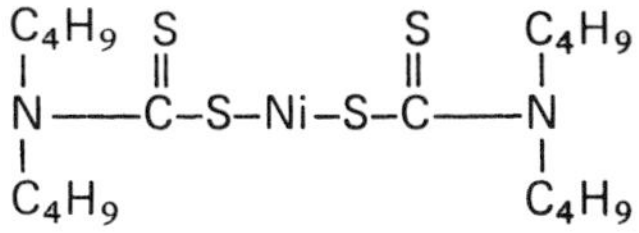

Carbon Black

This opaque material can be very effective even at low concentrations in shielding a polymer from U.V. degradation. It is also inexpensive. Carbon black is commonly used in PVC, PE, and PP to impart U.V. protection. The one problem with carbon black is that it can not be used in non-black color applications.

Mineral Fillers

These materials are also commonly used to help protect a given plastic from U.V. degradation. Rutile titanium dioxide is one of the best of this class. Also zinc oxide is considered effective. Calcium carbonate may afford some protection. On the other hand, not all mineral fillers are effective. For example, some talcs have reportedly hurt weatherability resistance.

ANTIOXIDANTS

Introduction

Articles made of plastic are subject to oxidative degradation which can shorten product service life. Oxidative degradation can result in discoloration, loss of tensile strength and impact strength as well as other property changes. The first step in the oxidative degradation process is the formation of macromolecular free radicals. These free radicals are brought about from the breaking of molecular chains during mixing or processing, exposure to U.V. radiation, residual catalysts or transition metal impurities present, ozone exposure, or just heat exposure.

$$\sim\!\sim\!\sim\!H \rightarrow \sim\!\sim\!\sim\!\cdot + H\cdot$$

$$\sim\!\sim\!\sim\!\sim\!\sim\!\sim \rightarrow \sim\!\sim\!\sim\!\cdot + \sim\!\sim\!\sim\!\cdot$$

These free radicals, in turn, will react with diatomic oxygen to form peroxide radicals and hydroperoxide species which propagate the formation of more free radicals.

$$\sim\!\sim\!\sim\!\cdot + O_2 \rightarrow \sim\!\sim\!\sim\!OO\cdot$$

$$\sim\!\sim\!\sim\!OO\cdot + \sim\!\sim\!\sim\!H \rightarrow \sim\!\sim\!\sim\!OOH + \sim\!\sim\!\sim\!\cdot$$

$$\sim\!\sim\!\sim\!OOH \rightarrow \sim\!\sim\!\sim\!O\cdot + \cdot OH$$

$$\sim\!\sim\!\sim\!O\cdot + \sim\!\sim\!\sim\!H \rightarrow \sim\!\sim\!\sim\!OH + \sim\!\sim\!\sim\!\cdot$$

$$\cdot OH + \sim\!\sim\!\sim\!H \rightarrow \sim\!\sim\!\sim\!\cdot + H_2O$$

This propagation will continue until a termination mechanism is reached where free radicals combine with one another to end the propagation. Well before this point, however, a considerable degree of alteration in the physical properties can result. Oxidative crosslinking and/or chain scission can result. Therefore, to prevent these changes from occurring, chemical additives called antioxidants are used.

There are two basic types of antioxidants, i.e., primary antioxidants and secondary antioxidants. Primary antioxidants work by donating a hydrogen (usually from either an OH or NH group) to a free radical. The resulting AO radical is usually stabilized by resonance which prevents future propagation. Examples of this type of AO are hindered phenols and aromatic amines.

$$\sim\sim\sim OO\cdot\ +\ AH\ \rightarrow\ \sim\sim\sim OOH\ +\ A\cdot$$

$$\sim\sim\sim\cdot\ \ \ +\ AH\ \rightarrow\ \sim\sim\sim H\ \ \ \ +\ A\cdot$$

$$\sim\sim\sim O\cdot\ \ +\ AH\ \rightarrow\ \sim\sim\sim OH\ \ \ +\ A\cdot$$

$$HO\cdot\ +\ AH\ \rightarrow\ H_2O\ +\ A\cdot$$

The secondary antioxidants protect the polymer from oxidative degradation by destroying the peroxidic radicals and in the process becoming oxidized. For example, dialkyl thioesters are oxidized to sulfones while phosphites are oxidized to phosphates.

In practice, it is very common for two or more different antioxidants to be used together in a plastic compound to achieve best results. In fact, using a primary antioxidant, such as hindered phenol, in combination with a secondary antioxidant, such as a phosphite or thioester, will result in a true synergism, that is, the resultant protection is greater than the protection that either AO could impart separately. Apparently the secondary AO can rejuvenate the primary AO in the following manner.

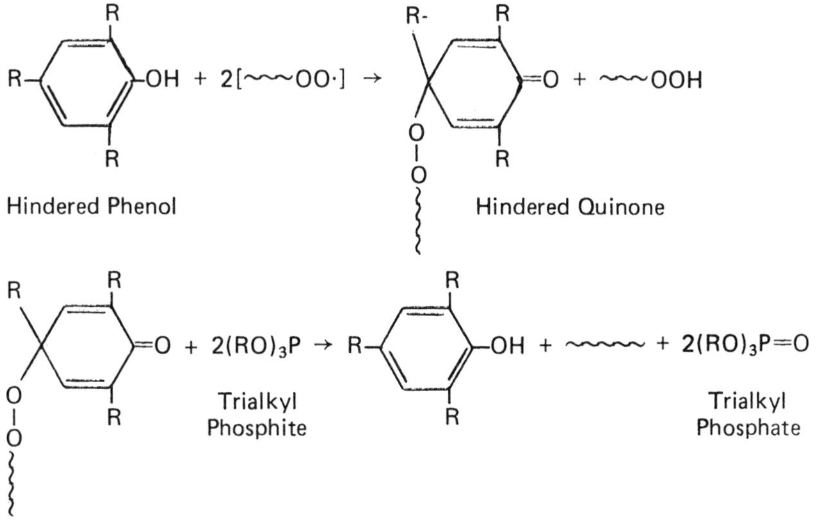

Hindered Phenol

Hindered Quinone

Trialkyl Phosphite

Trialkyl Phosphate

The susceptibility of a given polymer to oxidative attack is greatly dependant on its structure. Polymers containing unsaturated sites, such as rubber modified polystyrene, are most susceptible to oxidative attack while linear saturated polymers containing secondary carbons in their backbones (such as linear polyethylene) are least susceptible. Saturated polymers having tertiary carbons off the backbone, (such as poly-propylene), are more susceptible to oxidative attack than saturated linear polymers containing secondary carbons in the backbone, but less susceptible than unsaturated polymers to attack. Therefore, unsaturated polymers generally require the highest levels of AO while the saturated linear polymers without tertiary carbons require the least.

$$-CH_2- \qquad \begin{matrix} R \\ | \\ -CH- \end{matrix} \qquad -CH=CH-$$

Secondary Carbon   Tertiary Carbon   Unsaturated Carbons
(least susceptible   (most susceptible to
to O_2 attack) O_2 attack)

There is a wide variety of different antioxidants to select from because of many considerations that must be taken into account. Some of these considerations are given below.

Effectiveness

Cost

Discoloration

Staining

Odor

Volatility

Extractability

Toxicity

Compatibility (exudation)

Physical form (solid or liquid)

Possible synergisms

The following are important classes of antioxidants used in the plastics industry today.

Hindered Phenols

This is the most popular and widely used class of antioxidants. They work by allowing a hydrogen to be abstracted by a free radical leaving a stable, resonated antioxidant radical which will not propagate the degradation process further. Hindered phenols are generally non-staining; however, they can discolor if they are oxidized into quinoid structures. Probably the single most popular hindered phenol is butylated hydroxy toluene (BHT) which is used in a variety of polymers including PS, PP, and PE.

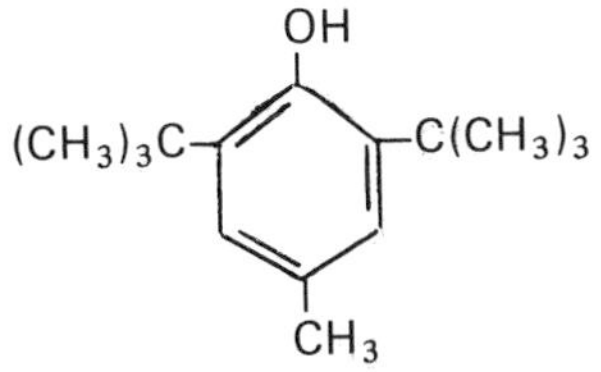

BHT

BHT is considered to be an effective AO that is low in toxicity. One problem with a low molecular weight antioxidant such as BHT is that it is somewhat volatile, which can prove a disadvantage. Therefore, in applications where greater permanence is needed, higher molecular weight hindered phenols such as tetrakis (methylene 3-(3'5'-di-tert-butyl 4'-hydroxyphenyl)propionate) methane are used.

Thiobisphenolics

This class is very similar in chemistry and function to the hindered phenol class just discussed. Thiobisphenolics have an advantage over hindered phenols in that they can be used with carbon black in a plastic compound while hindered phenols may not be as effective with carbon black. Generally thiobisphenolics are less volatile than hindered phenols.

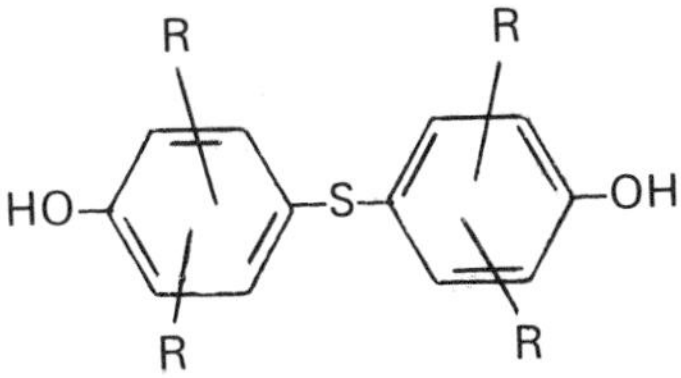

These anitoxidants are commonly used in wire and cable applications.

Aromatic Amines

This class functions in a similar manner to the hindered phenols in that these compounds donate a hydrogen from an NH group to neutralize a free radical. As with the hindered phenols, the aromatic amines upon loss of this labile hydrogen, form a stable resonated free radical which does not propagate the degradation further. Aromatic amines also have an important advantage of being able to destroy peroxides that are formed which render them more effective than the hindered phenols. On the other hand, unlike hindered phenols, many aromatic amines are very discoloring and staining which severely limits their application in the plastics industry. These amine antioxidants find greater application in the rubber industry. (See the chapter on the Rubber Industry for more details.)

Thioesters

These compounds are considered secondary antioxidants which work by destroying peroxidic radicals and in the process become oxidized into sulfones as shown below.

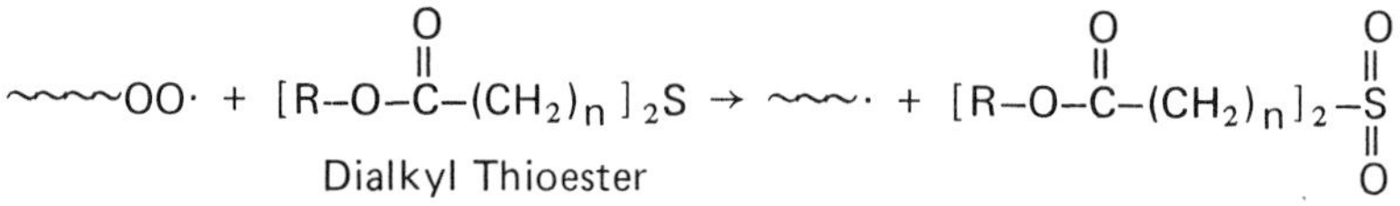

$$\sim\!\!\sim\!\!OO\cdot \;+\; [R\!-\!O\!-\!\overset{\overset{O}{\|}}{C}\!-\!(CH_2)_n]_2S \;\rightarrow\; \sim\!\!\sim\cdot \;+\; [R\!-\!O\!-\!\overset{\overset{O}{\|}}{C}\!-\!(CH_2)_n]_2\!-\!\overset{\overset{O}{\|}}{\underset{\underset{O}{\|}}{S}}$$

Dialkyl Thioester

These secondary antioxidants are usually used synergistically with hindered phenols. Thioesters are generally considered nondiscoloring but can impart odors to a product. An example of a compound commonly used from this class is shown below.

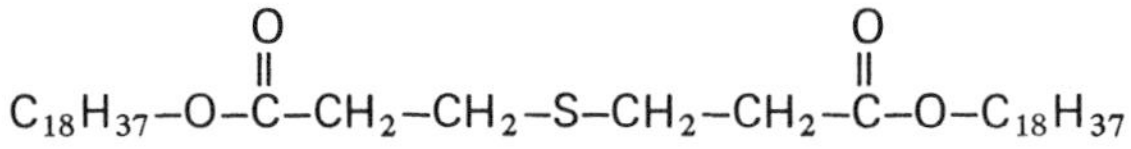

$$C_{18}H_{37}\!-\!O\!-\!\overset{\overset{O}{\|}}{C}\!-\!CH_2\!-\!CH_2\!-\!S\!-\!CH_2\!-\!CH_2\!-\!\overset{\overset{O}{\|}}{C}\!-\!O\!-\!C_{18}H_{37}$$

Distearyl Thiodipropionate

Phosphites

Again these compounds are considered secondary antioxidants which destroy peroxidic radicals and are oxidized to phosphates in the process as shown below.

$$\sim\!\!\sim\!\!OO\cdot \;+\; 2(RO)_3P \;\rightarrow\; \sim\!\!\sim\cdot \;+\; 2(RO)_3P\!=\!O$$

Trialkyl   Trialkyl
Phosphite   Phosphate

Again phosphites are commonly used in synergism with hindered phenols. These compounds are usually non-discoloring. One disadvantage to these compounds is that they are susceptible to hydrolysis. These compounds are commonly used in PP, PE, and PS. As discussed earlier, phosphites can also serve as heat stabilizers in some applications. Examples of phosphites that are used as a secondary AO are Tris(nonylphenyl) phosphite, and distearyl pentaerythritol diphosphite.

CURING AGENTS

These are chemicals that are used to promote or accelerate the crosslinking of resins.

Organic Peroxides

These unstable compounds are used to cure unsaturated polyesters and crosslink thermoplastics such as polyethylene. Organic peroxides achieve these effects through dissociating into free radicals as shown below.

$$ROOR \rightarrow 2RO\cdot$$

These free radicals in turn react with active hydrogens to form polymeric radicals which can be crosslinked.

$$RO\cdot + HR' \rightarrow ROH + \cdot R'$$

As a result, organic peroxides are used in the plastics industry to facilitate the styrene crosslinking of unsaturated polyesters (see previous discussion on unsaturated polyesters in the "Thermosets" section of this chapter) or, with other selected peroxides, to create carbon-carbon crosslinks in certain thermoplastic polymers, such as polyethylene, for improved dimensional stability. (See chapter on Rubber Industry for more information on peroxide curing of saturated polymers.) Examples of organic peroxides that are used in unsaturated polyester curing are given below. (Caution: some of these peroxides are too reactive for use with other thermoplastics.)

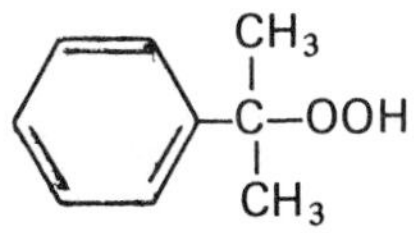

Cumene Hydroperoxide

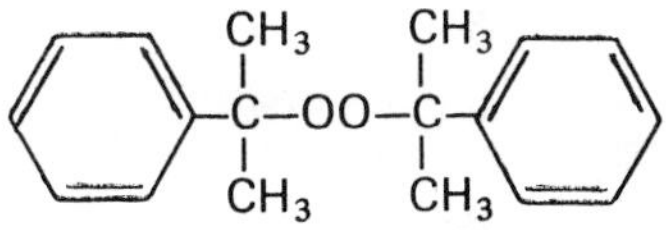

Dicumyl Peroxide

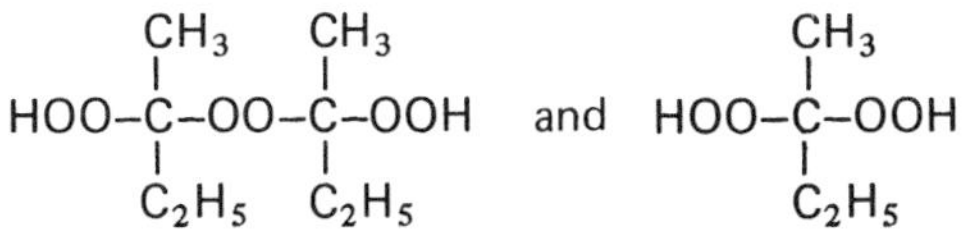

Dimer   Monomer   Tert-Butyl Perbenzoate

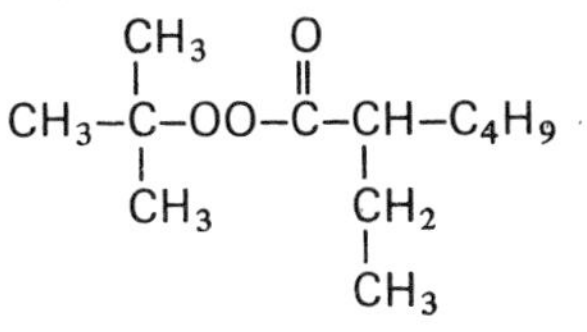

Methyl Ethyl Ketone Peroxide
(Mixed oligomers that are diluted
usually with dimethyl phthalate.)

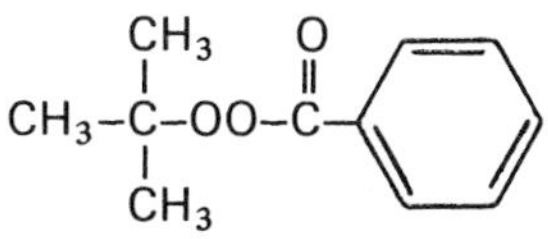

Tert-Butyl Peroctoate

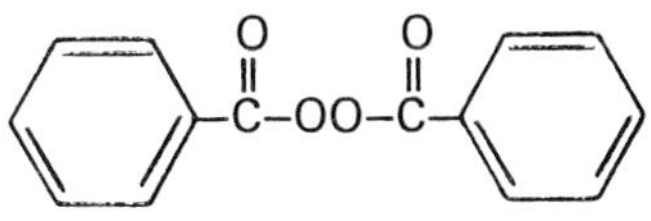

Benzoyl Peroxide

Many of these peroxides are so unstable that they are diluted by the manufacturers with dimethyl phthalate, dibutyl phthalate, tricresyl phosphate or some inert diluent which will reduce the fire and explosion risk somewhat. Many peroxides are stored under refrigeration. Organic peroxides can be very dangerous to work with if they are not handled properly. Some of them are extremely sensitive to shock, friction, heat, and foreign contamination which can cause them to explode or ignite. It is extremely important that these peroxides and promoters (discussed below) be kept completely apart and stored in separate areas of a plant. Organic peroxides and promoters must *never* be directly mixed together or an explosion or fire can occur. Always first thoroughly mix in the promoter into the resin, then carefully add the desired amount of peroxide catalyst to the resin. Also, precautions must be taken to avoid any worker contact in that exposure can cause severe throat irritation, lung damage, eye damage, skin burns, etc. There are many hazardous situations that can be encountered when handling organic peroxides which are beyond the scope of this book. A thorough review of all the safety procedures and precautions given by the various manufacturers, government agencies (such as OSHA and EPA), independent agencies (such as Manufacturing Chemists Association in Washington) and other published references is advised. These are dangerous materials to work with!

The many commercially available organic peroxides all differ in their heat sensitivity. One way this property is commonly determined is by half-life measurements. The half life of a selected organic peroxide at a specific temperature is defined as the time required to decompose one-half of the peroxide. However, most manufacturers use 10-hour half-life temperature to compare decomposition rates. (The ten hour half-life temperature is defined as the temperature at which one half of the specific organic peroxide will decompose in a ten hour time period.) Sometimes manufacturers will also give one-hour and one-minute half-life temperatures. Another important property of organic peroxides is their percent active oxygen content. The selection of an organic peroxide and concentration to be used is complicated and dependent on a variety of factors such as safety and handling considerations, the type of resin, curing temperatures, curing time constraints, thickness of end product, other additives in the compound, etc. Generally, only a small concentration of peroxide is required to cure a resin and this concentration is critical in determining cure rate.

Metal Salts and Amine Promoters

These additives are used in very small quantities to speed up the decomposition of peroxides usually through a redox type reaction thus making room temperature cures possible for unsaturated polyester resins. Without these metallic salts or amine additives, a higher cure temperature or longer cure time would be required. It is extremely important that these promoters and the organic peroxides (discussed above) be kept completely apart and stored in separate areas of a plant. Organic peroxides and these promoters must *never* be directly mixed together or an explosion and fire can occur. Always *thoroughly* mix the promoter in the resin *before* carefully adding the peroxide catalyst to the resin. As mentioned earlier, the appropriate safety precautions are beyond the scope of this book; however peroxides can explode or ignite on contact with promoters. Therefore the manner in which these promoters and peroxides are applied to a process is critical to safety and details of appropriate safety procedures are available through the manufacturers as well as government and independent sources.

Some examples of promoters commonly used to accelerate the cures of unsaturated polyesters are given below.

Cobalt naphthenate

Cobalt octoate

Manganese naphthenate

Dimethylaniline

Diethylaniline

The selection and concentration of the proper promoter is greatly dependent on which peroxide is chosen and at what concentration as well as other factors such as type of resin, cure temperature and time, and what other additives are present. The level of promoter used in a polyester cure is important to control of cure rate. Usually only a very small level of promoter is required.

Methylene Donor Hardeners

These agents are used to cure a reactive novolac phenol formaldehyde resin. As discussed previously, novolac PF resins are polymerized with a stoichiometric deficiency of formaldehyde in order to prevent gelation from occurring. This resin is then cooled and ground into a powder after which a powdered methylene donor hardening agent is blended with it. Upon reheating the resin, the methylene donor decomposes to evolve formaldehyde which gives the resin sufficient agent to crosslink three dimensionally and gel into a hard, cured thermoset.

The hardening agent most commonly used to cure novolac PF resins is hexamethylene tetramine (called "hexa"). Upon heating, the hexa decomposes into formaldehyde and ammonia. The formaldehyde contributes to crosslinking the novolac resin.

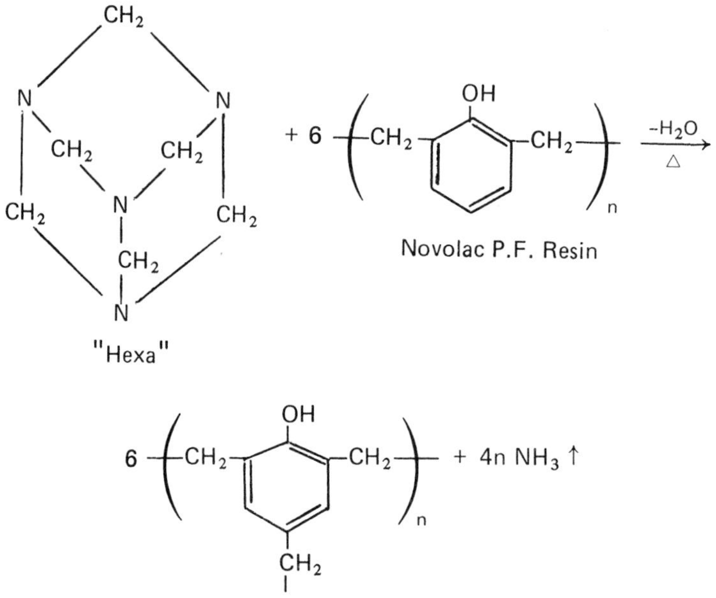

"Hexa"

Novolac P.F. Resin

COLORANTS

Introduction

Color is a very important concern for many plastic products. There are literally hundreds of colorants used in the plastics industry in order to achieve the desired color as well as other desired properties which shall be discussed. The art of color matching is a complicated skill, many times involving the selection of two or more colorants at a select concentration. The professional making these decisions is called a colorist. In selecting the proper colorant(s), the following are some of the properties that should be considered.

Hue—which color is to be used

Brightness—how bright a color does the colorant impart to the plastic

Opacity (or conversely transparency)—how much "hiding power" does the colorant impart; on the other hand, is the colorant transparent

Dispersibility—the ease with which a colorant disperses when added to a polymer is important. Poor dispersion can result in streaking or speckling in the product's appearance. Colorants of very small particle size are harder to disperse than colorants of larger particle size.

Chemical resistance—how resistant is the colorant to acids, alkalis, and oils

Migration resistance—This refers to the separation of the colorants from the plastic medium due to incompatibility usually resulting in a surface bloom on the product. This can result in a poor product appearance as well as build up of colorant on the mold surface (called "plate out").

Light stability—This property (sometimes called light-fastness) is the ability of a colorant to resist any color change resulting from prolonged exposure to light.

Heat stability—This property refers to the colorant's color stability after exposure to heat. This heat exposure can include heat resulting from processing. This property is sometimes referred to as "heat fastness".

Cost effectiveness—Some colorants are very expensive, but have such high tinctorial strength that only a very small concentration is required to achieve a desired color. Therefore, per pound cost should be considered along with performance.

Electrical properties—What effect does the colorant have on the plastic's electrical conductivity or resistivity.

Toxicity—Many of the colorants used in the plastics industry are extremely toxic and require special handling to assure that workers are not exposed to their hazards. Non-dusty forms should be used wherever possible. Follow all OSHA regulations when working with these colorants.

As can be seen from the cited considerations, the selection of the best colorant or colorant combination for a particular plastic application is not easy and many times requires a decision by an experienced colorist to make the best selection.

The following is a discussion concerning the six classes of colorants used in the plastic industry.

Dyes

These are lower molecular weight organic colorants that are soluble in the plastic medium. This class of colorant is not used extensively in the plastics industry. One reason for their limited use is their tendency to migrate because of their solubility. It is particularly important that a dye not be soluble in a plasticizer that is also being used in the compound. Dyes are generally not particularly known for their lightfastness properties either. One major advantage of dyes, however, is the extraordinary transparency that they can impart. Also their brilliance is usually quite good. A good example of the use of a dye is car tail-light lenses. Other examples of dyes used in plastics are given below. Some of these colorants are very toxic. Follow all OSHA regulations concerning their use.

> Azo dyes (dye that contains the -N = N- chromophore group)
>
> Diazo dyes
>
> Pyrazolones
>
> Quinolines
>
> Quinophthalones
>
> Anthraquinones
>
> Nigrosines

Organic Pigments

These colorants, unlike dyes, are for the most part basically insoluble in the plastic medium. Usually these organic pigments are of very fine particle size which impart degrees of opacifying power (dyes do not normally display opacity). Generally organic pigments give better heat stability and lightfastness, and are less likely to migrate than dyes. Some examples of classes of organic pigments used in the plastic industry are given below. Some of these colorants are very toxic. Use dustless forms where possible. Follow all OSHA regulations.

> Benzimidazolones (yellow, red, orange)
>
> Phthalocyanines (blue, green)
>
> Quinacridones (violet, red, orange)
>
> Dioxazines (violet)
>
> Isoindolinones (yellow, red, orange)
>
> Disazos (yellow, red)
>
> Pyrazalones (orange, red)
>
> Diarylides (yellow, orange)
>
> Dianisidines (orange)

Inorganic Pigments

These colorants are used extensively in plastics. Generally they do not have the brightness or tint strength that many organic pigments possess. On the other hand, inorganic pigments are generally less expensive and have superior heat stability and opacity when compared to organic pigments. Also inorganic pigments may display better lightfastness than organic pigments. Another typical advantage of inorganic pigments is their ease of dispersion compared to some organic pigments. This property is partially due to the larger average particle size of many inorganic pigments compared to organic ones. Lastly, inorganic pigments are more resistant to migration, extraction, and bleeding due to their insolubility. Some examples of inorganic colorants commonly used in the plastics industry are given below. Note that many of these colorants are extremely toxic. Special safety precautions must be applied in the use of pigments containing such heavy metals as lead, chromium, cobalt, cadmium, or nickel. Use dustless forms where possible. Follow all OSHA regulations when working with these colorants.

Titanium dioxide (white)

Lead chromates (yellow, orange)

Iron oxides (brown, red, maroon, yellow, black)

Chromium oxide (green)

Cadmium sulfoselenides (maroon, red, orange)

Lithopone (white)

Ultramarine blue (aluminosilicate complex with sulfur)

Nickel titanate (yellow)

Cobalt aluminate (blue)

Zinc chromate (yellow)

Lead molybdate (orange)

Cadmium sulfide (orange)

Lake Pigments

These are colorants which consist of a dye absorbed on an inorganic carrier such as alumina hydrate. These colorants are commonly used in such applications as refrigerator interiors and packaging.

Pearlescent Colorants

These are additives which impart a special "luster of pearl" appearance to the plastic product. This special effect is achieved by reflection and controlled light scattering by crystal particles of high refractive indices. Basic lead carbonate is an example of an additive which is used sometimes to achieve this effect; however it is highly toxic. Also mica particles coated with titanium dioxide can be used.

Daylight Fluorescent Colorants

These are unique colorants in that their molecules absorb light at short wavelengths and re-emit photons of light at specific longer wavelengths in accordance with the principles of quantum mechanics. Because light is not just reflected at many wavelengths, but rather re-emitted through fluorescence at specific wavelengths, the colors produced are extraordinarily bright and vivid. The colorant molecules emitting light for this effect are always excited by light exposure of higher energy (i.e., shorter wavelengths). This is why daylight fluorescent reds and oranges are brighter than say greens or blues (which are at the higher energy end of the light spectrum). These colorants are very sensitive to excessive heat exposure and can not be used in plastic processed at high temperatures. The chemical composition of many of these daylight fluorescent colorants is not disclosed by the manufacturers.

IMPACT MODIFIERS

Of course, plastics that are flexible are quite impact resistant in that they can easily absorb a shock without fracture. On the other hand, some rigid plastics with high glass transition values can be quite brittle and fracture on impact. These plastics can be rendered more impact resistant, however, by dispersing rubbery particles (impact modifiers) into the plastic medium to interrupt the propagation of cracks resulting from a sudden impact. These rubbery modifiers of course have low glass transition values. They render rigid plastics more fracture resistant without reducing rigidity.

The use of rubbery impact modifiers should result in a truly two-phase system. The solubility parameters of the rubbery impact modifier must be such that it will impart only marginal compatibility between the two polymers. The solubility characteristics of the impact modifier and rigid plastic must be different enough to assure a two phase system (a one phase system would result in loss of rigidity). On the other hand, the impact modifier must be similar enough to the rigid plastic to insure sufficient compatibility and adhesion at the rubbery particle interface with the plastic matrix.

For the best performance of impact modifiers, the loading level, particle size, and particle dispersion state must all be optimized. If good clarity is desired, then the refractive index of the impact modifier and plastic medium should be close and the average particle sizes should be quite small. The use of impact modifiers in rigid plastic can also affect color, appearance, heat stability, melt viscosity, chemical resistance, and weathering resistance.

The largest single application of impact modifiers is in rigid PVC. Also impact modifiers are used in polystyrene, polycarbonate, polysulfone, and polyester. Examples of impact modifiers commonly used are given below.

Chlorinated polyethylene

Ethylene vinyl acetate

Methacrylate-butadiene-styrene resin

Acrylonitrile-butadiene-styrene resin

Acrylonitrile-butadiene rubber

Acrylic resins

Polybutylene

NEW DIRECTIONS

As mentioned earlier, the plastics industry in the last 30 years has grown a phenomenal 12% per year because of effective, economical substitutions of easy-to-process plastic for traditional materials such as glass, ceramics, stone, wood, and metal. With the Arab Oil Embargo in 1973, this growth rate was reduced as was the growth rate of the general U.S. economy; however, plastics still commanded a growth rate significantly greater than that of the overall economy. In the 1980s, commodity plastics are expected to grow at a faster rate because of their special properties which allow them to be substituted for metals. On a volume basis, the total energy consumed to make a plastic product requires fewer BTU/in^3 than for traditional metal alloys, even with the alternate fuel value of the plastic feedstock included in the calculation. As the cost of energy rises, an extensive research effort is under way to improve plastic properties further so they can more effectively substitute for metals in more demanding applications. New composites of plastic resins and new high strength fibers are being combined for strength equivalent to metals yet lighter in weight. (Some of these new high strength fibers were discussed previously.) These new composites should display great strength, creep resistance, and dimensional stability. Also new "sandwich" composites of plastic bonded between two sheets of metal are being developed. These types of composites are lighter in weight than the all-metal counterparts, yet retain good strength qualities. These new composites may find use in automobile and aircraft construction where lighter weight means greater fuel economy.

Alloys

While new commercial plastics were developed in the 1950s and 1960s, the number of new plastic polymers that will be developed and marketed in the future will diminish. This reduction is due in part to restrictive environmental regulations regarding the introduction of new commercial materials and the rising cost of new capital equipment needed to produce new polymers. Therefore, in order to meet the future needs for new plastic properties, manufacturers are endeavoring to blend existing commercial polymers through proprietary techniques. These blends are called alloys. These alloys are commonly formed by melt mixing different polymers. An earlier example of this technique can be found in our discussion on blending rubber with rigid plastic for improved impact resistance. Today commercial alloys of the following combinations can be found.

> SAN/Polysulfone
>
> ABS/Polysulfone
>
> PVC/ABS
>
> Nylon/Ethylene-acrylate
>
> PET/Acrylic
>
> PVC/Chlorinated PE/PE
>
> PVC/NBR (rubber)
>
> PVC/EVA
>
> PP/EPR (rubber)
>
> PPO/PS

These different polymers must have enough compatibility in order that they do not phase separate completely either while processing or in the end product. If the two polymers are in solution with one another, then the combination is called a *homogeneous alloy* and displays only a single glass transition. More common, however, are the *heterogeneous alloys* which will consist of continuous and dispersed (discontinuous) phases with their own separate glass transition temperatures. Still, the heterogeneous alloys consist of polymers that are compatible enough to have secondary bonding between the different phases in the form of Van der Waal forces, dipole interaction, and/or hydrogen bonding. From these alloy combinations, improvements in cost, processing temperature, heat resistance, flame retardancy, impact resistance, and toughness can be achieved. Research efforts continue as companies try to find more effective combinations of existing commercial polymers to satisfy market needs without having to develop a new expensive polymer instead.

New Processes

New improvements in the manufacture and processing of plastics are also occurring. The new process for polymerizing linear low density polyethylene has been quite successful. This new process is significantly less costly than the conventional LDPE process (discussed previously). Growth in the manufacture of this new linear LDPE has been great over the first half of the 1980s. Some of the conventional LDPE capacity was displaced by this new process. Also many conventional processes for making other commodity plastics have become more efficient through the use of improved catalysts. In addition, polymerization reactors are being made larger and will be less expensive to operate per pound of output. Lastly, in relation to fabrication, the new reaction injection molding (RIM) techniques are emerging using such polymers as polyurethane and nylon. RIM technology will have a major impact on the plastics industry. RIM can be used to mold large plastic articles efficiently.

Ablative and Heat Resistant Plastics

A great deal of research is underway in the plastics industry to develop ablative and heat resistant plastics for special uses in aerospace applications and other areas. An ablative plastic survives tremendously high temperatures (up to 25,000°F) for short time periods through the sacrificial erosion of its surface caused by pyrolytic degradation from the heat exposure. An obvious application has been to protect reentry space vehicles from high temperatures upon reentry into the Earth's atmosphere.

In an effort to develop a polymer with very high heat resistance, polymer chemists have been synthesizing *ladder* polymers. Ladder polymers have structures consisting of two continuous strands connected together. Usually a ladder polymer structure is composed of a series of uninterrupted rings each connected with the other by two bonding atoms. For example, polyquinoxaline represents a true ladder polymer as shown below.

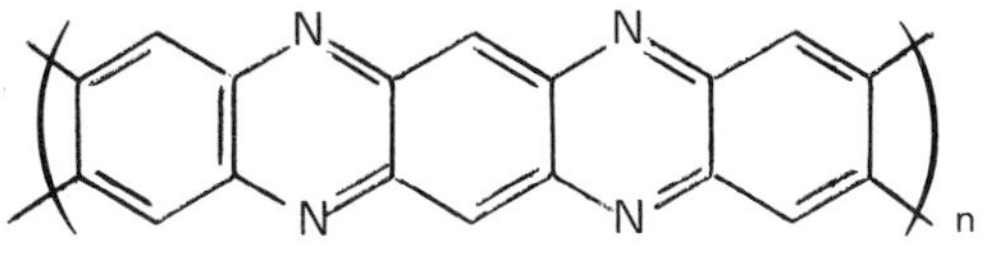

Polyquinoxaline

Semi-ladder polymers are similar to ladder polymers except the repeating ring sequence is interrupted by single bonds in the backbone. Because of this, semi-ladder polymers generally are not quite as stiff or heat resistant as ladder polymers. An example of a semi-ladder polymer is polybenzimidazole.

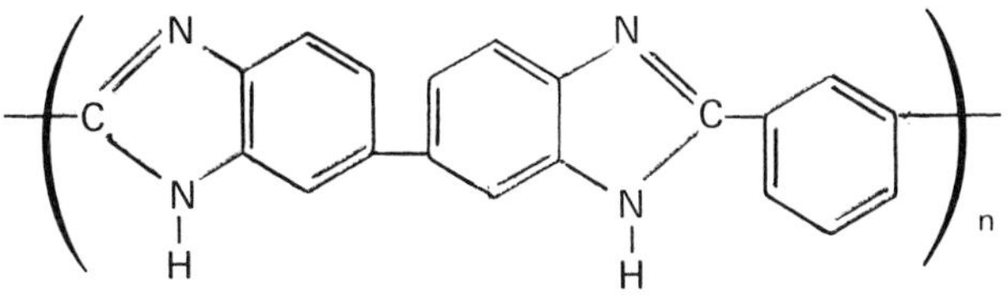

Polybenzimidazole

By having these "fused" rings contained in the chain's backbone, the polymer will be very stiff, extremely resistant to deformations, resistant to very high temperatures (because there are two strands that must be broken), and very hard to process. Because of the difficulty in processing these polymers, many of these ladder polymer structures are completed in situ by a cyclization reaction on a polymer intermediate that could be processed prior to cyclization. This can be illustrated from the cyclization of a polymer intermediate to produce polyimidazopyrrolone.

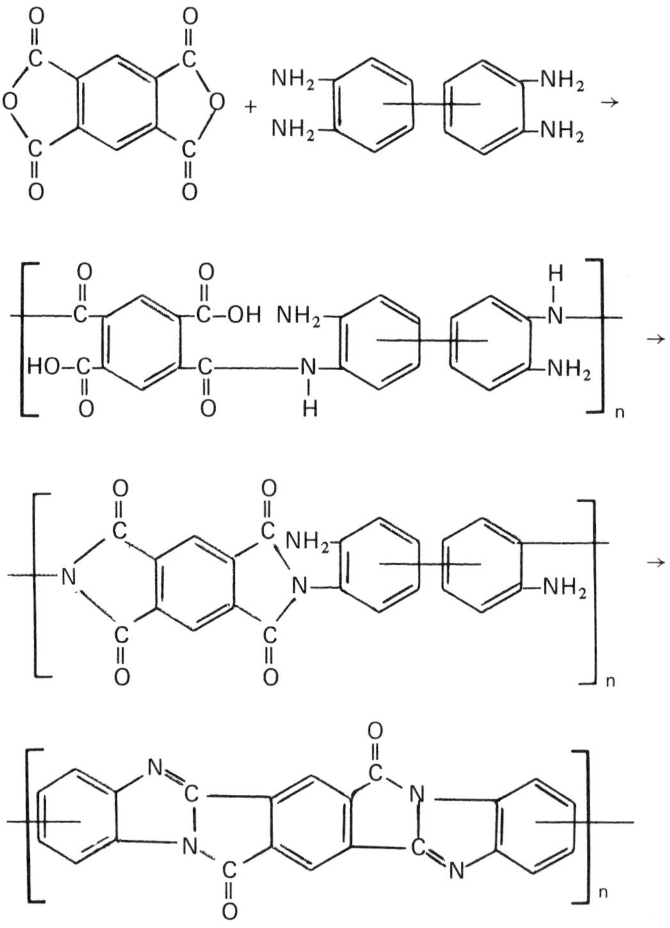

Imidazopyrrolone

Much research has been undertaken in this area. The following are examples of some of the ladder and semi-ladder polymers that have been reported in the literature.

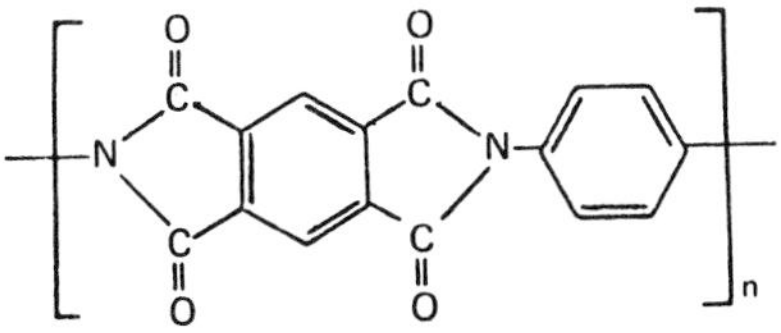

1,4-Phenylenepyromellitimide

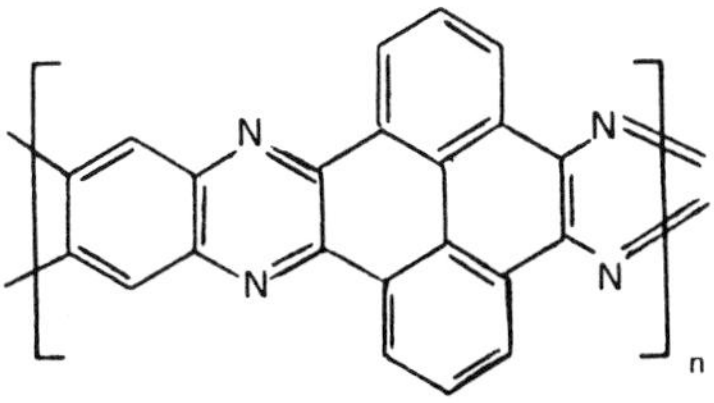

A Polyquinoxaline

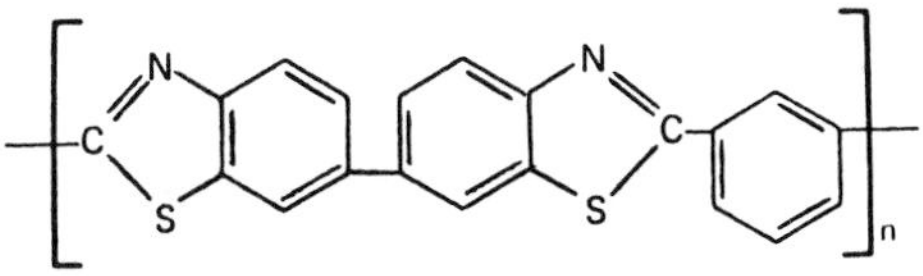

Polybenzothiazole

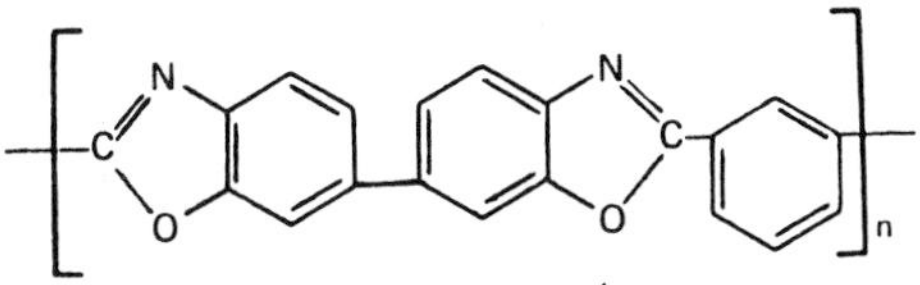

Polybenzoxazole

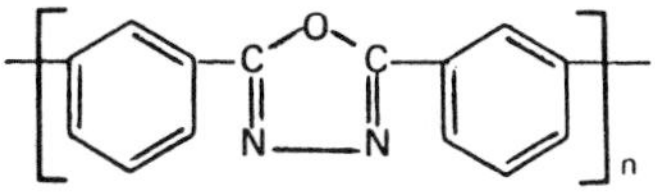

Polyoxadiazole

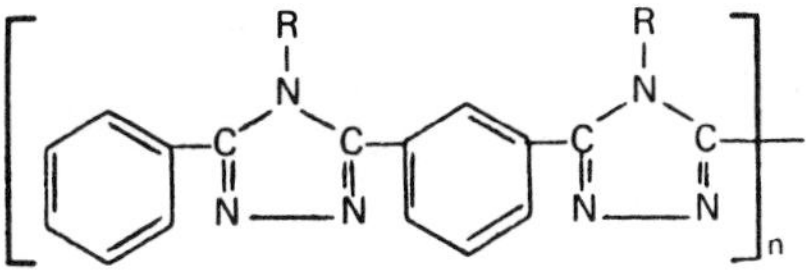

Polytriazole

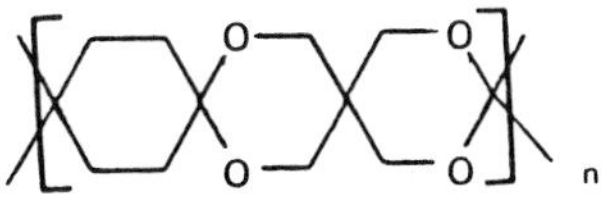

Spiroketal Polymer

These are just a few examples of new heat resistant polymers that have been developed. Extensive research continues in this area.

Electrical Conductance

As discussed earlier, many plastics display excellent electrical insulating properties and have been used for years in insulation of wire and cable. However, recent research in the plastics industry has discovered ways of rendering certain polymers electrically conducting. When such polymers as stretch-orientated polyacetylene or polyphenylene sulfide (PPS) have been "doped" with inorganic salts such as arsenic pentafluoride, they become conductors of electricity. To date, the highest conductivity values are still a long way from being equal to the conducting metals such as copper; however, research in this area is continuing. Potential future application for such polymers may be in making all-plastic batteries or solar cells.

SUGGESTED READING

Baird, Ronald J., *Industrial Plastics*, The Goodheart Willcox Co, Inc., South Holland, Ill., 1976.

Briston, J.H. Gosselin, C.C., *Introduction to Plastics*, Howlyn Publishing Group Ltd., Middlesex, 1968

Crosby, Edward G., Kochis, Stephen N., *Practical Guide to Plastic Applications*, Maple Press Co., N.Y., 1972

D'Alelio, G.F., Parker, John A., *Ablative Plastics*, Marcel Dekker, Inc., N.Y., 1971

DuBois, Harry John, Frederick W., *Plastics*, 6th ed., Reinhold Publishing, N.Y., 1981

Flick, E.W., *Industrial Synthetic Resins Handbook*, Noyes Publications, Park Ridge, N.J., 1985

Flick, E.W., *Plastics Additives—An Industrial Guide*, Noyes Publications, Park Ridge, N.J., 1986

Goodman, S.H., editor, *Handbook of Thermoset Plastics*, Noyes Publications, Park Ridge, N.J., 1986

Grandilli, Peter A., *Technicians Handbook of Plastics*, Von Nostrand Reinhold, N.Y., 1981

Harper, Charles A., *Handbook of Plastics*, McGraw Hill, N.Y., 1975

Katz, H.S., Milewski, J.V., *Handbook of Fillers and Reinforcements for Plastics*, Van Nostrand Reinhold, N.Y., 1978

Kresser, Theodore O.J., *Polyolefin Plastics*, Van Nostrand Reinhold, N.Y., 1969

Kresta, Jiri E., *Polymer Additives*, Plenum Press, N.Y., 1984

Kuz, Minskii, *The Aging and Stabilization of Polymers*, Elsevier Publ., N.Y., 1966

Mianes, Karl, *Plastics in Europe*, Chemical Publishing Co., N.Y., 1964

Milby, Robert V., *Plastics Technology*, McGraw Hill, N.Y., 1973

Modern Plastic Encyclopedia, McGraw Hill, N.Y., annual

Nass, Leonard I., *Encyclopedia of PVC*, Vol. 1,2,3, Marcel Dekker, Inc., N.Y., 1976–77

Patton, William J., *Plastics Technology, Theory, Design, and Manufacture*, Reston Publishing Co., Reston, Va., 1976

Pinner, S.H., *Modern Packaging Films*, Butterworth, London, 1967

Plueddemann, Edwin P., *Silane Coupling Agents*, Plenum Press, N.Y., 1982

Radian Corporation, *Plastics Processing*, Noyes Publications, Park Ridge, N.J., 1986

Radian Corporation, *Polymer Manufacturing*, Noyes Publications, Park Ridge, N.J., 1986

Ritchie, P.D., *Plasticizers, Stabilizers and Fillers*, Iliffe Book, London, 1972

Saechtling, Hansjurgen, *International Plastic Handbook*, Macmillan Publishing Co., N.Y., 1984

Sarvetnick, H.A., *Polyvinyl Chloride*, Plastics Applications Serv., 1977

Seymour, R.B., *Modern Plastics Technology*, Reston Publishing Co., A Prentice-Hall Company, Reston, Va., 1975

Seymour, Raymond, B., *Additives for Plastics*, Academic Press, N.Y., 1978

Simonds, H.R., *Source Book of the New Plastics*, Reinhold Publishing, N.Y., 1961

Titow, W.V., Lankam, B.J., *Reinforced Thermoplastics*, John Wiley & Sons, N.Y., 1975

2

The Rubber Industry

The rubber industry is a mature industry which really emerged as a commercial enterprise after Charles Goodyear discovered how to vulcanize rubber in 1839. From its early applications in shoe soles and bicycle tires, the industry has given birth to a wide variety of products which are given below.

Truck Tires

Farm Tires

Passenger Tires

Off-The-Road Tires

Hose

Conveyor Belts

Molded Mechanical Goods

Matting and Sheeting

Tank Lining

V-Belts

Shoe Heel and Soles

Sponge Rubber

Tires by far represent the largest segment of the rubber industry. In fact, tires alone consume approximately 70 percent of all rubber produced and represents over 12 billion dollars in sales. Next to tires, the combined product areas of conveyor belting and hose represent the second biggest dollar value in sales approaching two billion annually in the U.S.A.

Although there are hundreds of rubber fabricators in the United States, the largest segment of the industry, i.e. the tire industry, consists of only 12 firms. The four largest tire companies represent approximately 85% of domestic tire production. The reason for this high concentration in the tire sector of the industry appears to be attributed to the very large capital investments required to build a modern tire plant large enough in size to achieve "economies of scale" and be competitive. Today, a modern tire plant must be large enough to produce approximately 10,000 tires per day or greater in order to lower the per unit cost to a competitive level. The very high cost of building these plants and the relatively low return on investment has restricted entry of new firms into the tire sector.

Presently the top eight rubber companies are as follows:

Company	1985 Annual Sales (Millions)
Goodyear	9,896
Firestone	3,836
B.F. Goodrich*	3,250
Gen Corp.	3,020
Uniroyal*	2,115
Dayco	905
Armstrong	768
Cooper Tire	522

*Uniroyal and Goodrich tire divisions merged into one company in 1986.

These companies used to be exclusively rubber companies; but today they have all diversified into non-rubber enterprises. For example, rubber products represent only one-third of Gen Corp. (General Tires) total sales.

HISTORY

When natural rubber was first discovered in Central and South America by the European explorers, it was mostly a curiosity rather than of practical use because of its weak strength and tendency to become sticky in the summer and stiff in the winter. These problems were eliminated in 1839 when Charles Goodyear discovered vulcanization. Goodyear found that combining sulfur and white lead with natural rubber and heating it would remove the rubber's sticky qualities and strengthen it greatly, thus rendering it useful commercially. From that time on, the number of products fabricated from rubber expanded rapidly and a whole new industry emerged.

Still many problems such as long curing times and poor aging properties persisted. It was not until 1906 when George Oenslager and A.H. Marks of Diamond Rubber Company discovered that certain amines such as aniline would speed up the vulcanization process and improve aging properties. Then in 1921, it was discoverd at Goodyear Tire Company that thiazole derivatives would speed up the cure rate in vulcanization while not activating the curing process at the lower mixing and extruding temperatures. Also, it was Diamond Rubber in 1912 that discovered the significant improvements in abrasion resistance and other physical properties through the addition of lamp black to a rubber compound.

From these early key discoveries in the first quarter of this century, rubber chemistry has advanced incredibly in many directions including the development of many new synthetic rubbers, new catalysts to make stereospecific elastomers, more effective accelerators (better delayed scorch and cure rate), better antidegradants, development of better reinforcing carbon blacks, new methods of rubber adhesion to fabric and metal surfaces, improved scorch control, and new curing methods, to name a few.

MANUFACTURING PROCESS

Uncured rubber, whether natural or synthetic, behaves as a viscoelastic fluid under forced mixing conditions at elevated temperatures. Under these processing conditions, various rubber chemicals, fillers, and pigments can be added and mixed into the rubber to form what is called an uncured "rubber compound". These compounding ingredients are generally added to the rubber on one of two basic types of mixers, i.e. the two roll mill or the internal mixer.

The two roll mill consists of two heavy metal rolls which are jacketed with steam and water to control temperature. These rolls turn towards each other with a pre-set distance separating them to allow the rubber to pass through this opening with high shear mixing force. The rubber will generally form a "band" around one of the two rolls. Many times, in order to increase shear, the rolls are fixed to turn at different speeds toward each other. Mill mixing is the oldest method of rubber mixing dating back to the very beginning of the rubber industry; however, it is a relatively slow method and its batch sizes are limited. Internal mixers overcome this problem.

Internal mixers were first developed by Fernley H. Banbury in 1916. Even today Banbury mixers are the most commonly used mixers. They consist of two rotor blades turning toward each other in an enclosed metal cavity. The cavity is open to a loading chute through which rubber, fillers, and various chemicals are placed. Upon completion of the mix cycle, the mixed rubber stock is discharged through a door in the bottom of the mixer.

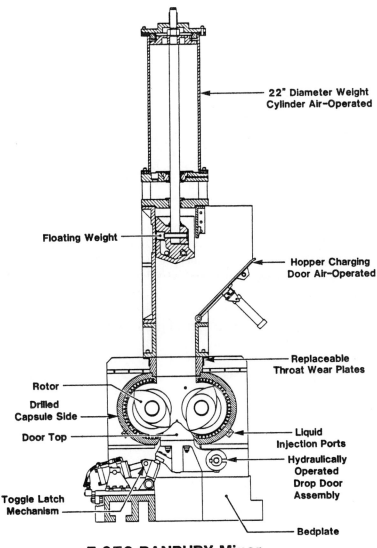

F 270 BANBURY Mixer

Farrel F Series Banbury® Mixer

The Banbury Mixer is an exclusive registered trademark of the
Farrel Corporation.

The Banbury mixing time is determined by the shape and size of the rotors, the rotor speed and horse power of the motor turning them. Over the years, Banbury rotor speeds have been getting faster and batch sizes larger. Today a No. 27 Banbury can handle batches in excess of a thousand pounds and in some cases completely mix a compound in less than 2 minutes. Of course, with so much energy being absorbed by the rubber stock being mixed, the batch temperature can rise well above 300°F before being dumped and cooled.

After the rubber stock is mixed, it generally is later remilled and fed into either a calender or extruder. A three or four roll calender is generally used when applying uncured rubber to fabric as a coat. An extruder is used when the uncured rubber stock is to be shaped into say a tire tread, a belt cover, or a hose tube for example. Generally, after calendering or extrusion, the uncured rubber parts are usually brought together in a building step. For example, the uncured tire components are put together by hand on a rotating tire building drum where the calendered carcass plies are placed over the innerliner; the belt, and sidewalls, over the carcass, etc. In a similar manner, conveyor belts are constructed on large building tables in which each calendered ply is mannually placed over the other followed by the laying of the extruded cover stock.

After the shaping and construction of the uncured rubber product, the "green" product must then be cured. With tires, the uncured constructed assembly is placed in a Bag-O-Matic press where it is shaped in a tire mold and cured under high temperature and pressure. Many other rubber products are also cured under high temperature and pressure. Other products are sometimes cured in autoclaves under steam pressure. Rubber cure temperatures can range from as low as 212°F to over 400°F. The higher the cure temperature, the shorter the cure cycle. Many rubber products are therefore cured at the higher temperature range. In addition, most rubber products are cured under pressure as well to avoid gas formation and porosity.

TESTING

The most basic properties of a cured rubber compound are those obtained from stress-strain measurements, mainly ultimate tensile, ultimate elongation, and modulus. These properties are obtained from measured stress resulting from a given rate of elongation of a cured rubber specimen usually in the shape of a dumbbell of known thickness as shown below.

This cured specimen is placed between the jaws of a Scott or Instron tester. The jaws separate the ends of the dumbbell at a given predetermined rate while the resulting pounds of stress are recorded. The pounds force per given crossectional area of the specimen (measured in megapascals or pounds per square inch) required to separate and rupture (break) the sample is called the *ultimate tensile strength*. The elongation at break is called the *ultimate elongation*, and the force per cross sectional area required to elongate the specimen a given percent elongation (say 300% elongation) is called *Modulus*. These three properties are the basic compound properties obtained from stress-strain testing and relate to the overall strength of the rubber compound.

Another basic test in the rubber industry is the *Mooney* Viscometer. A Mooney Viscometer measures a rubber sample's viscosity by the amount of torque required to turn a rotor at 2 r.p.m. in a heated metal cavity containing a preheated rubber sample.

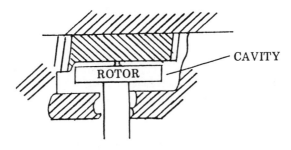

When a Mooney is run on raw rubber, not a rubber compound, the measured viscosity of that rubber after a given time period (usually 4 minutes) is called a *polymer Mooney viscosity* and relates roughly to the elastomer's average molecular weight. On the other hand, a Mooney run on mixed rubber stock is generally run long enough to measure the rise in viscosity due to incipient crosslinking. The time required at a given temperature for the compound to achieve a given torque rise (usually a 5 unit rise) is called *Time to Scorch* (TS). This quality relates to how quickly a rubber compound will scorch or set up from continued mixing or processing.

Today, compund Mooneys have been replaced in many cases by the oscillating disk rheometer, which not only measures scorch time, but also cure rate and state of cure. Unlike the Mooney viscometer, which measures the amount of torque required to rotate a rotor at a given speed, the ODR measures the amount of torque required to oscillate a disc through a predetermined arc (say 1° or 3°) in a heated cavity filled with a sample of the rubber compound. Thus the ODR can accurately measure torque increases of a compound fully cured where a Mooney rotor begins to slip as the compound starts to cure and lose its fluid quality. (See discussion on accelerators for more information).

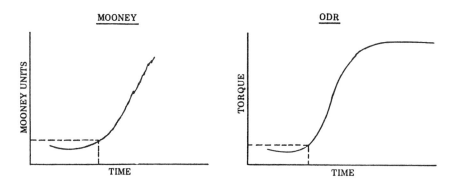

Further discussion on rubber tests is beyond the scope of this book; however, many of the more common tests are listed below. For more details, please make reference to volume 09.01 of the ASTM Standards.

Physical Test	ASTM No.
Goodrich Flexometer (hysteresis)	D 623
Goodyear-Healy Rebound (resilience)	D 1054
Durometer Hardness	D 2240
Tear (Dies A, B, and C)	D 624
Yersley Oscillograph	D 945
DeMattia Flex Test	D 430
Scott Flex Test	D 430A
Brittle Point (low temperature properties)	D 746
Gehman Freeze Test	D 1053
Compression Set	D 395
Pico Abrasion	D 2228
Static Ozone Testing	D 1149
Air Aging	D 573
Test Tube Aging	D 865
Oxygen Aging	D 572

COMPOUNDING

As indicated in the description of the rubber mixing step, a variety of different chemicals and fillers are added to the rubber being mixed. Basically, these additives are to enable the compound to be cured, improve physical properties, reduce cost, ease processing before cure, improve cured aging properties, impart color, or a variety of other functions. The basic groups of these additives that make up a rubber compound are as follows.

(1) Base Elastomer (s)

(2) Curatives (Activators, crosslinking agents and accelerators)

(3) Fillers and Reinforcing Agents

(4) Softeners

(5) Tackifiers

(6) Antioxidants

(7) Colorants

(8) Flame retardants (if needed)

(9) Blowing Agents (if needed)

A typical rubber compound recipe (for say a tire carcass compound) might be as follows.

Components	Parts*
Rubber	100
Carbon Black	35
Processing Oil	8
Antioxidant	1
Zinc Oxide	5
Stearic Acid	1
Accelerator	0.7
Sulfur	2.5

*Parts means "parts per hundred rubber" or PHR.

This typical formulation (called a rubber compound) could be based on natural rubber SBR, BR or say any blend combination of these elastomers (compounds are commonly based on blends of two or more different elastomers to achieve the best balance of properties). Also the parts given for each of the other ingredients are by accepted convention always expressed in "parts per hundred rubber" (PHR). The total parts of rubber, or sum of the parts of two or more rubbers to be blended, is always defined as 100. This way compounders can compare easily the levels of curatives, loadings, etc. between different compounds based on the same relative proportion of rubber hydrocarbon and do not have to recalculate a percentage for every component after adjusting levels of only one or two components. If the rubber being used is oil or black extended (a premade oil or black masterbatch), then the part level for the extended rubber is assigned a value sufficiently above 100 to adjust the calculated rubber hydrocarbon content to equal 100.

Of course, this is only one example of a rubber formulation for a specific application. Even for this one use, there are literally thousands of alternate combinations of components that would still provide acceptable properties for this particular application. Rubber formulations can be compounded to provide a wide variety of properties in their end use. Some of these techniques will be discussed.

ELASTOMERS

Introduction

Before we discuss the chemical additives used with the base rubber, an explanation of the different rubbers used today, (commonly referred to as elastomers) should be given. To convey some idea of the relative production volumes for each of the more common elastomers, the following data is provided below.

Elastomer	U.S. Production Capacity 1984 (Millions of Pounds)
Styrene-Butadiene (SBR)	1,977
Polybutadiene (BR)	675
Isoprene-Isobutylene (IIR)	223*
Ethylene-Propylene (EPR)	465
Polychloroprene (CR)	197*
Acrylonitrile-Butadiene (NBR)	145
Polyisoprene (IR)	60*

*1978 production because 1984 figures not reported by ITC.

Source: International Trade Commission.

Unlike the production of other polymers, the majority of the elastomer producers are also large users of these polymers. In fact, much of the productive capacity for elastomers is owned by tire manufacturers as shown below.

Company	. . % Share of U.S. Industrial Capacity . . .				
	SBR	BR	IR	EPDM	Butyl
Armstrong	—	4.84	—	—	—
Copolymer	6.89	—	—	14.86	—
Firestone	19.31	22.42	—	—	—
General Tire	6.90	9.90	—	—	—
B.F. Goodrich*	9.65	—	—	—	—
Goodyear	23.17	30.34	100	—	—
Uniroyal*	9.65	—	—	18.90	—
Other non-tire companies	24.35	32.35	0	66.24	100

*Goodrich and Uniroyal merged SBR production in 1986.

Unlike other non-rubber polymers which are rigid or non-resilient, elastomeric polymers have much lower glass transition temperatures (Tg) usually between -40° and -120°C. Another distinguishing feature of elastomers as compared to other polymers is that elastomers are far more amorphous in structure containing far less crystallinity than most rigid plastics. Generally there is far less attraction among macromolecules, such as hydrogen bonding or polar attraction.

Another important feature of most elastomers used today (with the exception of silicone rubber, EPM, Hydrin, and Hypalon), is that they generally have unsaturated bonds contained in the backbone of the polymer chain. This provides for active sites (alpha carbons) for crosslinking with conventional sulfur crosslinking agents (to be discussed).

Most of the synthetic elastomers used today are either made from the *emulsion* or *solution* polymerization process. Emulsion polymerization (involving a free radical mechanism in an aqueous emulsion system in which polymerization occurs in soap "micelles") usually produces an elastomer with broader molecular weight distribution which is easier to process (less "nervy" or resistant to thermoplastic behavior on the mill) than solution polymerized elastomers. On the other hand, solution polymerization is a process usually involving a stereospecific type catalyst. This ionic mechanism produces elastomers that are structurally purer and have a much narrower molecular weight distribution. This results in elastomers which provide better physical properties in the cured state, but are harder to process than emulsion polymerized polymers. The following chart shows which of the two processes are commercially used for commonly used elastomers.

	. . . . . .PZN Process.	
Elastomers Type	Emulsion	Solution
SBR	x	x
BR	x	x
CR	x	—
NBR	x	—
IR	—	x
EPDM	—	x

Natural Rubber

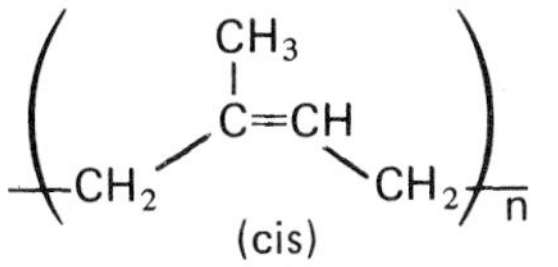

$$\left(-CH_2 \underset{(cis)}{\overset{\overset{\displaystyle CH_3}{\overset{\displaystyle |}{C}}=CH}{}} CH_2 \right)_n$$

This elastomer is made from the coagulation of natural rubber latex obtained from Hevea rubber trees grown mostly in Southeast Asia (85% of the world's rubber supply comes from this region). Before 1930, natural rubber was the only rubber available for commercial use. Today, it represents about one-third of the total elastomer consumption, two-thirds of which is used in tire production. Furthermore, this percent share of usage is not likely to decline significantly in the foreseeable future because of some unique advantages that natural rubber has over other synthetic rubbers. First of all, natural rubber possesses superior building tack over other synthetics. This is the "sticky" quality that is needed to build different rubber products from uncured rubber component parts.

Secondly, natural rubber is one of the very few elastomers that will crystallize when stretched. (Like most elastomers, natural rubber is mostly amorphous in the relaxed state). This property enables natural rubber gum compounds (compounds without carbon black or fillers) to have good tensile strength. Many synthetic elastomers, such as SBR, must have carbon black in the compound to have acceptable tensile strength and modulus.

A third important feature of natural rubber over other elastomers is its superior dynamic properties because of its extremely high resilience in the cured state. By being extremely resilient, less kinetic energy is lost as heat during stress deformation cycles. As a result, there is less heat buildup for products (such as tires) under constant deformation cycles. Thus products made of natural rubber under these severe dynamic conditions are less likely to fail than if made from any one of several other synthetic elastomers available.

Of course, there are important disadvantages to natural rubber use also. For one thing, natural rubber has inherently poor aging properties, whether heat aging, weathering, or ozone attack. On aging, a natural rubber compound softens, and cracks. An effective antidegradant must be used. Also natural rubber has poor resistance to oil attack.

Styrene-Butadiene Rubber (SBR)

$$\left[-CH_2-CH=CH-CH_2-\right]_x \left[-CH_2-CH-\right]_y$$

If any elastomer can be called "general purpose", SBR probably comes as close to that description as any other. It is the largest volume elastomer used in the world today. It is relatively inexpensive compared to other elastomers and is used more widely and in greater quantity than any other elastomer. Much SBR is manufactured from the emulsion process although some is also made in the solution process as well. SBR by itself has poor tensile strength; however when effectively compounded with carbon black and oil, its physical properties are greatly improved. Usually emulsion SBR contains 23% carbon bound styrene, but can range from 20 to 50% styrene. Higher bound styrene content reduces resiliency.

It should be noted that there are basically two types of "emulsion" polymerized SBR. One type is referred to as "hot," while the other is referred to as "cold". "Hot" SBR is derived from free radical emulsion polymerization carried out at 50°C while "cold" is emulsion polymerization carried out at 5°C using a more active free radical initiation system. "Hot" SBR was developed during World War II, but later in 1945, "cold" SBR was introduced as an improvement. "Cold" SBR generally has a higher average molecular weight, lower gel content, and a narrower molecular weight distribution than "hot" SBR. As a result, "cold" SBR gives better abrasion and wear resistance as well as higher tensile and modulus values than "hot" SBR. In some cases, a higher molecular weight can make it more difficult to process. As a result, "cold" SBR is commonly made in an oil extended masterbatch form (oil added to the latex before coagulation) in order to aid later factory processing. Also carbon black is sometimes added as a slurry before coagulation. The International Institute of Synthetic Rubber Producers, Inc. designates SBR by the following nomenclature.

SBR Series	Description
1000	Hot polymer
1100	Hot carbon black masterbatch
1500	Cold polymer
1600	Cold carbon black masterbatch
1700	Cold oil masterbatch
1800	Cold oil-black masterbatch

A third type of SBR used today is produced from the solution polymerization process commonly using a alkyl lithium catalyst. These polymers were developed commercially in the 1960s. They have been used to a more limited extent because of their higher price. Solution SBR, however, does have certain advantages over emulsion SBR. First of all, "solution" SBR has a much higher cis or vinyl content than emulsion SBR. (Emulsion SBR's cis content is generally only between 8 to 15%). Also "solution" SBR has a much narrower molecular weight distribution than either "hot" or "cold" emulsion SBR. Therefore, "solution" SBR can have superior physical properties, but emulsion SBR's process better on the mill. "Solution" SBR's having randomized styrene and butadiene units can be used in tire applications; however, block solution SBR cannot. Block solution SBR is commonly used in injection molding and adhesive applications.

Polybutadiene Rubber (BR)

$$-\!\!-\!\!(CH_2-CH\!=\!CH-CH_2)\!\!-\!\!-$$

This elastomer can be made by either emulsion polymerization or solution polymerization; however most BR today is made by the solution polymerization process using stereo specific catalysts (Ziegler Type) which provides a polymer which is greater than 95% pure cis-1,4-polybutadiene. Although this elastomer does not crystallize upon stretching to give good "gum" tensile strength as with natural rubber, it does have other characteristic advantages not found with other polymers. For one thing, it has a very low glass transition (Tg) value of -100°C (Considerably lower than most other commercial elastomers). This low Tg value means that, thermodynamically, BR will have higher resiliency at ambient temperatures than many other elastomers. This means that rubber products containing BR will generate less heat under continued dynamic compression cycles. This is why BR is commonly used today with other elastomers in tire compounds in order to reduce heat buildup from constant flexing. In addition, BR rubber, because of its low Tg, imparts improved cold temperature flexibility to a rubber compound. Also it has been found that BR displays superior abrasion resistance compared to other elastomers. Therefore, today it is not uncommon to find tire treads containing cis-BR blended with SBR. While SBR may improve traction, BR improves tire tread wear resistance as well as crack resistance.

cis-1,4-Polyisoprene (IR)

This is another solution polymerized elastomer using a Ziegler catalyst (alkyl aluminum). Chemically it has the same macromolecular structure that natural rubber possesses (see structure given for natural rubber previously). However, IR, unlike natural rubber, does not have the same high molecular weight that natural rubber has. Also IR does not contain other natural "impurities" normally found in NR such as proteins and fatty acids. As a result, IR does not have as good "green" (uncured) strength as natural rubber; however IR processes better than NR. Unlike natural rubber, IR does not require as much breakdown in the Banbury before other compounding ingredients can be added. IR is also more uniform, being derived from the petrochemical feedstock (isoprene). On the other hand, IR prices have been rising because of the price rise in petroleum. For this reason, IR has never significantly displaced NR and now only one company (Goodyear) still makes IR in the United States.

Ethylene-Propylene Rubber (EPR)

EPDM (using ethylidene norbornene as the 3rd monomer)

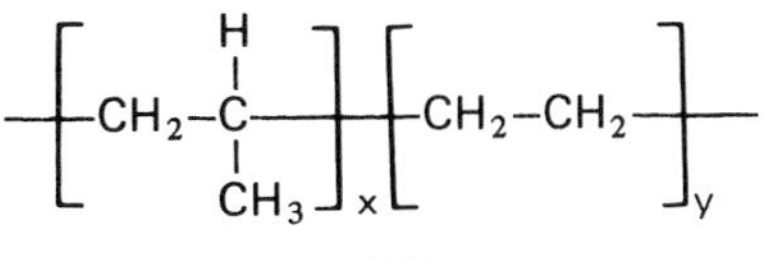

EPM

The EPR name actually denotes both the copolymer of ethylene and propylene, called EPM, and the terpolymer of ethylene, propylene and a third diene monomer, called EPDM. Unlike EPM, EPDM has unsaturated sites which enable it to be vulcanized with conventional sulfur curatives. In contrast, EPM must be peroxide cured. For this reason, EPDM is the more commonly used elastomer of the two. By positioning the unsaturated site on the pendant of the third monomer unit and not in the backbone of the macromolecule itself, EPDM has far superior aging properties over other conventional elastomers. (Unsaturated sites within the molecular backbone of a conventional elastomer are most receptive to oxidative attack resulting in chain scission or crosslinking). EPDM's resistance to aging, even at elevated temperatures as high as 300°F, is the reason it is used to make parts which are exposed to severe aging conditions whether at hot service temperatures or high exposure to ozone (commonly produced from electrostatic discharge). For these reasons, EPDM is used to make automobile parts, steam hose, washer and dryer parts and wire insulation to name a few uses.

It is important to note that not all EPDM elastomers are the same. First of all, the third diene monomers used to make EPDM can be any one of three monomers shown below.

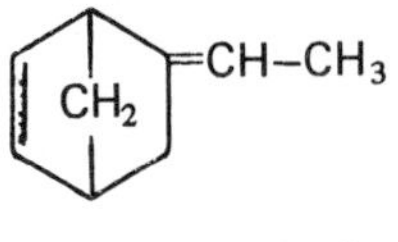

$CH_2=CH-CH_2-CH=CH-CH_3$

1,4-Hexadiene

5 Ethylidene-2-norbornene   Dicyclopentadiene

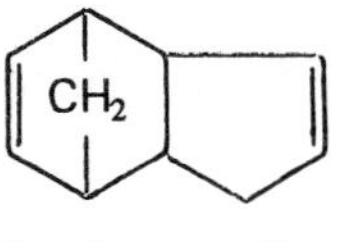

Each of these three dienes has a different degree of reactivity that must be considered. EPDM elastomers also differ in ratios of ethylene to propylene. Generally, the higher the ethylene content, the more thermoplastic behavior the EPDM will display during processing.

In addition to the advantages in using EPDM discussed earlier, another advantage is that this elastomer can be loaded to relatively high levels of inexpensive fillers and still give good rubber properties. This enables EPDM compound costs to be lowered.

Even though EPDM is considered a general purpose elastomer, it has some disadvantages that should be noted. One problem with EPDM is it has very poor oil resistance.

Also, another important disadvantage with EPDM compounds is that their building tack is very poor.

Acrylonitrile Butadiene (NBR)

$$\left[\!\! \begin{array}{c} CH_2-CH \\ \quad | \\ \quad CN \end{array} \!\!\right]_x \left[\!\! \begin{array}{c} CH_2-CH=CH-CH_2 \end{array} \!\!\right]_y$$

NBR

This polymer along with SBR is one of the two emulsion polymerized synthetic rubbers to be developed and used during World War II. NBR's key advantage over other elastomers is that it is very resistant to various organic solvents and petroleum oils, more so than most other elastomers. The higher the acrylonitrile content in NBR, the better this resistance, but correspondingly the lower the resiliency of the rubber and the poorer the cold temperature flexibility. NBR does not have a high resistance to aging; however, when PVC (Polyvinyl chloride) is blended with NBR, it improves NBR's aging properties.

NBR is commonly used as a specialty elastomer in such products as gasoline hose, hydraulic parts, oil gaskets, and linings.

Chloroprene Rubber (CR)

$$\left[\!\! \begin{array}{c} \quad\quad Cl \\ \quad\quad | \\ CH_2-C=CH-CH_2 \end{array} \!\!\right]_n$$

This elastomer was the first commercial synthetic elastomer produced by Dupont (under the trade name of Neoprene) by the emulsion polymerization of 2-chloro 1,3-butadiene. Neoprene has since been widely accepted in many applications because of its unique combination of good age and weather resistance properties and its moderate resistance to oil (although not as good as NBR for oil resistance). Just as with natural rubber, neoprene also crystallizes upon stretching to provide high tensile. Also being halogenated, neoprene is somewhat fire resistant. On the other hand, its relatively high price has prevented its acceptance as a general purpose elastomer. Neoprenes today are commonly used in such applications as V-belts, hose, and automotive parts where a combination of heat and oil resistance is needed.

There are two basic types of neoprene used today. One class of neoprene is designated the G-type by Dupont. These neoprenes use sulfur and tetramethylthiuram disulfide as modifying and stabilizing agents during emulsion polymerization. As a result, G-type neoprene can be cured using only zinc oxide as the curing agent. On the other hand,

Dupont's W- types do not use sulfur and thiuram in the emulsion poly-merization process. As a result, the W types must have an organic accelerator added to the compound before curing to ensure a satisfactory cure rate.

Chlorosulfonated Polyethylene (CSM)

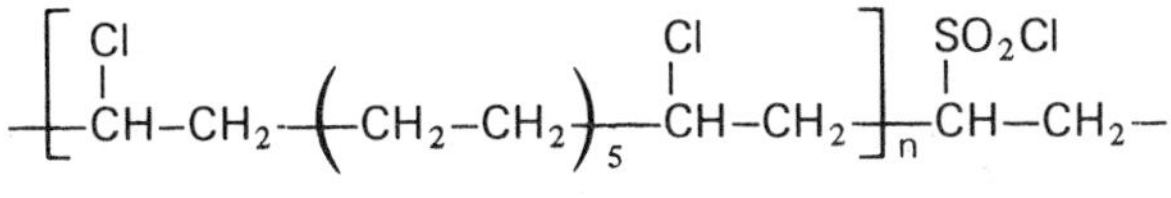

$$\left[\underset{\overset{|}{CH}}{\overset{Cl}{|}} - CH_2 \left(CH_2 - CH_2 \right)_5 \underset{\overset{|}{CH}}{\overset{Cl}{|}} - CH_2 \right]_n \underset{\overset{|}{CH}}{\overset{SO_2Cl}{|}} - CH_2 -$$

CSM

This elastomer is formed from the reaction of chlorine and sulfur dioxide with polyethylene. This polymer is sold under the trade name of Hypalon by Dupont as a specialty elastomer. Although its volume of usage is relatively small, Hypalon possesses the unique feature of supe-rior chemical resistance. Because of this property, Hypalon is used in making acid hose, chemical tank linings, and gaskets. In addition, the chlorine substitution along the polyethylene chain serves to reduce flam-mability (in a similar manner as the chlorine content of neoprene). Ox-idative resistance is also good because there is no unsaturation (Hypalon is cured using metallic oxides).

Butyl Rubber (IIR)

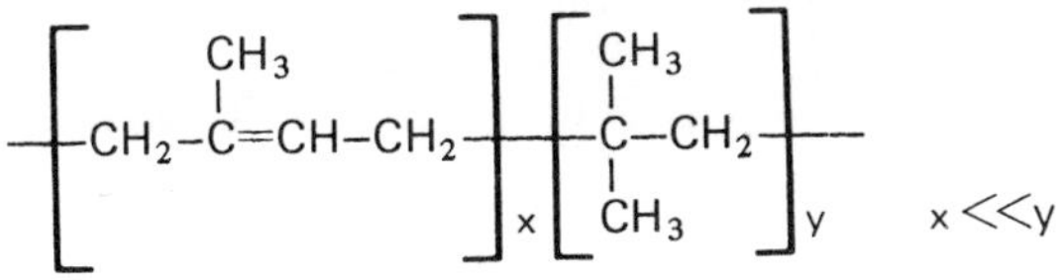

$$\left[CH_2 - \underset{\overset{|}{}}{\overset{CH_3}{C}} = CH - CH_2 \right]_x \left[\underset{\overset{|}{CH_3}}{\overset{CH_3}{C}} - CH_2 \right]_y \quad x \ll y$$

What is commonly called butyl rubber is a copolymer of isobutylene and isoprene. Only a small amount of isoprene (between 1.0 to 4.5%) is present to provide unsaturation for sulfur cures. Butyl is made from the cationic polymerization of isobutylene and isoprene in methyl chloride at -140°F.

This elastomer has very poor resiliency when compared to other elastomers. Its one important "claim to fame" is its extremely good resistance to air permeability which explains its main application in making tire inner tubes. Also, because of its low unsaturation content, butyl is sometimes used in electrical wiring due to its favorable resistance to ozone attack. Disadvantages of butyl rubber are its poor compatibility with other elastomers and its poor oil resistance.

Chlorobutyl Rubber (CIIR)

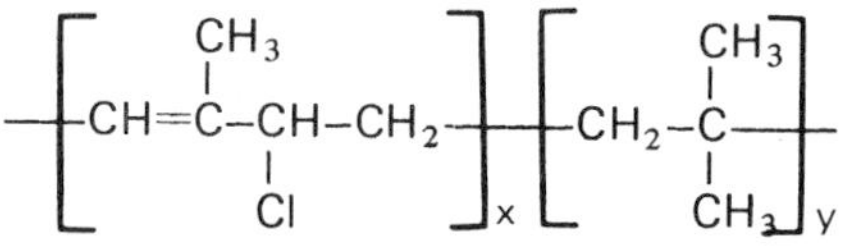

$$\left[\begin{array}{c} CH_3 \\ | \\ -CH=C-CH-CH_2- \\ | \\ Cl \end{array} \right]_x \left[\begin{array}{c} CH_3 \\ | \\ -CH_2-C- \\ | \\ CH_3 \end{array} \right]_y$$

Chlorobutyl Rubber

This elastomer is formed from the controlled chlorination of butyl rubber. The chlorine introduced onto the butyl backbone is usually about 1.2% by weight. The chlorine participates in an electrophilic substitution reaction resulting in the formation of an allylic chloride adjacent to the double bond.

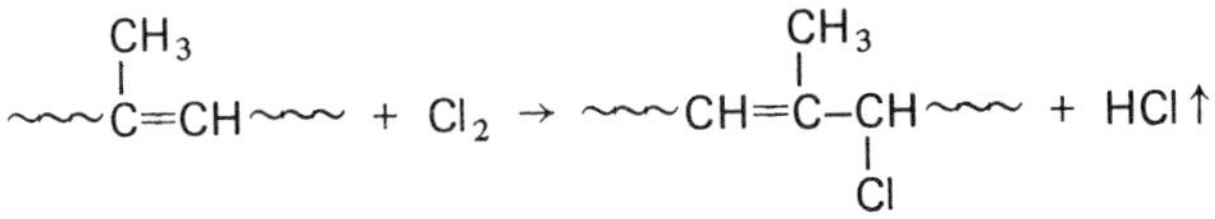

$$\underset{\sim\sim}{C}=CH\sim\sim + Cl_2 \rightarrow \sim\sim CH=\underset{|}{\overset{CH_3}{C}}-CH\sim\sim + HCl\uparrow$$

The allylic chloride greatly enhances the "curability" of the polymer, enabling chlorobutyl to be covulcanized with other conventional elastomers of higher unsaturation. Thus the chlorobutyl is more compatible with other elastomers than straight butyl. Therefore, chlorobutyl is commonly used in the innerliners for tubeless tires to help prevent air diffusions. Another halogenated butyl, called bromobutyl (BIIR), is also available for use in basically the same application as chlorobutyl.

Epichlorohydrin Rubber (CO and ECO)

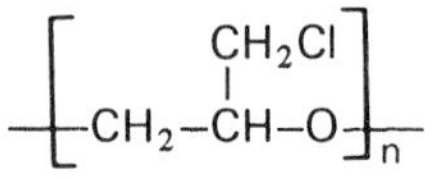

$$\left[\begin{array}{c} CH_2Cl \\ | \\ -CH_2-CH-O- \end{array} \right]_n$$

CO (homopolymer)

$$\left[-CH_2-CH_2-O- \right]_x \left[\begin{array}{c} CH_2Cl \\ | \\ -CH_2-CH-O- \end{array} \right]_y$$

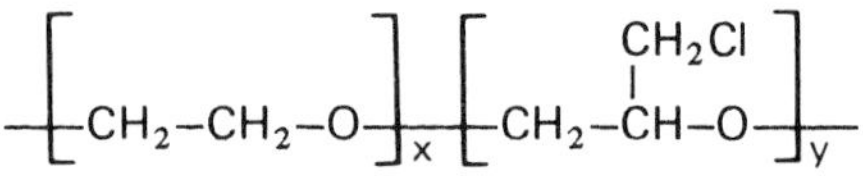

ECO (copolymer)

The term epichlorohydrin rubber (sold under the trade name of Hydrin® by B.F. Goodrich) refers to the homopolymer from the epichlorohydrin monomer and the copolymer of epichlorohydrin and ethylene oxide monomers.

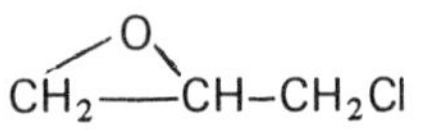

$$CH_2\overset{O}{\overbrace{\quad\quad}}CH-CH_2Cl$$

Epichlorohydrin
(monomer)

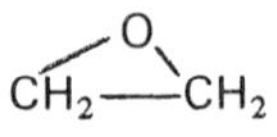

$$CH_2\overset{O}{\overbrace{\quad\quad}}CH_2$$

Ethylene Oxide
(monomer)

These elastomers are known for their resistance to solvents, oil, fire, and ozone attack. Also these polymers are known for their outstanding resistance to air permeability (twice as resistant as butyl rubber). For these reasons CO and ECO are used for gas tank liners and refrigerator hose. Also they are used for making gaskets, seals, gas regulators, and pump valve components. The copolymer containing ethylene oxide, has better low temperature properties than the homopolymer, however the copolymer has less resistance to fire, oil, and air permeability.

Basically, epichlorohydrin rubber costs more than general purpose elastomers; however, it is not nearly as expensive as silicone and fluorocarbon rubber (to be discussed). Still, this elastomer remains relatively low in production volume.

Fluorocarbon Rubber

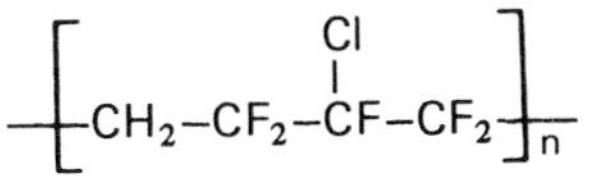

$$\left[CH_2-CF_2-\overset{\displaystyle Cl}{\underset{\displaystyle |}{C}F}-CF_2 \right]_n$$

These elastomers are sold under the trade names of Viton (by Dupont) and Fluorel (by 3M). Although there are several versions of this elastomer, one that is commonly used is the copolymer of chlorotrifluoroethylene ($ClCF = CF_2$) and vinylidene fluoride ($CF_2 = CH_2$). These elastomers are among the most expensive commercially available elastomers and are used in specialty applications where their superior resistance to attack by strong acids and alkali, various solvents, and ozone is needed. Also these polymers possess very good thermal stability up to 450°F and are fire resistant. These superior properties are largely due to the high fluorine content. Fluorocarbon rubbers are generally vulcanized through nonsulfur cure systems using peroxides, metal oxides or diamines. These polymers are used only to a very limited extent because of their very high cost, but they have found use in the aerospace industry as well as in the making of rolls, O-rings, and hose.

Polysulfide Rubber

$$\left[CH_2-CH_2-O-CH_2-O-CH_2-CH_2-S \frac{}{x} \right]_x \left[CH_2-CH_2-S \frac{}{x} \right]_y$$

This elastomer is unique in that it contains sulfur in its backbone. The structure of the polysulfide shown above is formed from the substitution reaction between sodium polysulfide and dichloroethyl-formal. The sulfur in the backbone of the resulting polymer enables this elastomer to display excellent cold temperature flexibility as well as excellent resistance to various solvents and ozone attack. Unfortunately, these polymers have low resilience as well as an unpleasant odor. Although polysulfide consumption is relatively low compared to other elastomers, it is used extensively in limited applications, such as hose lining for solvents or paints, gas meter diaphragms (because of its excellent combination of low temperature flexibility and oil resistance) and sealants.

Silicone Rubber

$$\left[\begin{array}{c} CH_3 \\ | \\ -Si-O- \\ | \\ CH_3 \end{array} \right]_n$$

This rubber is very expensive but has a unique advantage over other elastomers in that it can remain in service over the broadest temperature range of from -150°F to +600°F. This is because its polymeric backbone is made up of silicon and oxygen bonds which provide greater flexibility over a broader temperature range. Also silicone rubber is resistant to oxidative and water attack. Silicone rubbers are generally cured by the use of peroxides through the establishment of carbon-carbon crosslinks between methyl groups. The tensile strength of silicone rubber is relatively low and this elastomer is not compatible with other carbon based elastomers. Generally, *fumed silica* instead of carbon black is used as a reinforcing agent. Silicones are commonly used in Aerospace applications. Special *fluorinated silicone rubber* is also available which provides even a greater range of temperature use as well as, improved resistance to solvent attack.

CURATIVES

In order to render a rubber usable in a practical sense, it must be "vulcanized" which means the rubber molecules must be crosslinked. Rubber molecules are very long, flexible and unordered, one might say having the appearance analogous to spaghetti at the micro level. Rubber molecular chains are for the most part completely random and without crystallinity, i.e. amorphous in structure. Also, these string like molecules are entangled or coiled with one another so that when the rubber is stretched, these molecules want to return back to their original position when the force is relaxed, provided the period of time for extension is relatively short. When uncured rubber is in a stretched state for any prolonged period of time, the entangled molecular chains begin to slip past each other and untangle, thus reducing stress. So when the stretching force is relaxed, the uncured rubber can not return completely back to its original dimensions. The ability of the rubber to recover from the deformation is called the *elastic* quality of the rubber. The failure of the uncured rubber to return to its original shape is the *plastic* or viscous quality of rubber. The longer the time of deformation of the uncured rubber, the greater this viscous quality and the smaller the elastic quality we see. Some refer to this phenomenon as "cold flow" of uncured rubber.

To enhance the elastic quality and drastically reduce the "flow" quality of rubber, the rubber molecular chains are crosslinked to one another through the vulcanization process. This crosslinking prevents the rubber molecules from sliding past one another and disentangling or uncoiling even after long periods of stretching. Thus the vulcanized rubber returns back to its original shape when the stress is removed. Indeed, through establishing crosslinks among the rubber molecules, a macromolecular three dimensional network or matrix is established throughout the mass. The more crosslinks (the higher the crosslink density), the tighter the network and the higher the resulting modulus and hardness.

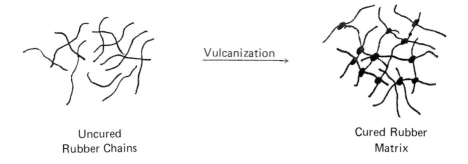

Vulcanization →

Uncured
Rubber Chains

Cured Rubber
Matrix

The crosslinking agent that is most commonly used throughout the rubber industry is sulfur. Indeed this is the crosslinking agent that Charles Goodyear discovered. Today the use of sulfur to vulcanize rubber in a reasonable time period requires the use of activators, as well as, accelerators which must be mixed into the rubber compound along with the sulfur.

Activators

The two activators used together in most all rubber compounds are rubber grade *stearic acid* and *zinc oxide*. The stearic acid is present mainly to react with zinc oxide and solubilize it (make the zinc ion available for the vulcanization process through the formation of zinc stearate). The zinc ions are necessary to activate the organic accelerators. Stearic acid is not the only fatty acid that will work. Other fatty acids, such as, lauric acid can also be used. Also zinc oxide is not the only metallic compound that will activate accelerators. Lead and cadmium compounds will also work. However, zinc oxide and stearic acid are by far the most common combination of activators used by the rubber industry today. Also, lead and cadmium compounds are toxic and pose a health hazard to workers. Therefore, the use of these toxic activators is discouraged.

Accelerators

Before the advent of organic rubber accelerators it used to require 3 hours to cure rubber with zinc oxide alone. With the discovery of the first organic rubber accelerators based on aniline at the turn of the century, cure time was reduced to one-half hour. With accelerators used today, cure times of only a few minutes are common.

Accelerators, which are almost always used with zinc oxide and stearic acid, are activated by the zinc ion present which complexes with the accelerators. Through a complicated mechanism which no one fully understands, the activated accelerator species are able to break up the eight membered rings of the elemental sulfur present and establish mono, di, and poly sulfidic crosslinks between alpha carbons of the unsaturated chains.

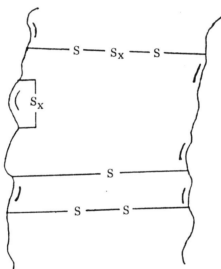

SULFUR CROSSLINKING BETWEEN
RUBBER MOLECULAR CHAINS

Accelerators have been improved over the years because they can give not only fast cure rates but also good delayed scorch. To understand this we must explain the concept of scorch. The term time-to-scorch for a rubber compound is the time required at a given cure temperature for incipient crosslinking to occur among the rubber molecules. After this scorch condition is reached, the rubber compound can no longer be processed well on a mill or calender and cannot be smoothly extruded. In other words, the rubber has started to lose its thermoplastic quality. As time progresses beyond this "scorch time" during cure, the rubber will become tougher (modulus increases) at a given rate. This rate of modulus rise is called the cure rate. Finally, the ultimate state of cure is reached in which rubber compound modulus and tensile reach their peaks.

These changes in modulus just discussed during the curing of a rubber compound can be measured by an instrument called the Oscillating Disk Rheometer (ODR) (Discussed earlier). The ODR chart shown below will illustrate the properties of time-to-scorch, cure rate, and state of cure.

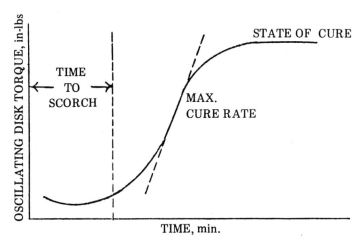

The point to be made from this illustration is that different rubber accelerators give different cure properties. The various accelerators used today all give different scorch times, cure rates, and states of cure.

The following are some of the accelerators commonly used by the rubber industry today.

Guanidines.

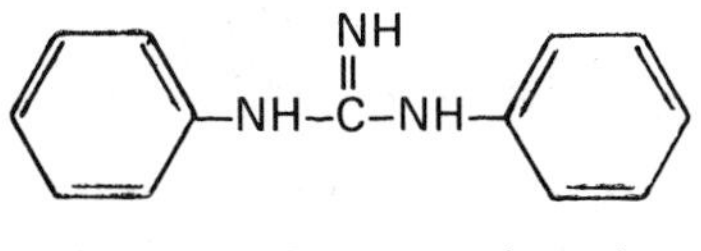

Diphenyl Guanidine (DPG)

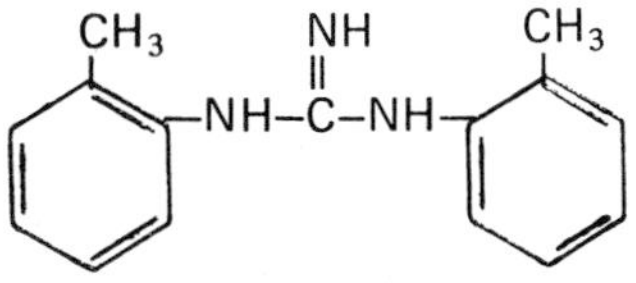

Di-o-tolylguanidine (DOTG)

Thiurams.

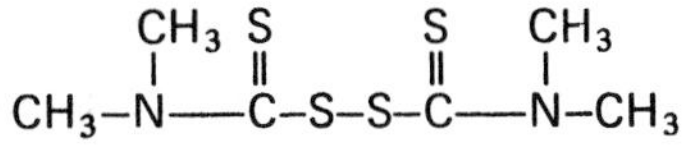

$CH_3-N \overset{\underset{|}{CH_3}}{\underset{}{}} \overset{\underset{\|}{S}}{C}-S-S-\overset{\underset{\|}{S}}{C} \overset{\underset{|}{CH_3}}{\underset{}{}} N-CH_3$

Tetramethyl Thiuram Disulfide
(TMTD)

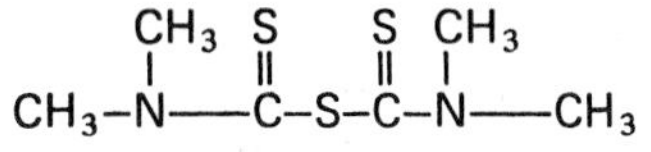

$CH_3-N \overset{\underset{|}{CH_3}}{\underset{}{}} \overset{\underset{\|}{S}}{C}-S-\overset{\underset{\|}{S}}{C} \overset{\underset{|}{CH_3}}{\underset{}{}} N-CH_3$

Tetramethyl Thiuram Monosulfide
(TMTM)

Thiazoles.

2-Mercaptobenzothiazole
(MBT)

Mercaptobenzothiazole Disulfide
(MBTS)

Sulfenamides.

N-Cyclohexyl-2-Benzothiazyl
Sulfenamide (CBTS)

tert-Butyl Benzothiazole
Sulfenamide (BBTS)

2-(Morpholinothio) Benzothiazole
Sulfenamide (MBS)

Dithiocarbamate Salts.

Zinc Dimethyldithio
Carbamate (ZMDC)

Zinc Dibutyldithio
Carbamate (ZBDC)

DPG was discovered to be a useful accelerator in 1921. Since then guanidines such as DPG and DOTG have been used extensively by the rubber industry; however that usage has declined greatly over the years because guanidines and early amine accelerators gave short scorch times (poor scorch protection). Today guanidine accelerators are mainly used as secondary accelerators (used to activate another primary accelerator) or in special compound applications where stock is to be air cured.

Instead, another accelerator called 2-mercaptobenzothiazole (MBT), also discovered in 1921, was found to have superior scorch protection (longer scorch time) while still imparting a fast cure rate. Later, mercaptobenzothiazole disulfide (MBTS) was found to be even better for delayed scorch. These two thiazole accelerators are widely used as primary accelerators; however they are not the largest volume accelerators used today. Instead, sulfenamide accelerators, which are amine derivatives of thiazoles, are today the more widely used accelerators throughout the rubber industry because they impart the best delayed scorch properties to allow the compound a longer time to be mixed, calendered, or extruded while still giving a very fast cure rate at vulcanization temperatures. CBTS was initially used; however BBTS and MBS are now the more commonly used accelerators in the rubber industry. BBTS is used when a faster cure rate is important with some delayed action on scorch while MBS is popular because it imparts the longer delayed scorch, yet provides a high state of cure after vulcanization. These sulfenamide accelerators are in a sense "blocked" forms of MBT. When mixed in a rubber compound, their scorch delay is caused by the time required for these accelerators to thermally decompose into MBT and their respective amines. The MBT intermediate then enters into the vulcanization mechanism as discussed earlier. Thiazoles and sulfenamide accelerators represent over 80% of the total of 135 million pounds of rubber accelerators produced in the United States annually.

While sulfenamide accelerators are the most common *primary* accelerators used today, they are many times used with *secondary* accelerators (called "kickers") which help activate the cure system. The thiuram and dithiocarbamate accelerators shown above are examples of accelerators which are used for this purpose. They are commonly used at relatively small levels with a primary accelerator to help activate the cure. Also they can be used alone in a rubber compound as the primary accelerator, although these compounds are somewhat scorchy.

Thioureas. Before leaving this discussion on rubber accelerators, one last group, the thioureas, should be mentioned.

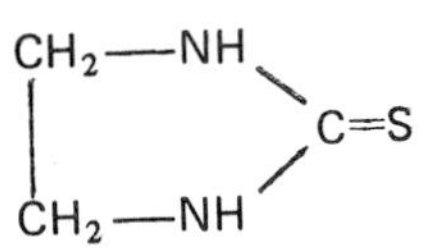

Ethylene Thiourea (ETU)

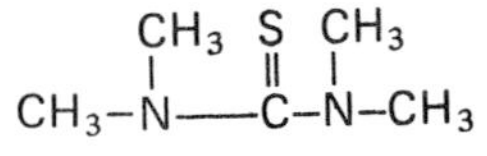

Tetramethyl Thiourea (TMTU)

Thioureas are used almost entirely as accelerators for neoprene. They are used with W type neoprene in particular because these types require such an accelerator to give the cured properties desired. As a result, ETU and, to a lesser extent, TMTU, are used in the industry as neoprene accelerators. Both accelerators give good scorch safety, fast cure rate, and high state of cure. TMTU gives more delayed scorch than ETU. There is a trend to shift away from using these thiourea accelerators in favor of other accelerators because they have been found to be carcinogenic. When ETU or TMTU are used today, they are usually used in an encapsulated elastomeric masterbatch form in order to minimize worker exposure.

Sulfur Donors

One thiuram accelerator, TMTD, not only is used as a primary and secondary accelerator in many different rubber compounds, but also it functions as a sulfur donor as well, supplying sulfur for crosslinking. Another chemical commonly used almost exclusively as a sulfur donor is dithiodimorpholine (DTDM).

Dithiodimorpholine

Although DTDM and other "sulfur donors" are far more expensive than elemental sulfur, they are used with other curatives in what are called "efficient vulcanization" (EV) cures in which a higher number of mono and disulfidic crosslinks are established with proportionately fewer polysulfidic crosslinks formed. This means that vulcanizates obtained with EV cures will display improved thermal stability as well as better aging resistance (mono and di-sulfidic crosslinks are more stable than polysulfidic crosslinks). On the other hand, conventional cures with a higher proportion of polysulfide crosslinks may provide better dynamic fatigue resistance.

Retarders and Inhibitors

Many times in rubber compounding, not only will a primary and secondary accelerator be used in a compound, but also a retarder or inhibitor will also be used. For years scorch retarders such as *phthalic anhydride, salicylic acid* or *benzoic acid* were used and still are used today. However, not only do these chemicals retard the scorch time, (providing more scorch safety), but they also slow the cure rate and reduce the ultimate state of cure.

One new chemical marketed by Monsanto in the early seventies, N-(cyclohexylthio)-phthalimide, not only delayed scorch, but did not significantly reduce the cure rate or the ultimate state of cure for sulfenamide cured vulcanizates. This chemical is called a pre-vulcanization-inhibitor or PVI by Monsanto.

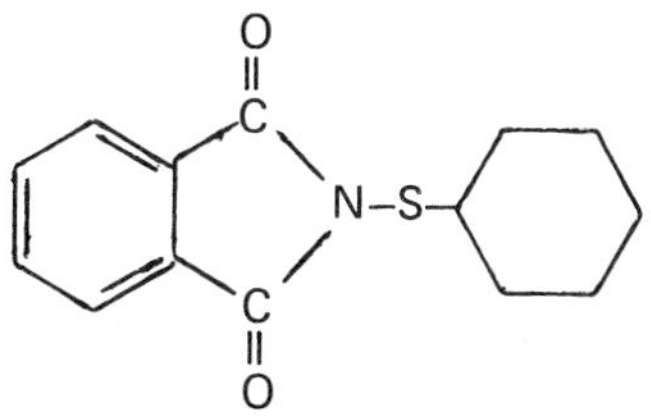

PVI

PVI works most effectively with sulfenamide accelerators. During cure, accelerators such as CBTS, BBTS, or MBS dissociate into their respective amines and MBT which, you may recall, is an effective thiazole acclerator. The process of the dissociation of the sulfenamide partially explains the improved delayed scorch of these accelerators compared to MBT and MBTS (thiazoles). PVI lengthens the delayed scorch of sulfenamides even further by reacting with the MBT rate determining intermediate to form cyclohexyldithiobenzothiazole (CDB). CDB is a good delayed action accelerator in its own right and dissociates to accelerate the cure, but with even more delay in scorch.

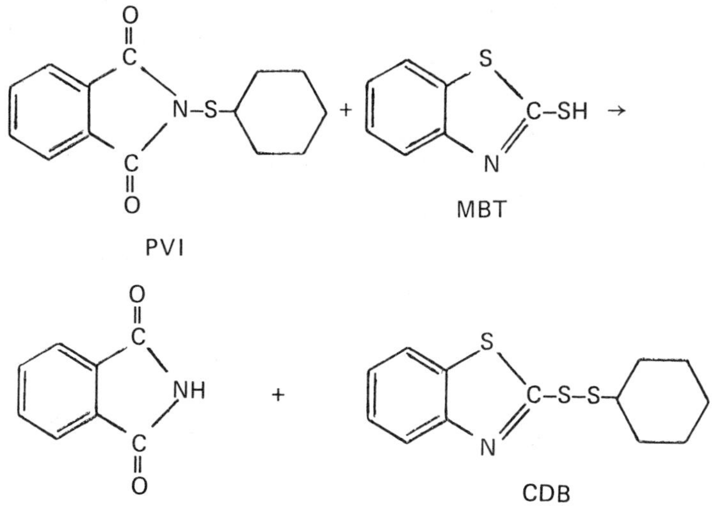

PVI   MBT

Phthalimide   CDB

Another chemical called N-nitrosodiphenylamine (NDPA) is also used as a retarder with sulfenamide accelerators. However, NDPA is not as effective or as expensive as PVI. Also, because of toxicity and environmental health risks (potential for forming carcinogenic nitrosamines), NDPA is declining in usage.

Non Sulfur Curing Agents

Although sulfur is the accepted crosslinking agent used in most all rubber compounds today, there are small limited uses for non-sulfur cures. One of the most noted non-sulfur cure systems involves the utilization of peroxides.

Shown below is the most commonly used organic peroxide, for curing rubber. It is usually available as a 40% active, white powder extended with a mineral filler.

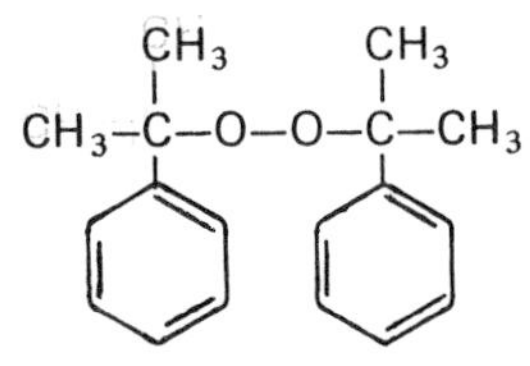

$$CH_3-\underset{\underset{\text{C}_6H_5}{|}}{\overset{\overset{CH_3}{|}}{C}}-O-O-\underset{\underset{\text{C}_6H_5}{|}}{\overset{\overset{CH_3}{|}}{C}}-CH_3$$

Dicumyl Peroxide

Organic peroxides can be highly unstable and even explosive in some cases. Some of them are extremely sensitive to shock, friction, heat, and foreign contamination, which can cause them to explode or ignite. Also, precautions must be taken in handling peroxides to avoid any worker contact in that exposure can cause severe throat irritation, lung damage, eye damage, skin burns, etc. Therefore, care should be exercised in using these chemicals.

The thermal instability of these organic peroxides and the resulting formation of free radicals partially explains how they can be useful in curing rubber.

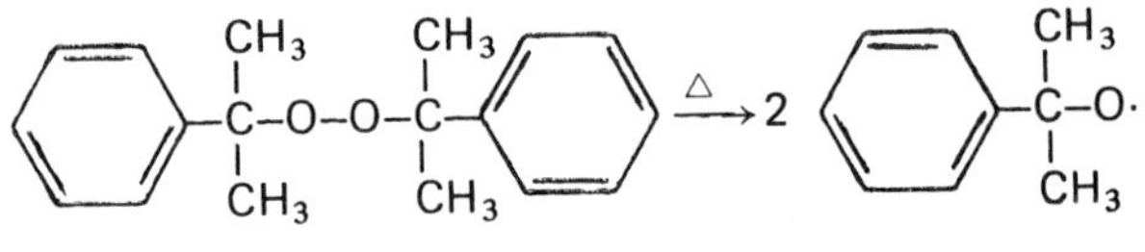

These free radicals in turn react with allylic hydrogens on the rubber chains forming polymeric free radicals which then react to form a carbon-carbon crosslink.

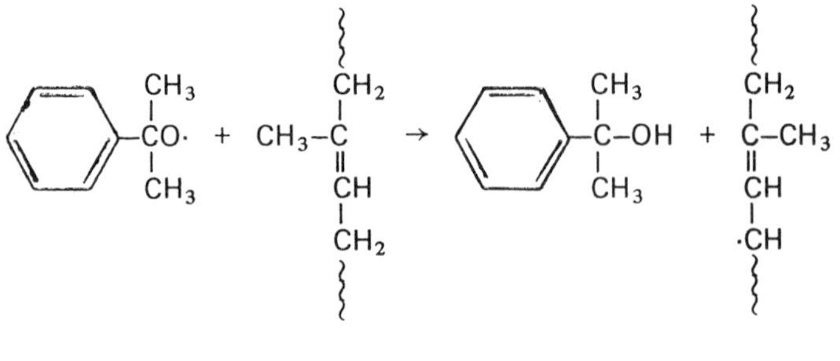

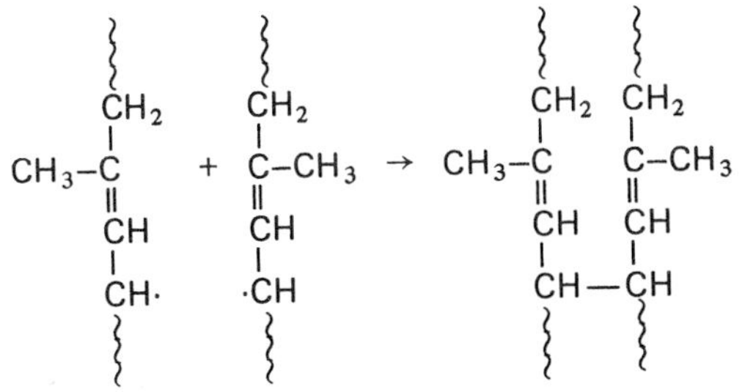

$$\begin{array}{ccccc}
\overset{\displaystyle\mid}{CH_2} & & \overset{\displaystyle\mid}{CH_2} & & \overset{\displaystyle\mid}{CH_2}\quad \overset{\displaystyle\mid}{CH_2} \\
\mid & & \mid & & \mid\qquad\mid \\
CH_3{-}C & + & C{-}CH_3 & \rightarrow & CH_3{-}C\qquad C{-}CH_3 \\
\parallel & & \parallel & & \parallel\qquad\parallel \\
CH & & CH & & CH\quad CH \\
\mid & & \mid & & \mid\qquad\mid \\
CH\cdot & & \cdot CH & & CH{-}CH \\
\end{array}$$

Unsaturated elastomers such as natural rubber, SBR, BR and NBR can be peroxide cured through the mechanism shown above.

In addition, not only can unsaturated elastomers, but more importantly, saturated elastomers can also be cured through the use of organic peroxide as shown below.

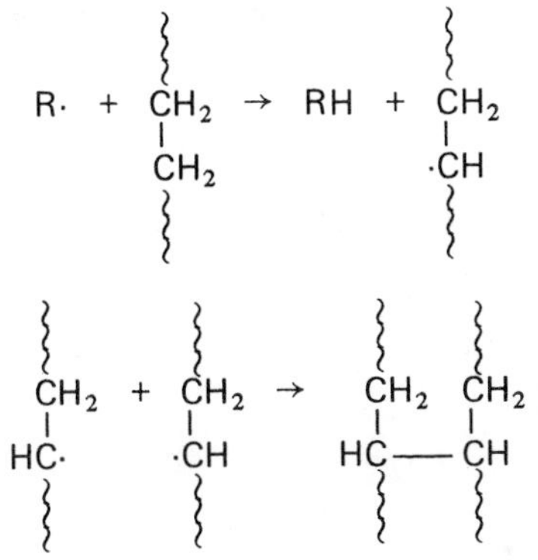

$$R\cdot \;+\; \begin{array}{c}CH_2\\\mid\\CH_2\end{array} \;\rightarrow\; RH \;+\; \begin{array}{c}CH_2\\\mid\\\cdot CH\end{array}$$

$$\begin{array}{c}CH_2\\\mid\\HC\cdot\end{array} \;+\; \begin{array}{c}CH_2\\\mid\\\cdot CH\end{array} \;\rightarrow\; \begin{array}{cc}CH_2 & CH_2\\\mid & \mid\\HC & {-}\!{-}\; CH\end{array}$$

Saturated polymers such as EPM, silicone rubber, and even polyethylene can be crosslinked through this mechanism. One important exception is butyl rubber which can not be cured using peroxides because they cause chain scission and depolymerization to occur.

Dioxime cures are another method of vulcanizing elastomers without the aid of sulfur. The crosslinking agent to be discussed here is *para-quinone dioxime*. Although this chemical can be used to cure olefin rubber, it is primarily used to cure butyl rubber. Generally p-quinone dioxime is used with an oxidizing agent such as MBTS to cure rubber in the following manner.

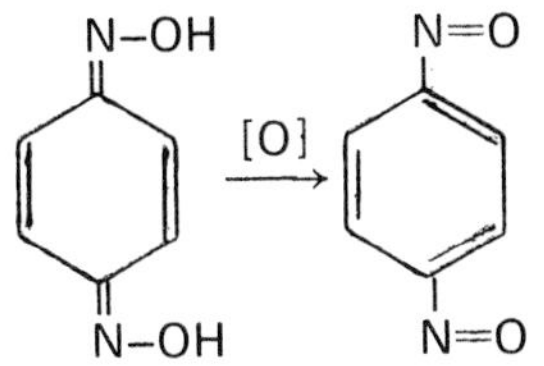

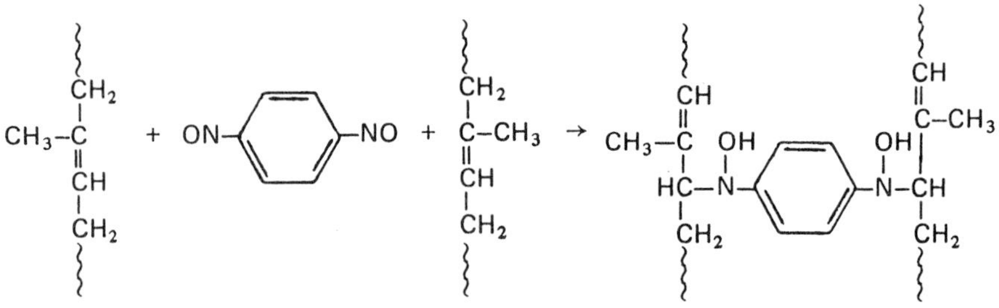

$$CH_3-\underset{\underset{CH_2}{\overset{CH_2}{|}}}{\overset{CH_2}{\underset{|}{C}}} + ON-\langle \rangle-NO + \underset{\underset{CH_2}{\overset{CH_2}{|}}}{\overset{CH_2}{\underset{|}{C}}}-CH_3 \rightarrow CH_3-\underset{\underset{CH_2}{\overset{CH}{|}}}{\overset{CH}{\underset{|}{C}}} \begin{matrix} OH \\ HC-N \end{matrix} \langle \rangle \begin{matrix} OH \\ N-CH \end{matrix} \underset{CH_2}{\overset{C-CH_3}{\underset{|}{\overset{CH}{|}}}}$$

In addition, other non-sulfur curatives such as methylol terminated phenol formaldehyde resin with stannous chloride can also be used to cure butyl rubber.

Lastly, metallic oxide cures such as zinc oxide or lead oxide (highly toxic) can act as a crosslinking agent for halogenated elastomers such as neoprene or hypalon. In the situation illustrated below, zinc oxide reacts with the small percentage of allylic chlorine groups of neoprene to form an oxygen crosslink between the two active sites.

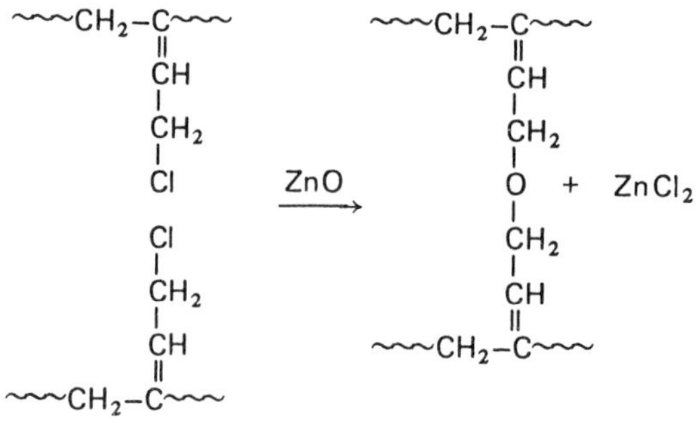

$$\begin{matrix} \sim\sim CH_2-C\sim\sim \\ \| \\ CH \\ | \\ CH_2 \\ | \\ Cl \\ \\ Cl \\ | \\ CH_2 \\ | \\ CH \\ \| \\ \sim\sim CH_2-C\sim\sim \end{matrix} \quad \xrightarrow{ZnO} \quad \begin{matrix} \sim\sim CH_2-C\sim\sim \\ \| \\ CH \\ | \\ CH_2 \\ | \\ O \\ | \\ CH_2 \\ | \\ CH \\ \| \\ \sim\sim CH_2-C\sim\sim \end{matrix} + \ ZnCl_2$$

FILLERS AND REINFORCING AGENTS

Essentially the "loadings" used in the rubber industry are dry filler or reinforcing agents such as carbon black, clays, silicas, and calcium carbonates. Also softeners such as petroleum processing oils and synthetic ester plasticizers are used. By far the most commonly used loadings are a combination of carbon black and petroleum oil to extend a rubber compound.

Unlike the use of fillers in plastics which are many times used strictly to reduce compound cost, fillers in rubber are not only used for this purpose but also to improve the compound's physical properties. For example, the use of carbon black in rubber compounds is often essential to achieve acceptable compound properties such as tensile strength, wear resistance, hardness, and modulus. On the other hand, softeners such as petroleum processing oils are essential in many rubber compounds to improve the processing qualities of an uncured compound—such as the ease in which the compound can be extruded. Many times a predetermined combination of carbon black and petroleum processing oil will be used in a rubber compound to impart the desired combination of physical properties such as hardness. Too much carbon black without oil might make the cured compound too hard and stiff. Conversely, too much oil without carbon black might make the compound too soft. Therefore, many times a balance is needed between oil and carbon black.

In reviewing dry fillers such as carbon blacks and other non-black fillers, it is important to keep in mind that there are three important properties of these fillers which determine the qualities they will impart to cured rubber. These three properties are as follows:

(1) Particle size

(2) Structure (Morphology)

(3) Compatibility

Average particle size is one of the most important properties in predicting the properties that a given filler will impart to cured rubber. The rule to remember here is that the smaller the average particle size of a filler, the higher the degree of reinforcement imparted to the vulcanizate assuming all other factors remain constant. Increased reinforcement means improved tensile, modulus, abrasion resistance, tear resistance, etc. Differences in the average particle size of fillers are illustrated below.

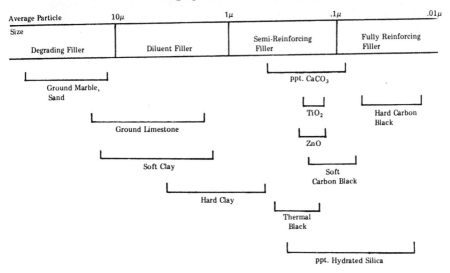

As can be seen, fillers with an average particle size greater than 5 or 10 microns are considered degrading fillers. That is the filler actually has a negative effect on the physical properties of the rubber. Very large particles here may cause severe cracking. Fillers having average particle sizes in the diluent filler range (1 to 5 microns), do not significantly improve or have a negative effect on the physical properties of the vulcanizate provided the loadings are not too high. Generally these diluents may affect compound hardness somewhat. Diluent fillers are primarily used to reduce the per pound cost of the rubber compound. Those loadings with an average particle size between 1.0 and 0.1 micron are considered semi-reinforcing and fully reinforcing if under 0.1 micron. Reinforcing fillers give improvement to those vulcanizate properties mentioned earlier.

The structure or morphology of a filler is an important quality also. Structure relates to the aggregation of particles. Carbon blacks tend to have much higher structure than inorganic fillers. The structure of most inorganic fillers is lower than carbon black and tends to break up more easily in mixing.

Particle Structure of Carbon Black

Carbon blacks with the same particle size but different structures will give different physical properties in rubber. A higher structure carbon black in a rubber compound will impart less extrudate shrinkage, swell less on extrusion and impart a higher cured compound modulus than a lower structured carbon black.

Lastly, rubber compatibility is an important consideration. While some precipitated silicas and carbon blacks have the same small average particle size, it is the peculiar organic surface chemistry of the carbon blacks that enable them to be much more compatible with rubber and impart much more reinforcement to the cured compound than silica. Carbon black particle surfaces have quinone, lactone, carboxyl, and other organic functional groups which improve the interaction between the carbon black and rubber. It is believed that the surface of the carbon black particle actually forms chemical bonds with the rubber matrix. Most inorganic fillers will not form these bonds. Inorganic filler particle surfaces tend to be "rubberphobic" while carbon black surfaces are "rubberphilic".

Carbon Black

This material is by far the most important and widely used dry loading in the rubber industry. Compounds not containing carbon black are the exception rather than the rule. The only rubber compounds that carbon black is generally not used in are either "gum rubber" and latex compounds or non-black colored or white rubber articles. Carbon blacks are used so widely because of their unique surface chemistry (which enhances compatibility with rubber) and particular particle size and structural properties. Also they are relatively low in cost for the degree of reinforcement they impart to rubber.

Today the United States has a four-billion pound annual capacity in carbon black production. The rubber industry accounts for 94% of the carbon black consumption, the balance going into plastics and printing ink. Sixty-five percent of all carbon black consumption goes into tire production. There are 6 U.S. producers of carbon black, listed below in approximate decreasing order of market share.

(1) Cabot Corporation

(2) Columbian Company

(3) Ashland Company

(4) J.M. Huber

(5) Sid Richardson

(6) Witco

Carbon blacks are produced by either the thermal decomposition of heavy aromatic petroleum oil feedstocks in horizontal furnaces (these blacks —N 100 to N700 series—are called "furnace blacks") or the pyrolysis of natural gas in larger vertical furnaces (these larger particle sized blacks being called "thermal blacks"—N800 and N900 series). The following are the nine basic types of carbon black used today which are classified by average particle size.

Old Classification	ASTM Classification	Average Particle Size (nm)
SAF (Super Abrasion Furnace)	N100 Series	11-19
ISAF (Intermediate Super Abrasion)	N200	20-25
HAF (High Abrasion Furnace)	N300	26-30
FF (Fine Furnace)	N400	31-39
FEF (Fast Extruding Furnace)	N500	40-48
GPF (General Purpose Furnace)	N600	49-60
SRF (Semireinforcing Furnace)	N700	61-100
FT (Fine Thermal)	N800	101-200
MT (Medium Thermal)	N900	201-500

As one can see, a wide range of carbon blacks is available to rubber compounders today to meet their specific product needs. In addition, there are several different carbon blacks available within each of these nine classes. These different carbon blacks within the same class differ primarily in structure (usually measured by DBP absorption number) as well as other properties. As mentioned earlier, particle size is most important in determining the degree of reinforcement a carbon black will impart (very small particles give better tensile strength and abrasion resistance while very large particles give better resilience). However in many cases structure is equally important when selecting a black in that this property greatly affects such properties as ease of extrusion and compound modulus values.

Furnace carbon blacks such as SAF, ISAF, and HAF are called "tread" blacks by tire technologists. They are also called "hard" blacks and have very fine particle sizes which enable them to provide a very high degree of rubber reinforcement to the rubber compound as seen by superior tread wear resistance in tire applications.

On the other hand, furnace blacks such as GPF and SRF blacks with larger particle sizes are used in compound applications where more resiliency and flexibility are needed. These so called "soft" blacks are commonly used in tire carcass and sidewall compounds where this flexing is important. Thus, these carbon blacks are many times called "carcass" blacks.

Lastly, the FT and MT blacks are thermal carbon blacks (blacks made from natural gas instead of petroleum aromatic oil feedstock). These blacks have relatively large particle sizes and are used where very high resilience is needed in a compound application. These carbon blacks have been declining in usage because of shortages and price controls of natural gas.

Precipitated Silicas

These fillers have average particle sizes as small as hard carbon blacks; however, because of the inorganic surface chemistry, lower structure, poorer dispersion characteristics, and a retarding effect on cures, these fillers have never been accepted as equivalent to carbon black. As a result, precipitated hydrated silicas (commonly sold under the trade name of Hi-Sil) are many times used today in specialty applications such as adhesion promotion of rubber to textile cord or brass.

One method has been found to improve the compatibility of silicas to rubber through the use of mercaptosilane coupling agents. These agents consist of inorganic silicon alkoxy groups which hydrolyze to interact with the siliceous surface of the filler particle. On the other hand, these coupling agent molecules also have organic mercapto functional groups which interact in the vulcanization process and may become bonded to the rubber chains themselves. These mercapto silane coupling agents are similar to the aminosilane used in making polyester fiberglass composites.

(See the "Plastics" chapter for more details). One of the most commonly used mercapto silane in the rubber industry is gamma-mercaptopropyltrimethoxy silane.

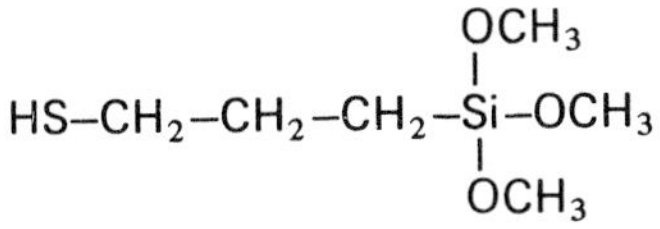

$$HS-CH_2-CH_2-CH_2-\underset{\underset{OCH_3}{|}}{\overset{\overset{OCH_3}{|}}{Si}}-OCH_3$$

This coupling agent hydrolyzes in the presence of water as shown giving off methanol (a toxic alcohol).

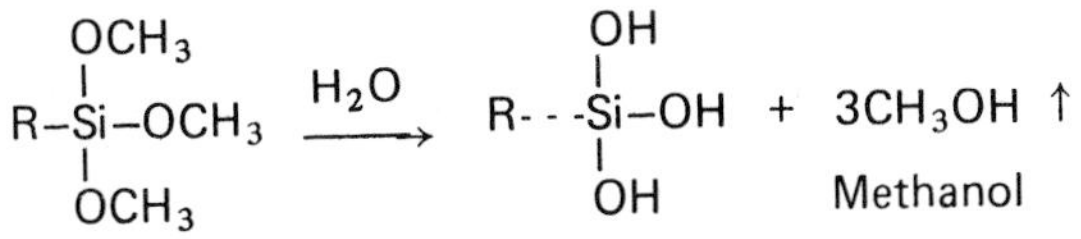

$$R-\underset{\underset{OCH_3}{|}}{\overset{\overset{OCH_3}{|}}{Si}}-OCH_3 \quad \xrightarrow{H_2O} \quad R--\underset{\underset{OH}{|}}{\overset{\overset{OH}{|}}{Si}}-OH \; + \; 3CH_3OH \uparrow$$
$$\text{Methanol}$$

Then the silane is free to react with silanol groups on the particle surface of the siliceous filler.

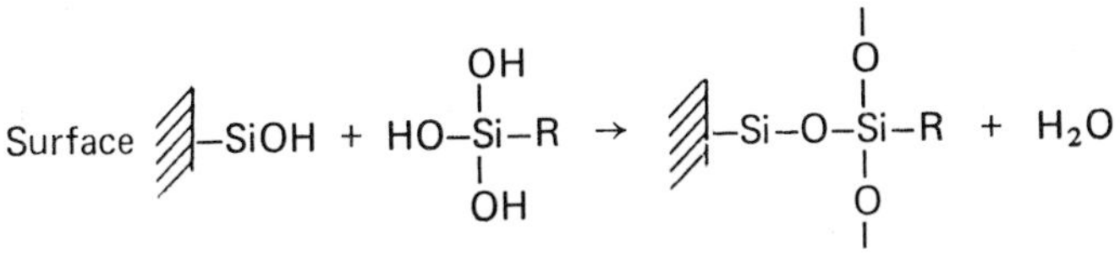

$$\text{Surface} \; \diagup\!\!\!\!\diagup -SiOH \; + \; HO-\underset{\underset{OH}{|}}{\overset{\overset{OH}{|}}{Si}}-R \; \rightarrow \; \diagup\!\!\!\!\diagup -Si-O-\underset{\underset{O}{|}}{\overset{\overset{|}{O}}{Si}}-R \; + \; H_2O$$

The next step occurs during cure when the mercapto group enters into the vulcanization reaction.

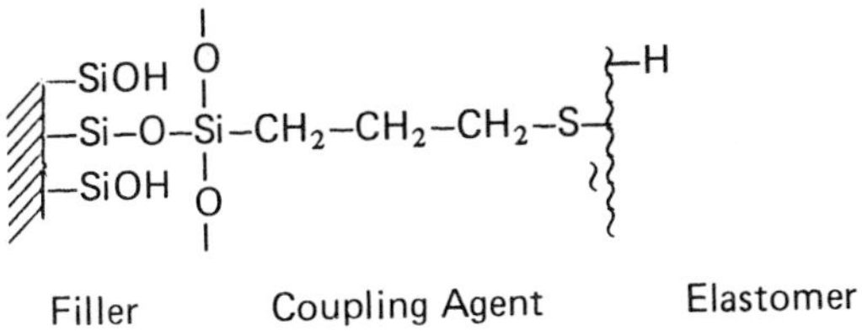

$$\diagup\!\!\!\!\diagup \begin{matrix} -SiOH \\ -Si-O-\overset{\overset{|}{O}}{\underset{\underset{O}{|}}{Si}}-CH_2-CH_2-CH_2-S \\ -SiOH \end{matrix} \begin{matrix} -H \\ \\ \end{matrix}$$

Filler   Coupling Agent   Elastomer

Nevertheless, the use of mercaptosilane with precipitated silicas has not been completely satisfactory as a replacement for carbon black to say the least. More research work in this area needs to be undertaken. Commonly, siliceous fillers such as precipitated silicas are pretreated with silane coupling agents by the manufacturer.

Clays

Use of these fillers are one of the most effective ways of reducing rubber compound costs. Clays are siliceous fillers consisting of a variety of aluminum silicates. Kaolin clays, mined in Georgia and South Carolina, are the most commonly used clays in the rubber industry. These clays are available as "hard clays" (smaller particle size) and "soft clays" (larger particle size). While the soft clays impart less stiffness to a compound, the hard clays give better physical reinforcement.

Clays should certainly not be considered as replacements for hard carbon blacks such as HAF. However, clays have an advantage over whitings such as ground limestone in that clays will impart a higher modulus to a given rubber compound. This greater stiffness is due in part to the flat plate-like shape of the clay particles. Another important aspect of clay is that because it is a siliceous filler with acidic surface chemistry, it has a retarding effect on rubber cures. Just as discussed previously with silica, mercapto silane coupling agents can and are being used with clays to improve their reinforcing qualities in rubber. It should be noted, however, that the treatment of clay with silanes does increase the per pound cost dramatically. For worker protection, avoid using clays which contain unsafe levels of silica impurities.

Calcium Carbonates

These fillers are available as ground or precipitated materials. The ground forms usually come from grinding limestone. The precipitated forms are purified and precipitated by carbonating aqueous calcium hydroxide under carefully controlled conditions. The ground versions are used as diluent fillers in a similar manner that clays are used in rubber. These ground calcium carbonates have an advantage over clays in imparting better resilience in rubber. Also calcium carbonates are whiter than clays in color. On the other hand, clays impart higher modulus values. Just as with clays, ground calcium carbonates are relatively inexpensive.

The precipitated calcium carbonates have particle sizes only one-tenth the size of many ground forms. They are used as semi-reinforcing fillers and as a whitener.

Titanium Dioxide

This relatively expensive filler is semi-reinforcing and used principally as a white coloring agent. The refractive index of titanium dioxide is 2.5 for the anatase crystalline form and 2.7 for the rutile form. These values are higher than any other white fillers used in rubber. Generally, the greater the difference between the refractive index of the filler and the refractive index of the rubber medium, the greater the opacity. Also, most titanium dioxides used in rubber today have particle sizes at approximately one-half the wave length of light, the optimum for light scattering. For these reasons, titanium dioxide imparts to rubber compounds the highest degree of "hiding power" or opacity. For example, titanium dioxide is commonly used by tire manufacturers in white sidewall compounds to assure a very white appearance.

Anatase is the softer form that is commonly used in some compound applications because it tends to chalk. On the other hand, rutile imparts greater hiding power (higher refractive index) and resists chalking.

Zinc Oxide

As discussed earlier, zinc oxide is used in most all rubber compounds at a 3 to 5 part level as an activator for vulcanization accelerators. Occasionally, however, zinc oxide is used at higher loading levels even though it is relatively expensive, because of its superior thermal conductivity. Also higher loadings impart better aging properties.

SOFTENERS

Processing and Extender Oils

When carbon black is added to a rubber batch being mixed in a Banbury, the mixing temperature goes up very rapidly. Processing oil is usually added to the Banbury mix to reduce the mixing viscosity and improve the carbon black dispersion. This is just one important reason that oils are added to rubber compounds. As mentioned earlier, they are also used in balance with carbon black as extenders to provide a rubber compound at a predetermined hardness and low cost. Oils also serve as processing aids for better extrusion and calendering. In the processing of many compounds, the use of an oil as a physical plasticizer is essential in order to increase the plastic behavior of the rubber compound on the mill or calender and reduce the "nerve" to allow smooth processing.

When petroleum oils are added into the Banbury, they are called *processing oils*. They are believed to improve the thermoplastic behavior of an elastomer during mixing by interfering with the Van der Waal forces between rubber molecular chains thus allowing these chains to slide past one another. On the other hand, when these oils are included with the raw polymer as say in an oil extended SBR or BR masterbatch, then the oil used for this purpose is called an *extender oil*. Many times the same oils are used as processing oils and extender oils.

Petroleum extender oils and processing oils are grouped in the following ASTM classification.

ASTM Designation	Description	Percent Saturates	Percent Polars
101	highly aromatic	20 max.	25 max.
102	aromatic	20.1–35.0	12 max.
103	naphthenic	35.1–65.0	6 max.
104	paraffinic	65 min.	1 max.

To understand the oils used in the rubber industry, we have to understand the important oil properties which relate to how well the oils will behave in a rubber compound. These important properties are as follows.

(1) Structure

(2) Average Molecular Weight

(3) Percent Polar Compounds

First of all, the structure of a petroleum oil is very complicated. Below is just an example of an oil molecule of which there could literally be millions of alternate variations.

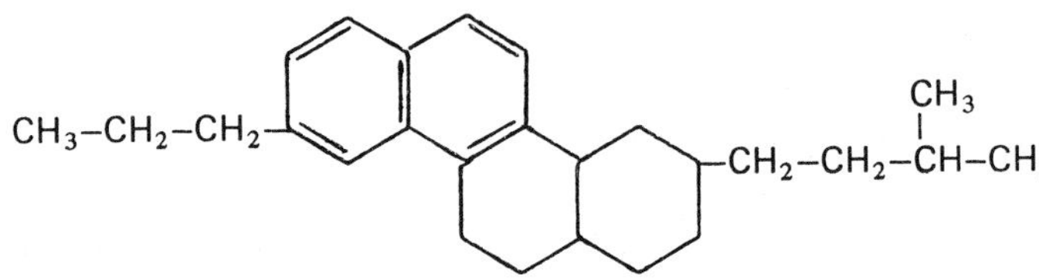

As can be seen, these structures are quite complicated; however, the basic units of these oil structures can be broken down into aromatic, naphthenic, and paraffinic structures. The relative amounts of each of these structures dictate the behavior of a given process oil in a rubber compound. For example, a highly aromatic oil will give better processing characteristics in an SBR compound than a paraffinic oil will. On the other hand, the paraffinic oil will give better cold temperature properties to the rubber compound. The preponderance of one of these structures over the others will reflect the oil's compatibility with the elastomer it is to be used in. Aromatic oils are more compatible with SBR while paraffinic oils are more compatible with butyl. This relates to the old rule that "likes dissolve likes".

Two important methods are commonly used for determining the structure of an oil. One method is to analyze the oil by separating it into its respective saturate, aromatic, and polar fractions through column chromatography in accordance with ASTM method D2007. In fact the ASTM classification listed earlier is based on this test which is called "clay-gel analysis".

The second common method used to classify these oils regarding structure is to calculate the oil's viscosity-gravity constant (VGC) from the oil's measured Saybolt viscosity at 210°F and specific gravity at 60°F.

$$VGC = \frac{g - 0.1244 \; Log \; (V_{210} - 31)}{0.9255 - 0.0979 \; Log \; (V_{210} - 31)} - 0.0839$$

where g = specific gravity @ 60°/60°F

V_{210} = saybolt universal viscosity @ 210°F

Oils that are very paraffinic will have VGC values around 0.80 while highly aromatic oils may have VGC values as high as 0.98. Oils around say 0.87 are considered naphthenic in nature.

VGC Scale 0.8	0.82	0.85	0.9	0.95	1.0
Paraffinic Oil	Relatively Naphthenic	Naphthenic Oil	Relatively Aromatic	Aromatic	

Average molecular weight is another important property of petroleum oils. A low average molecular weight oil will have a lower viscosity and perhaps be more effective in plasticizing a rubber than an oil of higher viscosity. Also oils with different viscosities may impart somewhat different physical properties to the vulcanizate.

Another important consideration is the amount of polar compounds present in an oil. Petroleum oils will many times contain heterocyclic compounds based on oxygen, sulfur, and nitrogen. These so called "polar compounds" are somewhat thermally unstable and have reportedly affected scorch properties of a rubber compound.

Generally, there are two basic types of oils commonly used in the rubber industry. The first type of oils are the staining aromatic oils such as the ATSM 101 and 102 types. These oils are used in applications such as extending rubber compounds where staining and discoloration are no problem. Oils such as these are usually dark extract oils which are a byproduct from lube oil manufacture. The second general type is the non-staining oil which is usually either naphthenic or paraffinic in nature. This type of oil is commonly used in applications where staining and discoloration are critical such as in non-black compounds. These oils can be raffinate oils from lube oil plant operations in which aromatics, polars, and waxes have been removed by extraction, hydrotreating and/or a dewaxing process. These raffinate oils are generally more expensive than the extract oils.

Synthetic Ester Plasticizers

There are rubber compounds in which petroleum oils will not work. For example, highly polar elastomers such as NBR may require a more expensive synthetic ester plasticizer rather than petroleum oil because a polar elastomer favors a polar plasticizer for compatibility. Also synthetic plasticizers give better cold temperature properties than petroleum oils do. Polychloroprene, acrylonitrile-butadiene rubber, epichlorohydrin, polyurethane, and polysulfides commonly use synthetic ester plasticizer. Some of these commonly used synthetic plasticizers are given below.

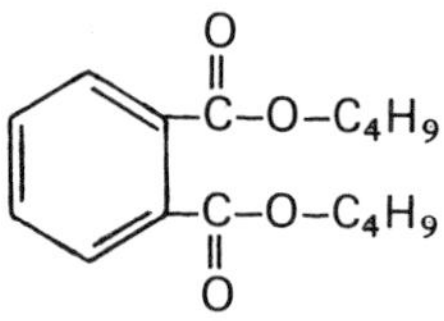

Dibutyl Phthalate (DBP)

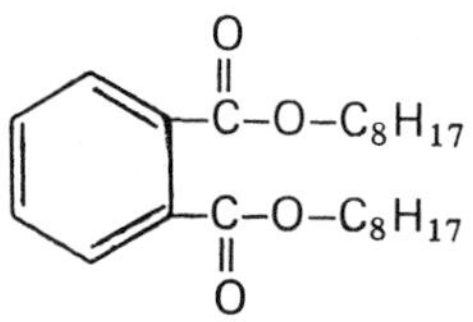

Dioctyl Phthalate (DOP)

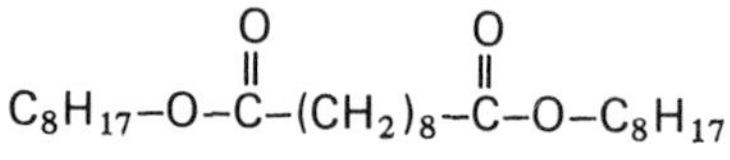

$C_8H_{17}-O-\overset{O}{\overset{\|}{C}}-(CH_2)_8-\overset{O}{\overset{\|}{C}}-O-C_8H_{17}$

Dioctyl Sebacate (DOS)

$C_8H_{17}-O-\overset{O}{\overset{\|}{C}}-(CH_2)_4-\overset{O}{\overset{\|}{C}}-O-C_8H_{17}$

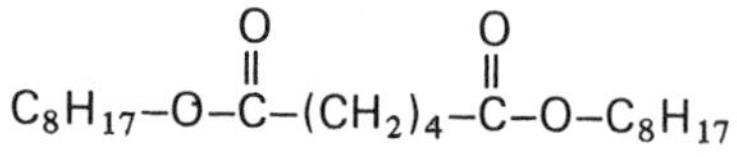

Dioctyl Adipate (DOA)

DBP and DOP both have very good plasticizing power; however DOP possesses more "permanence" in a compound (less loss with aging) than DBP. On the other hand, DBP is more polar and may give better plasticizing properties in certain compounds. DOS and DOA have advantages over the phthalates in having superior cold temperature properties.

These ester plasticizers just discussed are just examples of some of the many synthetic plasticizers that are used in specialty elastomer compounds. Other examples are alkyl and aryl phosphates, caprylate-caprates, azelates, glutarates, epoxidized oils, and low molecular weight liquid polyester.

Peptizers

The softeners discussed so far are all called "physical plasticizers" in that their plasticizing effect on rubber compounds is strictly a physical effect, not chemical. On the other hand, peptizers are "chemical plasticizers" in that they work on a completely different principle. Peptizers aid in the breakdown of rubber during mixing by reacting with the free radical chain ends resulting from chain scission during mixing, thus preventing these chain ends from recombining. These chemicals are usually added at relatively small quantities at the start of the mix cycle for best results. They work best in plasticizing tough natural rubber, however they can also be used with synthetic rubber as well. Examples of two common peptizers used today in the rubber industry are shown below.

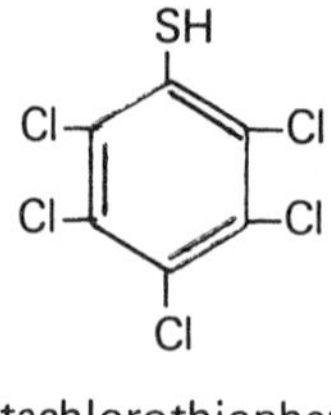

Pentachlorothiophenol

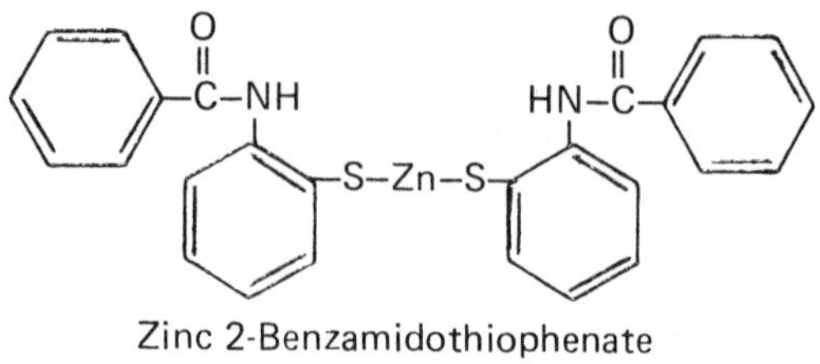

Zinc 2-Benzamidothiophenate

ANTIDEGRADANTS

There are approximately 190 million pounds of rubber antidegradant chemicals produced in the United States annually. Of this production, about 38% of these antidegradants are of the phenylenediamine class and 11% are of the phenolic class. These two types and other classes of antidegradants will be discussed.

As shown earlier, most elastomers used today consist of long entangled molecular chains with unsaturated sites along the backbone. Although a small portion of these unsaturated sites promote sulfur crosslinking during cure, they are also very susceptible to oxidative attack. Rubber oxidation is initiated by the formation of free radicals usually produced from the energy of light or heat. The free radicals in turn react with diatomic oxygen to generate peroxidic radicals in the following manner.

$$R \cdot + O_2 \rightarrow ROO \cdot$$

These unstable peroxidic radicals take hydrogen from other macromolecules in the rubber compound matrix.

$$ROO \cdot + RH \rightarrow ROOH + R \cdot$$

This chain reaction would terminate when two radicals combined; however the hydroperoxides formed by these reactions are unstable themselves and decompose to catalytically form progressively more free radicals in the following manner.

$$ROOH \rightarrow RO \cdot + \cdot OH$$
$$RO \cdot + RH \rightarrow ROH + R \cdot$$
$$\cdot OH + RH \rightarrow H_2O + R \cdot$$

This propagation will accelerate the formation of free radicals. Rising radical formation will degrade the rubber compound by causing scission of the polymer chains or oxidative cross linking between the chains. Chain scission, which is the dominant occurrence in natural rubber, causes the rubber to become softer and weaker in tensile strength. However, oxidative crosslinking is the dominant outcome from the oxidative attack of elastomers such as SBR, NBR, and neoprene, causing their cured compounds to become harder and crack. To help stop these oxidative degradation processes, a free radical "trap" is needed to absorb the free radicals as soon as they are formed in order to prevent the propagation cited above. This is the role that antioxidants play. They serve as propagation "chain stoppers". These antioxidants react with the free radicals to produce "inert" radicals that are stabilized by resonance. These "stabilized" free radicals will not react with oxygen to propagate the reaction further.

$$ROO\cdot + HA \rightarrow ROOH + A\cdot$$

Substituted Phenols and Amines

Two basic classes of antioxidants will be considered which can serve as "free radical absorbers". One class consists of substituted phenols and the other class is made up of secondary amine derivatives. Some examples of antioxidants from each of these classes are given below.

Substituted Phenols

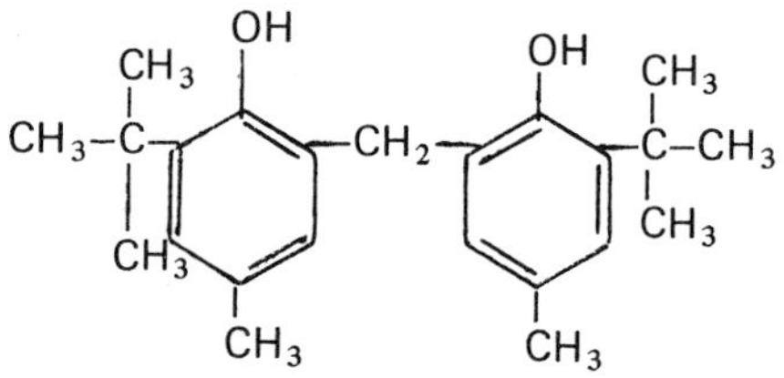

2,2'-Methylenebis(4-methyl-6-tert-butylphenol)

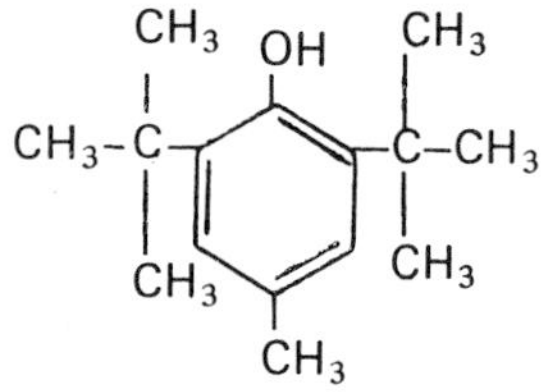

2,6-Di-tert-butyl-4- Methylphenol (BHT)

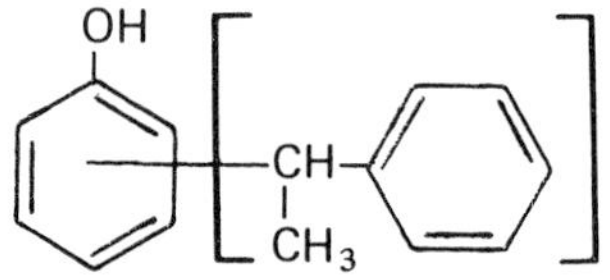

Styrenated Phenols

Substituted Amines

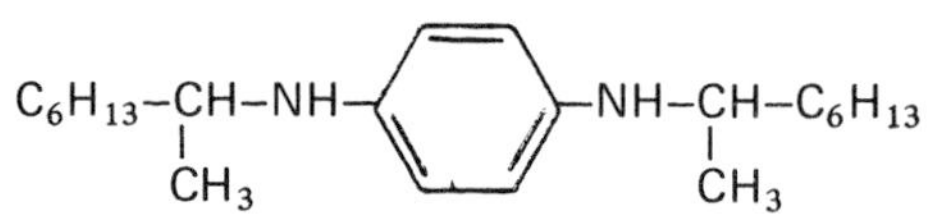

N,N'-Bis(1-methylheptyl)-p-phenylenediamine

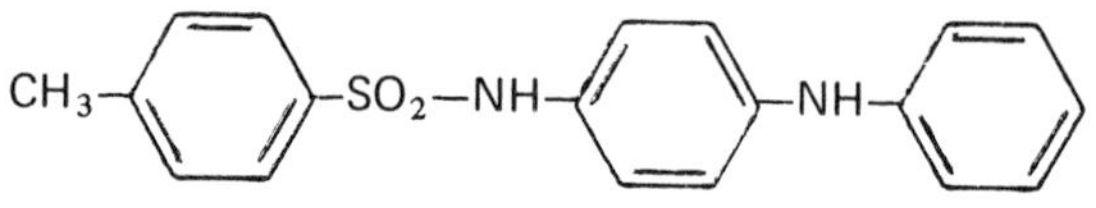

N-Phenyl-N'-(p-toluenesulfonyl)-p-phenylenediamine

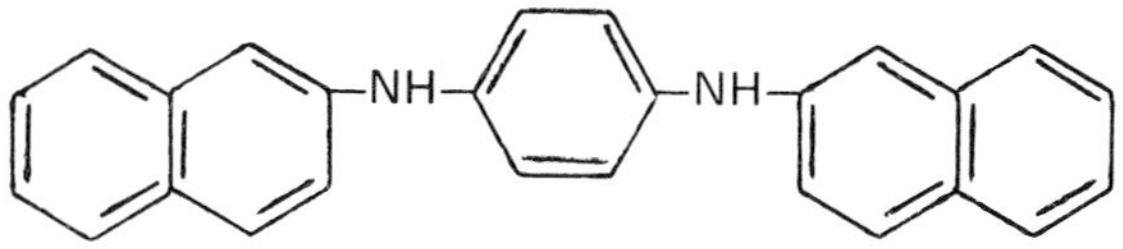

N,N'-Di-β-naphthyl-p-phenylenediamine (DNP)

N,N'-Diphenyl-p-phenylenediamine (DPPD)

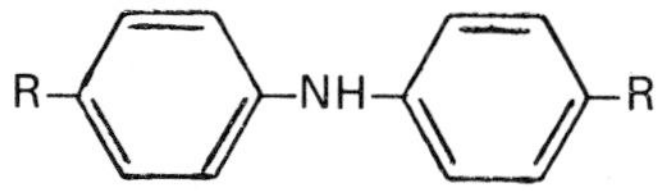

Alkylated diphenylamine

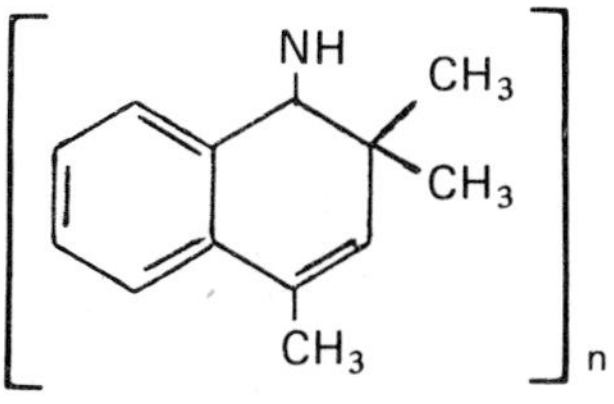

Poly-2,2,4-trimethyl-1,2-dihydroquinoline

Substituted phenols have been found to be only about half as effective as amine antioxidants in stopping oxidative degradation. However their main advantage is that they do not generally decompose into oxidized colored products which will contribute to the discoloration of the rubber compound or staining. This non-discoloring and non-staining characteristic of substituted phenols is the main reason for their use especially in light colored or white rubber products. The phenols used are substituted because straight phenol would be too acidic (retarding the cure). These AO's do not work well in black loaded compounds.

Substituted amines, on the other hand, are much more effective antioxidants than substituted phenols because the amines not only react with the free radicals which propagate the degradation process, but they also decompose the peroxides formed as well. Unfortunately, amines also form colored oxidation products while in service which limits their use to conventional black rubber compounds where discoloration and staining are not a problem. Also, care has to be exercised that the use of amine antioxidants in a particular compound will not migrate into an adjacent white or light colored compound of the rubber product in question.

Many of the substituted amines used as antioxidants are phenylene-diamines. These phenylenediamines vary in their respective solubility in different rubber compounds. Some are relatively insoluble in a given compound and may separate out, that is "bloom" to the surface of the compound. Other phenylenediamines may "bloom" to the surface at a much slower rate. This "blooming" property of phenylenediamine is very important in imparting surface protection properties to the cured compound against ozone attack.

Antiozonants

Ozone is generally present in the air at concentrations between 1 to 6 parts per hundred million (PPHM). In some regions, such as Los Angeles County, the concentration can be as high as 25 PPHM. Unsaturated elastomers are extremely sensitive to ozone attack even when at low concentration. Unprotected rubber surfaces are very receptive to ozone attack across the double bonds of the polymeric backbone which results in chain scission and cracking on the rubber surface thus leading ultimately to product failure.

$$
\begin{array}{c}
R' \\
| \\
R-C=CH-R + O_3
\end{array}
\rightarrow
\begin{array}{c}
R' \\
| \\
R-C\!-\!-\!-\!-CH-R \\
| \quad\quad | \\
O \quad\quad O \\
\diagdown \; O \; \diagup
\end{array}
\rightarrow
\begin{array}{c}
R' \\
| \quad (-) \\
R-C-O-O \\
(+)
\end{array}
+ RCH{=}O
$$

If a vulcanizate contains a given phenylenediamine which blooms to the surface of the compound continuously, it will protect that rubber compound against ozone attack on the surface. This is why many phenyl-enediamine antioxidants are also known as *antiozonants*. The rate of bloom for a given phenylenediamine is not only determined by the elastomer base and compound composition in general, but also by the alkyl and aryl substitutions of the particular phenylenediamine. Different combinations of alkyl/alkyl, aryl/aryl and alkyl/aryl substitutions along the phenylenediamine structure result in different solubility properties for the particular antiozonant which will determine its rate of bloom. Ideally, for a given rubber compound, the antiozonant should bloom at a sufficiently fast rate to provide effective antiozonant protection while still having this rate slow enough to assure good long term protection throughout the life of the product.

Protective Wax. Another important antiozonant to be discussed is petroleum wax. A blend of petroleum waxes is commonly used in rubber compounds to provide antiozonant protection. Petroleum waxes bloom to the surface of the rubber in a similar manner described for the phenylene-diamines; however, unlike the phenylenediamines which react with the ozone present, the petroleum wax bloom forms an inert protective barrier through which the ozone can not penetrate.

These protective waxes that are commercially available for rubber compounding usually consist of blends of paraffins and microcrystalline waxes. Different amounts of each of these components provide a characteristic rate of bloom and flexibility.

TACKIFIERS

Building tack is the term generally used to describe the ability of an uncured rubber part to stick to another uncured rubber part with only a moderate amount of hand pressure applied for a short duration. Building tack is very important in the rubber industry when a rubber product is being built by placing rubber parts one on top of another, which is the case in the construction of tires or conveyor belts. Without building tack, the rubber parts would not "stay put" and the assembled uncured product might even fall apart before it could be cured. Some base elastomers, such as natural rubber, many times provide sufficient building tack without the need for tackifiers. However, many synthetic elastomers such as SBR and particularly EPDM, require tackifiers in order to have sufficient building tack in the green state.

Tackifiers used today are usually either natural or synthetic resinous materials. The earliest tackifiers were the so called "naval stores" or rosins obtained from wood. There are basically three types of rosins used in rubber. Rosins obtained from the destructive distillation of turpentine oil are called *gum rosins*. Rosins extracted from old pine stumps are called *wood rosins*. And finally, rosins obtained from steam distillation of tall oil (a byproduct from paper manufacture) are called *tall oil rosins*. All three rosins are considered abietic acid derivatives.

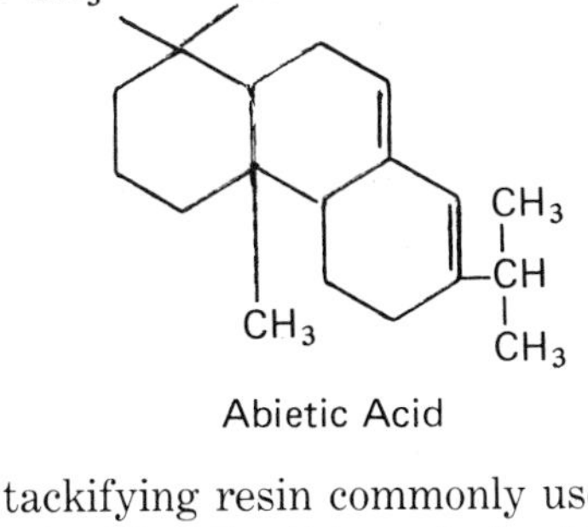

Abietic Acid

Another type of tackifying resin commonly used in rubber is the so called hydrocarbon resin. Usually, this resin is obtained as a byproduct from petroleum oil refining, although still a small amount is derived from coal tar. A common hydrocarbon resin used in rubber is the polyindene resin, a highly aromatic resin. The polyindene resins are also referred to as coumarone-indene resins even though some contain very little coumarone.

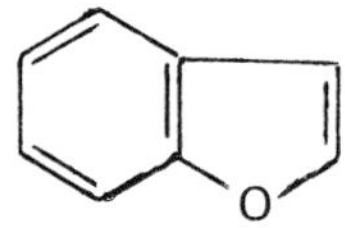

Coumarone

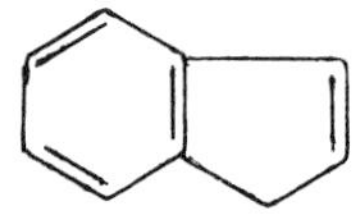

Indene

Phenol formaldehyde resins are a third type of resin normally used to tackify rubber compounds. One important point to note is that these phenol formaldehyde resins are non-heat reactive. That means they will not crosslink or gel on further heating. This is because these P.F. resins are alkylated (usually with tert-butyl or octyl groups) at the para positions to prevent gelling.

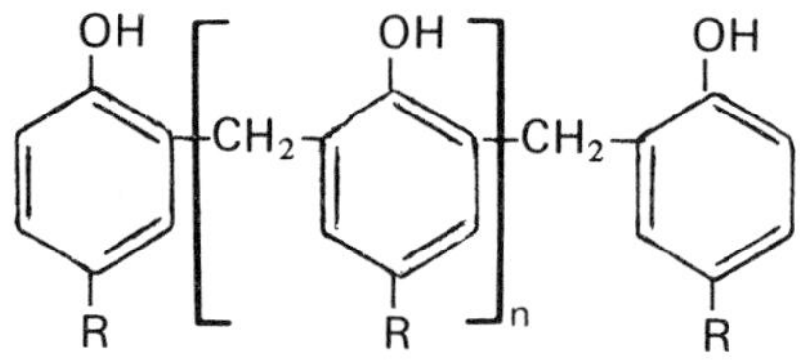

p-Alkyl substituted phenol-formaldehyde resin

These P.F. tackifying resins are considered more effective as rubber tackifiers than the coumarone-indene resins. On a per pound basis, P.F. resins impart a higher degree of tack than the others discussed. On the other hand, P.F. resins may be more expensive than the other two resin types.

Lastly, a fourth type of tackifying resin used to impart tack to rubber compounds is the polyterpene resin. This resin is polymerized from pinene, a constituent of turpentine oil.

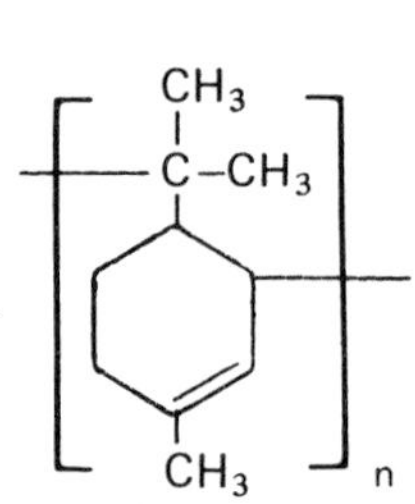

Polyterpene

FLAME RETARDANTS

For some product service applications, the rubber product must be flame resistant or self extingishing. Most common elastomers do not display this quality. Therefore, retarding additives must be included in a compound formulation when flame retardance is desired. *Antimony oxide* is one of the most effective and commonly used flame retardants in rubber compounds. For antimony oxide to be effective, it generally must be used with a chlorine donor in order for antimony chloride to be formed on combustion, thus self extingishing the flame in the vapor phase. *Chlorinated paraffin hydrocarbons* as well as other chlorine compounds are used for this purpose. Another commonly used class of flame retardant is phosphate ester plasticizers such as, triaryl phosphate (TAP). All the flame retardants we have just mentioned have different toxic properties which should be reviewed before selection.

COLORANTS

The majority of all rubber compounds mixed are just one color—black. Of course, the black color is from the carbon black loading, the main filler and reinforcing agent used in the rubber industry. Not only does carbon black reinforce and reduce the rubber compound's per pound cost, it also serves as a U.V. absorber for polymer stability. Even gum rubber compounds which do not require carbon black for reinforcement may still contain a small amount of carbon black just to impart a black color. This may be done by some manufacturers to avoid the need of maintaining a separate mixer just for non-black compounds. Such batches can not be mixed routinely in a Banbury that has previously mixed black batches because of the free carbon black remaining that is costly to remove and will contaminate non-black stocks.

Non-black rubber compounds, while representing only a small percentage of the total rubber mixed, are still quite common in many different product lines. White sidewall compounds for tires are probably the single highest volume application for non-black compounds. *Titanium dioxide* is a commonly used whiting agent for this application. The advantages of titanium dioxide over other whiting agents were discussed earlier under "fillers".

Colored compounds are even less common than white compounds. Some inorganic colorants are *red iron oxide, yellow iron oxide, green chromium oxide*, and *ultramarine blue* (made from heating clay, charcoal, calcium carbonate, and sulfur). Inorganic colorants have good thermal stability; however, organic colorants have gained in usage even though they are far more costly per pound than inorganics. Usually a much smaller concentration of organic colorant is required to achieve the same color intensity that a comparable inorganic will give. As a result, organic colorants such as *indigo* and *phthalocyanine* derivatives are commonly used. To cure some organic colorant containing compounds, lower cure temperatures must be utilized in order to avoid thermal degradation of the colorant. The dust from many of these colorants is toxic and appropriate worker protection is required.

BLOWING AGENTS

Blowing agents are another family of chemicals sometimes used in rubber compounding. On heating, these chemicals decompose emitting a gaseous product in the rubber which creates pores or cavities in the rubber mass thus rendering it sponge-like or microporous in texture. Such products are used in rug underlays, upholstery, etc. Some blowing agents used are simply sodium bicarbonate which emits carbon dioxide. However, organic blowing agents are commonly used in rubber also.

In making a urethane cellular product, the blowing agent here can be simply water. Water reacts with isocyanate present to produce carbon dioxide gas as shown below.

$$RNCO + H_2O \rightarrow RNH_2 + CO_2 \uparrow$$

Isocyanate

Polyurethane chemistry will be discussed later.

Cellular rubber can consist of open cells (such as sponge rubber) in which the cavities are all interconnecting, or closed cells (such as some forms of foam rubber) in which most of the cavities are isolated. Different production conditions determine which type results.

NEW DIRECTIONS

Presently natural and synthetic rubbers are now produced with very high average molecular weights only to be broken down in an energy intensive Banbury mixer in order to be impregnated with curing agents, fillers, and other compounding ingredients as well as further processed and shaped for future curing. Intuitively, it would seem logical that a better approach would be to conduct a polymerization reaction to build high molecular weight macromolecules in the mold itself and eliminate the depolymerizing mix step which is quite costly in terms of tying up capital equipment and using energy. Indeed, this may be the trend in the future through polyurethane liquid component systems utilizing liquid injection molding (LIM) and reaction injection molding (RIM) systems. Some advantages of such a system are given below.

(1) Lower production cost—a LIM system would eliminate the Banbury mixers, calenders, extruders, building machines, etc. Therefore only one-fifth of the factory space might be required and capital investment might be 20 to 40% below that of the conventional manufacturing method.

(2) Smaller economies of scale—for example, a tire plant could be economical at only 1,500 tires per day production with LIM technology as compared to the 10,000 tires per day minimum production level required today.

(3) Superior uniformity—instead of making a product from assembly of many components made of rubber and fabric, the product could be molded without fabric provided the polymer possessed sufficient strength. Also splices would be eliminated.

(4) Less energy intensive process—a conventional rubber product will consume approximately 1500 BTU's/ lb. An LIM system would require far less.

Polyurethane Systems

The present state of the art of LIM technology is based mainly on urethane components. Polyurethanes are formed in the mold from the reaction of diols and difunctional amines with isocyanate terminated prepolymers.

$$n[O=C=N-R-N=C=O] + n[HO-R'-OH] \rightarrow \left[\overset{O}{\overset{\|}{C}}-NH-R-NH-\overset{O}{\overset{\|}{C}}-O-R'-O \right]_n$$

Difunctional isocyanate   Diol   Polyurethane

Also another coupling reaction is possible between an isocyanate and an amine.

$$R-N=C=O + R'NH_2 \rightarrow RNH-\overset{O}{\overset{\|}{C}}-NHR$$

Isocyanate   Amine   Urea Linkage

Urethane reactions become more complicated when it is realized that other reactions also occur with the active hydrogens of the urethane and urea links to provide active sites for crosslinking.

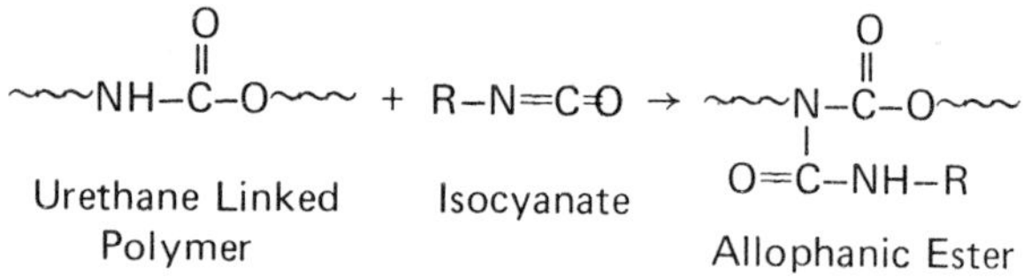

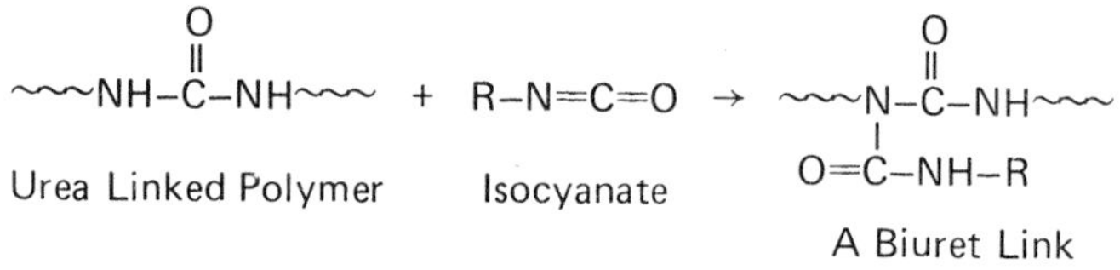

$$\underset{\substack{\text{Urethane Linked} \\ \text{Polymer}}}{\sim\sim\sim\text{NH–}\overset{\displaystyle O}{\overset{\|}{\text{C}}}\text{–O}\sim\sim} + \underset{\text{Isocyanate}}{\text{R–N}=\text{C=O}} \rightarrow \underset{\substack{\text{Allophanic Ester}}}{\sim\sim\sim\text{N–}\overset{\displaystyle O}{\overset{\|}{\text{C}}}\text{–O}\sim\sim}$$

$$O=\overset{\displaystyle}{\underset{\displaystyle}{\text{C}}}\text{–NH–R}$$

$$\underset{\text{Urea Linked Polymer}}{\sim\sim\sim\text{NH–}\overset{\displaystyle O}{\overset{\|}{\text{C}}}\text{–NH}\sim\sim} + \underset{\text{Isocyanate}}{\text{R–N}=\text{C=O}} \rightarrow \underset{\substack{\text{A Biuret Link}}}{\sim\sim\sim\text{N–}\overset{\displaystyle O}{\overset{\|}{\text{C}}}\text{–NH}\sim\sim}$$

$$O=\overset{\displaystyle}{\underset{\displaystyle}{\text{C}}}\text{–NH–R}$$

There are basically three types of urethane components used in urethane LIM technology. These three classes of components are given below.

Difunctional isocyanates. These components, in a strictly closed system, react with the polyols to form prepolymers which in turn can be crosslinked with the chain extenders (to be discussed). Some commonly used diisocyanates are shown below.

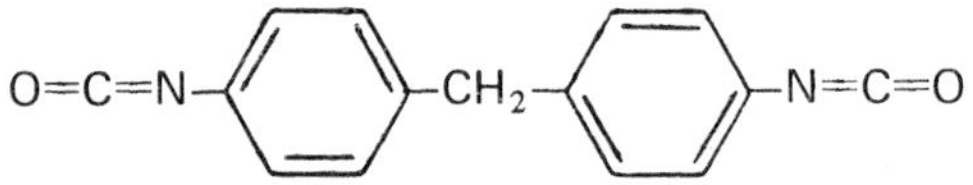

Diphenylmethane-4,4'-diisocyanate (MDI)

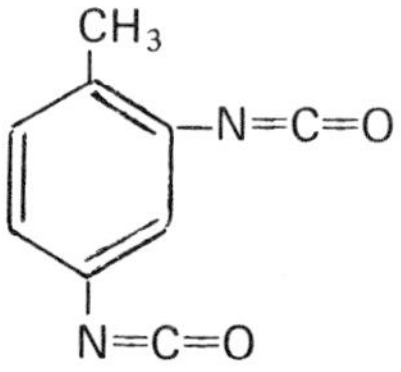

2,4-Tolylene diisocyanate (TDI)

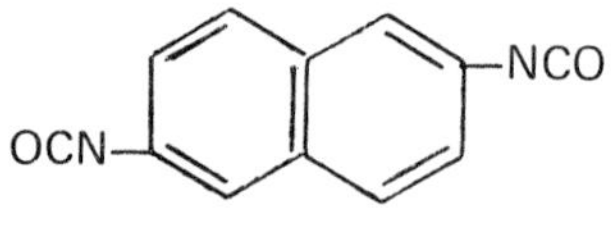

Naphthalene-1,5-diisocyanate (NDI)

Generally aromatic diisocyanates are used because they are faster reacting; however more expensive aliphatic diisocyanates (such as 1,6 hexamethylene diisocyanate are used when resistance to discoloration is required. All diisocyanates should only be used in closed systems to prevent worker exposure. Isocyanates are toxic and present a hazard to workers.

Polyols. These are liquid polymers made of a polyester or polyether backbone structure and hydroxyl terminated. These polymers usually have an average molecular weight around 2000.

$$HO-(CH_2)_2 \left[O-\overset{O}{\overset{\|}{C}}-(CH_2)_4-\overset{O}{\overset{\|}{C}}-O-(CH_2)_2 \right]_n -OH$$

Adipic acid-ethylene glycol polyester

$$HO \left[CH_2-\underset{CH_3}{CH}-O \right]_n CH_2-\underset{CH_3}{CH}-OH$$

Poly(propylene glycol)

$$HO \left[CH_2-CH=CH-CH_2 \right]_n OH$$

Hydroxyl-terminated polybutadiene

In many systems, instead of mixing isocyanates, polyols, and chain extenders together in what is called a "one shot" method, commonly the polyols will be reacted in an earlier step with the difunctional isocyanates under heat and an inert atmosphere to form an isocyanate terminated *prepolymer.* In many cases these so called prepolymers can be bought direct from the manufacturer in order to save a step and avoid the additional toxicity risk associated with diisocyanates.

$$HO\sim\sim OH + 2(OCN-R'-NCO) \rightarrow OCN-R'-NH-\overset{O}{\overset{\|}{C}}-O\sim\sim O-\overset{O}{\overset{\|}{C}}-NH-R'-NCO$$

Polyol   Diisocyanate   Prepolymer

The selection of the type of polyol or prepolymer to be used in the LIM system is very important in that this selection greatly determines the polyurethanes properties in the end product. The properties of strength, resistance to water, and flexibility at low temperatures are all determined by which polyol the urethane is based on.

Chain extending agents. These are diols, triols, or diamines. They are used to cure the isocyanate terminated prepolymer in the mold by serving as chain extenders and crosslinking agents. Examples of commonly used chain extenders are given below.

$$HO-CH_2-CH_2-CH_2-CH_2-OH$$

1,4-Butanediol

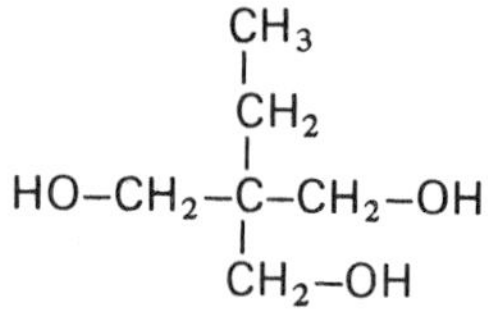

$$HO-CH_2-\underset{\underset{CH_2-OH}{|}}{\overset{\overset{CH_3}{|}}{\underset{|}{C}}}-CH_2-OH$$

Trimethylol propane

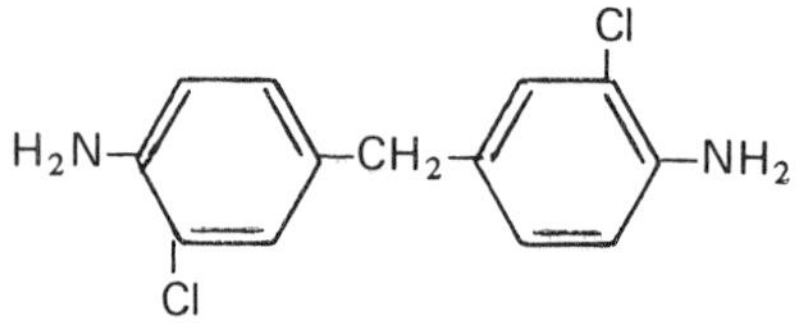

MOCA
4,4'-Methylenebis(o-chloroaniline)

MOCA has been found by the Federal Government to be carcinogenic; however, because of its unique reactivity, it is still used with the appropriate mandated handling precautions to prevent exposure to workers.

As noted, these di and tri functional compounds are used to extend and crosslink the isocyanate terminated prepolymers and thus "cure" the prepolymer.

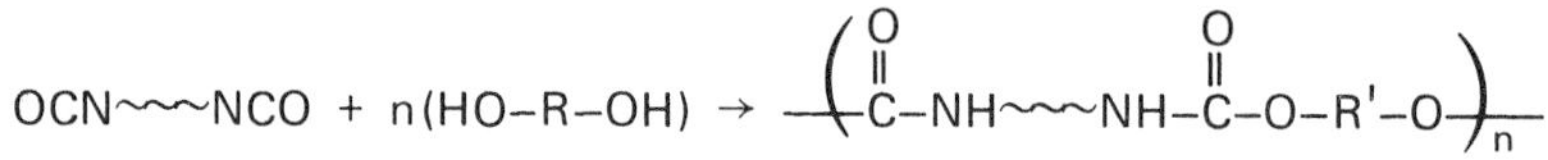

OCN〜〜NCO	+ n(HO-R-OH)	→	Polyurethane
Prepolymer	Diol		

Also

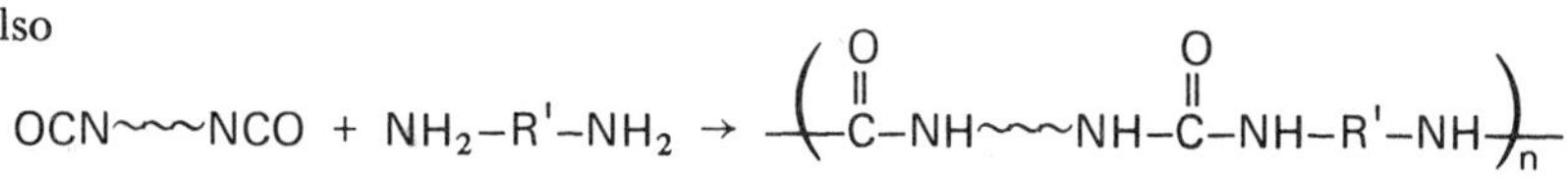

OCN〜〜NCO	+ NH₂-R'-NH₂	→	Polyurethane
Prepolymer	Diamine		

Polyurethanes display some very good properties. They have ultimate tensile values exceeding natural rubber. Polyurethanes have good low temperature properties, and good resistance to oil, tearing, abrasion, weathering, and ozone attack. Also there are the advantages of a simplified manufacturing process discussed earlier. So why haven't urethanes replaced most of the general purpose elastomers used today? The reason is that polyurethane by nature is a polycondensation polymer, meaning it contains functional groups as part of the polymer's backbone. Under severe dynamic stress cycle deformations which some rubber products may experience, these functional groups interfere with the recovery of the macromolecules to their original positions thus generating heat. This heat generation ultimately causes the polyurethanes to depolymerize and fail. On the other hand, most conventional elastomers used today are polymerized from an addition polymerization reaction in which no functional groups are actually in the polymer backbone. Therefore, these conventional polymers are more resistant to chain scission. Also many conventional elastomers are less expensive than polyurethane.

This explanation is perhaps overly simplistic in that many new polyurethane systems have been developed which have replaced conventional elastomers in use. In fact, polyurethane's markets have been growing rapidly and demand for polyurethane is expected to grow 10 percent annually. Work continues throughout the rubber industry to find reactive liquid components that can be used in a LIM system and provide improved performance and durability under severe dynamic conditions and high temperature service.

Of course the ultimate goal of urethane producers is to obtain acceptance of polyurethane by the tire manufacturers. Much work has been reported in the literature on just such an effort. So far, some of the problems reported in public press releases are as follows:

(1) Poor wet traction

(2) Softening under extreme heat

(3) Higher frequency of in-service "blowouts"

(4) Poor tread wear

Work continues in the industry to find solutions to these problems; however a successful "castable tire" is still quite a long ways from reality.

Thermoplastic Elastomers

Another relatively new class of elastomer that may grow in future use is thermoplastic elastomers. Referring back to discussions concerning plastics in the previous chapter, there were basically two plastic types, i.e., thermoplastic and thermosetting. Conventional elastomer compounds are analogous to thermosetting plastics in that these rubber compounds must have crosslinking agents added and be cured under heat and pressure in order to obtain acceptable strength and resilience properties. In a similar way thermoplastic elastomers are analogous to rigid thermoplastics such as polystyrene and polyethylene. Rigid thermoplastics already have acceptable strength properties at room temperature (without the need for hardeners to be added). However, when these plastics are heated above their crystalline melt point (Tm), they display flow properties that enable them to be easily processed on mills, calenders, extruders, etc. Likewise, thermoplastic elastomers also display good usable elastomeric properties (such as high tensile, modulus, and resiliency) at room temperature without the need for a curing step or the addition of curatives. On the other hand, upon heating the thermoplastic elastomer above its melt point, it behaves in a thermoplastic manner, lending itself quite well to processing by mill rolls, calenders, or extruders.

There is an explanation for the behavior just described for these "themoplastic elastomers". These thermoplastic elastomers are actually block polymers of plastic and rubber blocks. Each macromolecule consists of a center block of units that are rubbery in nature (either isoprene or butadiene units) flanked on each side by a blocked region of plastic units (usually styrene units). Thus each macromolecule has a linear plastic-rubber-plastic structure

AAAABBBBBBBBBBBBBBBAAAA

A = Glassy or crystalline units at service temperature
B = Elastomeric units

The hard plastic "glassy" ends of the polymer chain are joined together with the plastic ends of other molecular chains in large aggregates called "domains". These domains serve as crosslinks to establish a physical network structure just as sulfur crosslinks do in conventional rubber. However unlike conventional rubber, when the thermoplastic elastomer is heated to a temperature above the M.P. of these plastic domains, they melt resulting in thermoplastic behavior for good processibility on a calender or in an extruder just as an ordinary non-setting plastic.

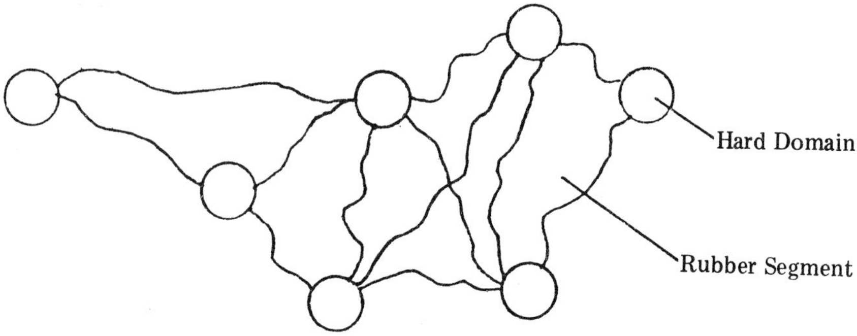

Thermoplastic elastomers commercially available today are either styrene-isoprene-styrene (SIS) or styrene-butadiene-styrene (SBS) block polymers. Both Shell Chemical Co. and Phillips Petroleum Co., make these polymers under the tradenames of Kraton and Solprene, respectively. Monsanto also makes a completely different thermoplastic elastomer called Santoprene. This new TPE consists of crosslinked EPDM rubber dispersed in either polyethylene or polypropylene. Lastly, Dupont markets high durometer TPE which is a copolyester block polymer.

Obviously the advantages to the use of these polymers over conventional elastomers are that no curatives need be added in a mixing cycle (the polymer is ready for use "as is"), no scorch problems can occur, and press cycles during molding are relatively short (since no curing occurs in the mold). In fact, these polymers can be processed on either rubber processing equipment or plastic processing equipment (such as injection molding). Thus the problem of going through a complicated mixing step with the necessity of adding sulfur, accelerators, zinc oxide, stearic acid, etc. is avoided.

On the other hand, there are certain disadvantages to these blocked polymers as well. First of all, mill temperatures must be set higher in temperature than normally used to process conventional rubber compounds because more heat is needed to melt the plastic domains (serving as cross-links). Another disadvantage of these thermoplastic elastomers is the poor thermal stability of the plastic crosslinks (domains). As the service temperature of these polymers begins to approach the melting point of these plastic domains, the "crosslinks" become more labile and fail with temperature rise. Thus severe dynamic service conditions which give high heat generation, will cause thermoplastic elastomers to fail. In these applications only conventional rubber will work. In this respect, thermoplastic elastomers have a similar disadvantage to that of polyurethane. In fact, there are special polyurethanes on the market that can be used as a thermoplastic.

Reclaimed Rubber

Reclaimed rubber is certainly not new. Rubber has been reclaimed from scrap rubber products since the "heater process" received a patent in 1858. What is new about reclaim may be an increased emphasis by the government to encourage rubber companies to use more of it in future products. One reason for a possible increase in future use is the emphasis being placed on "cyclic" economies rather than "through-put" economies. Now with the rising price of energy and the realization that natural resources are finite, future government planners would like to see more recycling of materials used in commerce. Also the cost of petroleum may also create economic inducements to use more reclaim in the future.

Reclaimed rubber is obtained basically from the grinding, heating, and/or chemical treatment of scrap rubber products (mostly whole tires). In many of the reclaim operations, methods of defibering these products have been developed. Chemically, the cured rubber is depolymerized, breaking up the vulcanization network previously formed. It is erroneous to say that reclaimed rubber is "devulcanized" in that the sulfur remains chemically bonded. As discussed earlier, much unsaturation remains after vulcanization so reclaimed rubber can easily be vulcanized again. In fact, one advantage of using reclaim rubber in a compound is that it increases the cure rate. Also, reclaim provides other processing advantages to a compound such as less nerve and extrusion shrinkage, faster extrusion rates, and better calendering.

Thus with the emphasis on reclaim conservation, reclaim use may well increase in the years ahead.

SUGGESTED READING

Alphen, J. Van, *Rubber Chemicals*, D. Reidel Publishing Co., Dordrecht, Holland, 1973

Babbit, R. O., *The Vanderbilt Rubber Handbook*, R. T. Vanderbilt Co., N.Y., Norwalk, Conn., 1978

Blow, C. M., Hepburn, C., *Rubber Technology and Manufacture*, Butterworth and Co., London, 1982

Eirich, Frederick R., *Science and Technology of Rubber*, Academic Press, N.Y., 1978

Evans, Colin W., *Practical Rubber Compounding and Processing*, Applied Science Publishers, Englewood, N.J., 1981

Freakley, P.T.K., *Rubber Processing & Production Organization*, Plenum Press, N.Y., 1985

Manual for the Rubber Industry, Bayer AG, Leverkusen, Germany, 1970

Morton, Maurice, *Rubber Technology*, 2nd Edition, Van Nostrand Reinhold Co., N.Y., 1973

Odian, George, *Principles of Polymerization*, 2nd Edition, McGraw-Hill Book Co., N.Y., 1981

Polysar Butyl Handbook, Ryerson Press, Toronto, 1966

Rubberworld Blue Book, Materials, Compounding Ingredients, and Machinery for Rubber. Rubber World Magazine, Bill Communications N.Y., Annual.

3

The Adhesive Industry

Although the adhesive industry is not nearly as large as other industries such as plastics and rubber, it is absolutely essential to the well being of our industrialized society. It is quite literally the "glue" that holds our technological state together. Adhesives are used today extensively in the construction of houses, automobiles, airplanes, electrical equipment, boxes, bags, shoes, tires, books, clothes, and even roads.

The adhesives industry generates approximately 4 billion dollars in sales annually. It is not a concentrated industry but is made up of over 800 companies of which the largest 8 firms only account for approximately 30% of all adhesives sales.

MARKETS

Adhesives are used in widely diverse markets. The largest single market for adhesives is building construction and wood products. This area accounts for about one-third of the total dollar value of adhesives produced. Very large quantities of phenolics and urea-formaldehyde adhesives are used in making such wood products as plywood and particle board. Resorcinol-phenol-formaldehyde adhesives are used in laminated lumber. Elastomer adhesives such as SBR, natural rubber, and neoprene are used to glue floors and wood frames. Adhesives based on polyvinyl acetate or starch are commonly used to glue gypsum board or wallpaper. Carpet is installed with elastomeric adhesive or hot melts such as polyethylene. These are just a few of the many applications that adhesives have in the construction industry.

Another large adhesive market is in the area of packaging and paper.

While packaging adhesive consumption does not represent even half the dollar value of adhesives sold in the construction market, on a pound basis, it represents over one-third of the total poundage of all adhesives sold. Historically, this area has been dominated by the naturally derived glues based on starches, dextrin, animal by-products, and sodium silicate. Recently, however, within the last twenty years there has been a trend with some package adhesive applicators to go over to hot melt adhesives based on synthetic polymers.

There are, of course, many other markets in which adhesives are used as well. Some examples of these are automobiles, which uses a variety of elastomer based adhesives; shoes, which use elastomeric adhesive as well as polyester and polyamide hot melts; electrical, which uses a wide variety of synthetic adhesives including epoxies, phenolics, polyester, PVC plastisols, silicones, etc.; and aircraft metal-to-metal construction adhesives which use so called adhesive "alloys" of epoxy, nitrile phenolics, and nylon components.

Adhesives End-Use Pattern

Construction	35%
Packaging	25%
Textiles	9%
Transportation	6%
Other	25%

HISTORY

The use of adhesives by man is known to date back to the ancient Egyptians who used them in applications such as bonding papyrus reeds or veneering furniture. Most of the adhesives used during this time were either an animal protein or starch base. Gum arabic is an example of a glue the Egyptians commonly used.

Even though the first commercial adhesive factory was built in England as early as 1700, little changed in adhesive technology till the arrival of synthetic polymers at the start of the twentieth century. Since then, the technological history of adhesives has been closely tied to the history of synthetic polymer development. Phenolic resins began to be used as adhesives in making plywood in 1912. Cellulose ester adhesives were commercialized in the decade of the 1920s. Goodrich developed cyclized rubber adhesives in 1927. Polychloroprene was used in adhesives in the early nineteen thirties. Urea-formaldehyde adhesives began in 1930. Carbide and Carbon Chemicals Co. developed polyvinyl acetate adhesives in 1939. Chlorinated rubber adhesives came along in 1940. Melamine-formaldehyde resin adhesives started in the 1940s. After

World War II, adhesive developments proliferated with the development of polyester, polyurethane, silicone, and epoxy adhesives by 1950. In a further proliferation in developments, by 1965 polyethylene, ethylene-vinyl acetate, polyvinyl ether, and cyanoacrylate were introduced commercially for adhesive applications. Today, new adhesives are developed and introduced almost yearly.

Because of the lower commercial volumes used in the adhesives industry compared to other polymer industries, many times chemicals and polymers used as raw materials for making adhesives are taken from raw material lines developed originally for use in such high volume industries as plastics, rubber, and paints. Therefore, many of the technical advancements that occur in adhesives have their origin in these other polymer industries.

MANUFACTURING PROCESS

The economies of scale involved in compounding adhesive formulations themselves are relatively small. In operations that formulate solvent or emulsion adhesives, the equipment can be quite simple. If one is compounding an emulsion (latex) adhesive, just simple mixers are needed of the desired capacity (say from 5 to 5,000 gallon). Also preblenders may be needed to make aqueous suspension of powder additives, in that all additives to the base latex have to be in a pourable liquid state when added to the mixer.

On the other hand, if one is compounding solvent based adhesives, more equipment may be necessary. For example, mixers for solvent based adhesives require special adaptations to assure that no accidental fires or explosions occur during their use. Also, such an operation will require solvent tanks and pipe lines as well as metering equipment and proper air ventilation to protect workers from excessive exposure to solvent vapors. Furthermore, if elastomers are to be used as the base polymer, then a rubber mill may be required to masticate the rubber (break it down) before it is added to the adhesive mixer (freshly masticated rubber generally requires less time to dissolve in a solvent). All these requirements add to the cost of an adhesive plant; however, still the cost is relatively low compared to some other industrial operations. Therefore, adhesive compounding is a competitive business which is not dominated by any one company or collection of companies.

On the other hand, the economies of scale and capital requirements for producing some of the synthetic base polymers used to make these adhesive formulations can be quite large. Relatively large economies of scale are characteristic of chemical plants carrying out polymerization reactions on a large scale to minimize per pound cost. As a result, the number of producers for a given synthetic polymer might be relatively few. In some cases, where a particular polymer has a very limited market, there may in fact be only one producer.

Adhesive plant laboratories are relatively small, usually consisting of equipment to measure the force required to pull apart two test substrates adhered together by the adhesive being tested. The American Society for Testing and Materials (ASTM) under committee D-14 gives a variety of standard adhesion tests used today. These include a variety of tests for lap shear, tensile, peel, cleavage, creep, fatigue, and impact. Two common tests are illustrated below.

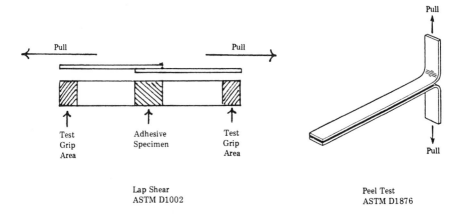

Lap Shear
ASTM D1002

Peel Test
ASTM D1876

BASIC PRINCIPLES

Adhesion is defined as the state of holding together two surfaces of bodies (called adherends) by interfacial forces which can be valence forces or mechanical interlocking or a combination of both.

Of course, in our discussion on adhesives, we will be most concerned with these valence forces through which adhesives work. Valence forces which are responsible for the surface attraction, can be either primary valence bonds or secondary valence bonds.

Primary valence bonds are the well known covalent, coordinate, electrovalent, or metallic bonds which are formed when atoms or molecules either share or transfer electrons. Their bonding forces range from 10–100 kcal/mole.

Far more important from the standpoint of adhesives, however, are the weaker secondary bonds known as Van der Waal forces (2–4 kcal/mol). These forces are responsible for the cohesion of non-polar liquid molecules, for example. Also, these weak forces can provide good adhesion between two surfaces; however, in order for these forces to act, molecules must be between 3 to 10 Å. Beyond 10 Å, these forces are no longer effective in that they decrease at a rate equal to the sixth power of the intermolecular distance. Even the smoothest of two solid surfaces cannot have enough surface area at intimate contact to achieve any adhesion between the two solids. This is because the apparent smooth surfaces are not at all smooth at the microscopic level as shown below.

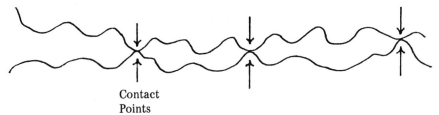

Contact
Points

This is where the role of the adhesive comes in. Most all adhesives, whether they are hot melts, evaporation types, chemical reactive types, etc, are applied to the adherend in a liquid state. The application as a liquid is necessary in order to sufficiently "wet" the surface of the adherend. This allows the adhesive molecules to get in close proximity to the surface molecules of the adherend in order to establish valence forces. Therefore, it is imperative that the adhesive being applied have sufficiently low viscosity to allow the adherend surface to be wetted properly. If the viscosity of the adhesive is too high, then it will not penetrate properly into the surface pores and crevices. This is important in that if gas bubbles or voids occur between the glue line and the adherend, stress will be concentrated more in a smaller area, thus weakening the strength of the bond. It is common practice in adhesive technology to clean and roughen the surface of an adherend before the application of adhesive. This is done not only to remove oxides from a metallic surface, but because a rough surface provides greater joint strength as well as helping to prevent crack propagation along the glue line by isolating air pockets that may have formed from incomplete wetting. The roughness provides "teeth" to the bonded joint. While the adhesive is still a liquid, it displays no cohesive strength to the bond. With cooling, vaporization of a carrier, or through a polymerization reaction, the adhesive will later become a solid which will then provide the cohesive strength to the bond.

In order for a given adhesive to effectively wet an adherend surface, that adhesive must be compatible with the adherend if a lasting bond is to result. Thermodynamically, for a bond between an adherend and an adhesive to be strong, there must be a resulting decrease in free energy (ΔG) from the combination of adhesive and substrate. In other words, the ΔG must be negative in the following equation.

$$\Delta G = \Delta H - T\Delta S$$

where

ΔG = Change in free energy
ΔH = Change in heat of mixing
ΔS = Change in entropy
T = Temperature

In general, when an adhesive and adherend meet, entropy is increased (more disorder), so entropy only contributes to the loss of free energy. Therefore, only the heat from mixing (ΔH), can prevent a negative ΔG from resulting. Thus, provided that the heat of mixing is not too great, the adhesive and adherend will combine. This is even more likely at elevated temperatures. Another way to look at it, is to say for the adhesive to work, the adhesive attraction between the liquid adhesive and the adherend must be greater than the cohesive forces within the liquid adhesive being applied. The following are examples of a "good" and "bad" adhesive bond.

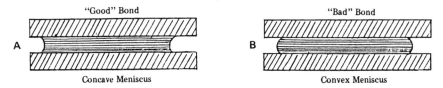

In figure A, the adhesive forces exceed the cohesive forces of the wet adhesive. Therefore, the adhesive spreads over the adherends and "wets" the adherend surface. In figure B, the cohesive forces of the wet adhesive are greater than adhesive forces between the adhesive and adherend. Therefore, in figure B, the adhesive is not compatible to the adherend material.

One method to help determine the compatibility between an adhesive and a given substrate was explored by W. A. Zisman of the Naval Research Laboratory in Washington, D.C. Zisman and his colleagues measured the critical surface tensions (γ_c) of different selected polymer surfaces. This critical surface tension measures the "wettability" of a given solid surface. γ_c gives the maximum surface tension that an adhesive can have. If the adhesive's surface tension exceeds γ_c, it will not be able to wet the surface of the given substrate. Zisman was able to measure γ_c of many surfaces by measuring the changes in contact angles of different selected liquids that were applied to the surface. All these liquids he used for his measurements already had known predetermined surface tensions. Therefore, the critical surface tension could be determined.

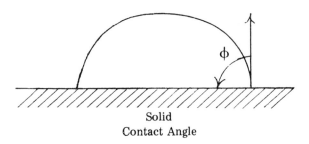

Solid
Contact Angle

As the contact angle approaches zero, the drop will cover a larger and larger area. When $\phi = 0$, complete spreading of the liquid has occured across the surfaces. Therefore, ϕ is inversely a measure of wettability.

In general, it can be noted that polymers that are highly polar in molecular structure give relatively high critical surface tensions. For example, nylon 6,6 (a polyamide) has $\gamma_c = 42$ and polyester (poly ethylene terephthalate) has a $\gamma_c = 43$. Conversely, the γ_c for silicone rubber is 22 and for teflon (polytetrafluoroethylene) it is only 18. The difference we see among these polymers is due to differences in the dispersing force component of intermolecular attraction.

Another method of measuring compatibility is by using Hildebrand solubility parameters. This solubility parameter δ is equal to the square root of the cohesive energy density (internal pressure) as shown below.

$$\delta = \sqrt{\Delta E / v}$$

where
$$\Delta E = \text{Change in energy of vaporization}$$
$$v = \text{Molar volume}$$

The energy of vaporization for low molecular weight liquids is easily measured. Fluorocarbon and hydrocarbon liquids are well known to have lower energy of vaporation values per cc compared to polar liquids of the same molecular weights such as alcohols or water. The energy of vaporation cannot be measured directly for polymers because they are obviously non-volatile. Therefore, indirect methods must be used to measure the solubility parameter δ of polymers. These values can be calculated, but the most common method is to simply compare the solubility properties of a given polymer to different selected solvents of known values in order to find the best solubility. A strong positive correlation has been reported between the critical surface tension (γ_c) and the Hildebrand solubility parameter (δ) for a wide range of different selected polymers. Although this correlation is by no means perfect, it does show that both γ_c or δ can be used to help predict the compatibility of an adhesive with a select adherend.

In general, different solids have different specific surface free energies (ergs/cm^2). Solids with strong intermolecular forces, such as carbides, metal oxides, metals, etc, have high energy surfaces. On the other hand, very low melting organic solids of weak intermolecular forces have low energy surfaces. Glasses and some salts are considered intermediate. Lastly, most all liquids are considered to have low energy surfaces. Therefore, liquids generally will easily spread over the high energy surface solids. For low energy surfaces such as Teflon, this is not the case. A high energy surface can also be converted to a low energy surface by the treatment with say polymethylsiloxane (silicone) to form a monomolecular layer across the surface. From this treatment, a liquid that previously would have spread quite readily is repelled instead. This is why silicone is used as a mold release agent in many applications.

So far discussed are those conditions necessary to obtain a good adhesive bond. But there is still one other condition that should be mentioned concerning the nature of the glue line itself. As mentioned earlier, an adhesive is applied in a liquid state, but later hardens to provide the necessary cohesive strength to the bond. The adhesive will harden by cooling, if it is a hot melt; by loss of a volatile carrier, if it is a solvent based or latex adhesive; or by polymerization, if it is a reactive adhesive. All these methods can cause the applied adhesive (glue line) to shrink, because of thermal contraction, loss of carrier mass, or loss of a low molecular weight by-product from a polymer condensation reaction. If this shrinkage is not controlled, it can create strains within the glue line which can result in an adhesive failure. Such methods to control this contraction are as follows.

(1) Use fillers in compounding the adhesive to reduce the contraction.

(2) Select an adhesive that either does not greatly contract or is somewhat flexible in its hardened state.

(3) Try not to make the glue line too thick.

(4) If the adherend is impervious, and solvent based adhesive is being applied, allow the greater part of the solvent to be lost after application before joining the two adherends to form the bond.

FUNCTIONAL USES

Before considering the functional uses of various chemicals and polymers used in compounding adhesives, we should first discuss the basic types of adhesives. These types are listed below.

(1) Chemically reactive types

(2) Evaporation types

(3) Hot melt types

(4) Pressure sensitive types

The chemically reactive types of adhesives usually involve the polymerization of low molecular weight liquid components upon application to form a polymerized glue line which will have good cohesive strength. There are two basic types of chemically reactive adhesives. One type consists of those reactive adhesives which give off a low molecular weight by-product (usually water) during a polymer condensation reaction. Examples of this type are the phenol-formaldehyde and resorcinol-formaldehyde adhesives. The other type of reactive adhesive polymerizes without the formation of a low molecular weight by-product sometimes through addition polymerization. Examples of this type are epoxies, cyanoacrylates, urethanes, and polyesters.

Evaporation type adhesives consist of a polymeric material being carried in a liquid state by a liquid carrier whether it is as an aqueous colloidal dispersion (latex) or as a solution in an organic solvent. Of course after application, the liquid carrier must be lost by evaporation or by diffusion into a porous substrate in order to obtain a solid glue line and good cohesive strength. The organic solution adhesives are faster drying than the latex adhesives; on the other hand, for the glue line to have good cohesive strength may require a higher molecular weight polymer. Generally, the higher the molecular weight, the higher the solution viscosity. But if the viscosity becomes too high, the adhesive may not wet properly. Therefore, there is a limit of say 30% or so in the total solid content of a solution system. A latex adhesive, on the other hand, has an advantage in that it can have a higher total solids content without raising the viscosity to excessively high levels. At levels above about 50% polymer concentration, latex particle agglomeration may occur.

Elastomeric adhesives are commonly used in both solvent solution form and in latex form. One complicating factor in classifying adhesives is that either of these adhesive forms may or may not contain rubber curatives. The rubber curatives are sometimes added to initiate rubber vulcanization and make the glue line stronger than it would be otherwise. Then these two types of elastomer adhesives with certain additions can be classified as both vaporation type and reactive type adhesives.

The third type of adhesive is called a hot melt adhesive. It consists of a thermoplastic polymer and usually other additives. This adhesive when applied hot (above melting) to a given substrate, effectively wets the surface; however, on cooling, the molten polymer returns to its solid form providing good cohesive strength to the bond. The advantage to hot melt adhesive types is that they set very quickly after applied.

Lastly, we have the pressure sensitive type adhesives. As you recall, the adhesives we have been discussing were all applied to a substrate in a liquid state in order to "wet" the substrate. Then through various means, that liquid is converted into a solid form. In the case of a pressure sensitive adhesive, this physical transformation does not occur. A pressure sensitive adhesive is, in a way, both in the liquid stage (for wetting) and in the solid stage (for cohesive strength) at the same time. A pressure sensitive adhesive consists of a viscoelastic material which is fluid enough to wet a surface under a slight applied external pressure, yet cohesive enough to provide some moderate strength in holding the adherends together after external pressure is released. Typically, these pressure sensitive adhesives are based on various elastomers which have been made sticky with the use of resin additives. The applications in providing adhesion for tape and labels are well known. One obvious disadvantage of a pressure sensitive adhesive is its relatively poor cohesive strength compared to other adhesive types. All these adhesives must only be applied when there is adequate ventilation for their use.

REACTIVE TYPE BASE POLYMERS

Epoxy Resins

These resins are usually the product of a reaction between epichlorohydrin and a diol which is generally either bisphenol A, bisphenol of formaldehyde or glycerin. The following shows the synthesis of one of the most common epoxy resins (the reaction product from epichlorohydrin and bisphenol A (BPA).

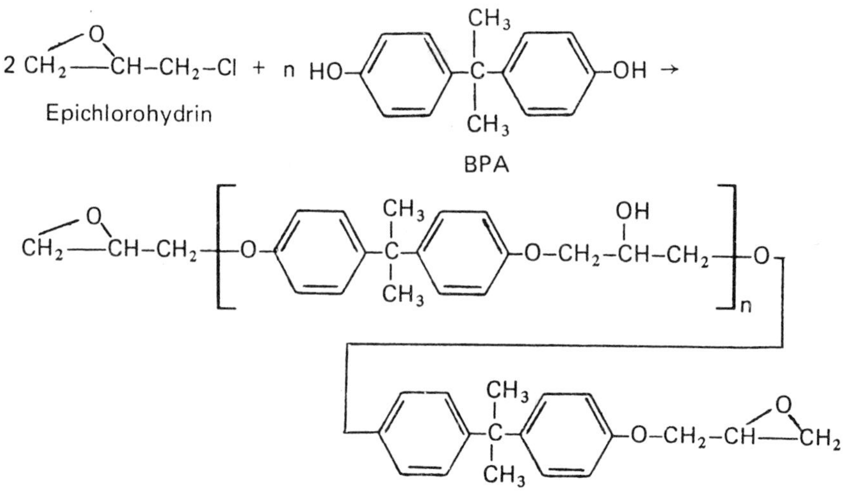

These epoxy resins are generally used in adhesive applications in a two part system in which a hardener must be added to the resin just before application. The hardener (usually an amine or anhydride) forms crosslinks between the epoxy resin molecules, thus converting the liquid resin into a hard matrix. This curing is achieved through an epoxy ring opening reaction with primary or secondary amines or acid anhydrides as hardeners.

Epoxies, because of their polar groups, display very high adhesive strength to metals, ceramics, glass, and other polymers. Epoxies are famous for their versatility. They can be used to bond most any two surfaces no matter how different. Also epoxies, after reacting with a hardener, give very high cohesive strength as well. Other advantages of epoxies are that they are highly reactive and do not give off high levels of volatile reaction by-products (some amine hardeners are volatile, however, and do flash evaporate during curing). Also epoxies shrink very little during cure. Their shrinkage is much less than other reactive polymers such as polyester, for example. The addition of fillers to the adhesive also reduces shrinkage even further. Because of the toxic vapors which may be given off when working with epoxies, all work with these resins should only be done in areas having adequate ventilation.

When epoxy adhesives set, they form a very hard, strong and tough glue line. In fact, for some applications, epoxies may be too hard. Unmodified epoxies lack flexibility and have poor peel strength. In order to improve flexibility, peel strength and ability to hold strength through wide temperature variations, special epoxy "alloys" are formulated with NBR, urethane, acrylics, and nylon. These special "alloy" formulations are now used extensively in making aircraft construction adhesives. For epoxy adhesive use, this area is a relatively large market. Although the cost of these adhesives is high, they meet the tough service requirements and specifications demanded for aircraft construction applications.

Polyurethanes

This polymer is also a product of a reaction of two different components. One is a polyhydroxy material. The other component is a di or polyisocyanate. (All precautions should be taken when working with these isocyanates because of their toxicity. They can also cause workers to have respiratory reactions. Proper ventilation is particularly important.)

A urethane is formed from the reaction of an isocyanate with an alcohol as shown below.

$$O=C=NRN=C=O + 2[HO\sim\sim OH] \rightarrow HO\sim\sim O\overset{\overset{\displaystyle O}{\|}}{C}NHRNH\overset{\overset{\displaystyle O}{\|}}{C}O\sim\sim OH$$

Just as with the epoxy curing reaction described earlier, no volatile by product is produced during this reaction. In a similar fashion urethane as an adhesive is very compatible to most substrates and can be used to bond very dissimilar materials. On the other hand, polyurethanes, unlike epoxies, are flexible and rubber like, usually being only lightly crosslinked (cured epoxies are very rigid). This flexibility enables urethane adhesives to be used in the bonding together of flexible plastics. Urethane's flexibility provides superior peel strength. Also urethane adhesives have an advantage over epoxies in providing good adhesion at cryogenic temperatures. (This superiority applies also to the epoxy-nylon "alloy" adhesives as well).

Polyurethanes can also be polymerized and crosslinked further by the use of amines or water (both forming polyureide bonds).

$$RN{=}C{=}O + R'NH_2 \rightarrow RNHCONHR'$$

$$2RN{=}C{=}O + H_2O \rightarrow RNHCONHR + CO_2$$

It should also be noted that some isocyanates are available in a blocked form in which the reactive isocyanate groups are not made available for reaction until they are heated to a high temperature. Some examples of these blocked isocyanates are given below.

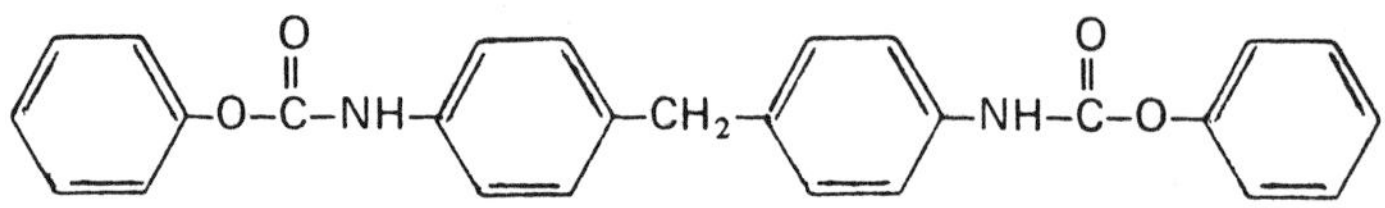

Methylenebis phenyl (4-phenyl carbamate)–A blocked MDI

(Gives off toxic phenol and MDI on thermal decomposition)

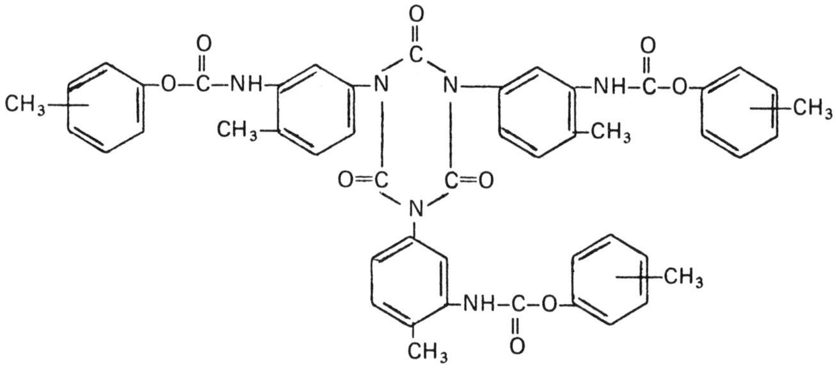

A Cresol Blocked TDI

(Gives off toxic cresol and TDI on thermal decomposition)

It is obvious that through the use of blocked isocyanates, it is possible to have a one-component urethane adhesive. A polyhydroxy material can be blended with the blocked isocyanate without reaction until the blend is heated to approximately 300° F.

Although the markets for polyurethanes have grown rapidly, their use in adhesive applications is not nearly as extensive as epoxies.

Cyanoacrylates

These reactive polymers are used in one part adhesive systems which are cured quickly from contact with moisture in the air.

Eastman Kodak developed the first commercial cyanoacrylate adhesive in 1958 under the trade name of Eastman 910. Eastman used methyl 2-Cyanoacrylate and ethyl 2-cyanoacrylate in producing most of the cyanoacrylate adhesives used today.

The principle for the fast curing of these adhesives is that these alkyl 2-cyanoacrylates can be rapidly polymerized when exposed to a weak base such as water. This polymerization is by a carbanion mechanism shown below.

$$B: + CH_2=\underset{\underset{COOCH_3}{|}}{\overset{\overset{CN}{|}}{C}}-COOCH_3 \rightarrow B CH_2\underset{\underset{COOCH_3}{|}}{\overset{\overset{CN}{|}}{C}}:^- \xrightarrow{CH_2=\overset{\overset{CN}{|}}{C}-COOR} B CH_2\underset{\underset{COOCH_3}{|}}{\overset{\overset{CN}{|}}{C}}-CH_2-\underset{\underset{COOCH_3}{|}}{\overset{\overset{CN}{|}}{C}}:^-$$

The very strong electronegative groups (CN and COOR) create conditions which permit a weak base (water) to initiate polymerization.

The advantages of cyanoacrylate adhesives are as follows:

(1) They are a one-step adhesive system.

(2) They cure at room temperature.

(3) They can bond very different substrates.

(4) They form strong bonds.

Because cyanoacrylates are one component systems which cure very fast at room temperature, they have been used in many manufacturing assembly like operations requiring fast cures. On the other hand, these adhesives have disadvantages in regard to moisture resistance. Again, as with polyurethanes and epoxies, no volatile by-product is given off from the setting of the cyanoacrylates.

Phenol-Formaldehyde Resins

These resins are formed from the controlled polycondensation reaction of phenol and formaldehyde. On a pound volume basis, these resins represent the largest volume usage in the production of synthetic adhesives. By far the largest single application for phenolic adhesives is in the production of exterior plywood. These phenol formaldehyde adhesives are spread on wood veneer by using roller coating to make the plywood. Phenolics have advantages in this application in that they are relatively inexpensive and water resistant. Other important areas of application include the use of phenolic resin in brake lining composite, adhering foundry sand together, contact adhesive for shoes and furniture, and in aircraft construction adhesive alloys.

There are basically two types of phenol-formaldehyde resins used (in either powder or liquid forms). One of these types is called a *two step novolac*. The novolac phenolic resin is produced with an acid catalyst from the polycondensation of phenol and formaldehyde with a stoichiometric excess of phenol in order to prevent the formed polymer from reaching a gel-point (crosslinking three-dimensionally). In order to cure a novolac phenolic resin, a methylene donor such as hexamethylene tetramine is added to the resin as a powder (these agents will be discussed more in detail under the "hardener" section). Upon reheating, the resin will then set up (gel) in its intended adhesive application.

The other type of thermosetting phenolic resin is called a *resole* resin. This resin is produced from the polycondensation of phenol and formaldehyde using an alkaline catalyst and a higher level of formaldehyde than used with novolacs. To prevent it from gelling, the resin polymerization is short stopped by cooling before the gel point is reached. Therefore, resole resins do not require a methylene donor additive (second step) to be used in adhesive applications. Instead just reheating will bring them to gelation. Thus this resin is called a *one step resin*. Resoles are commonly used in aqueous or alcoholic dispersions. (Dispersions are ideal for penetration in plywood applications).

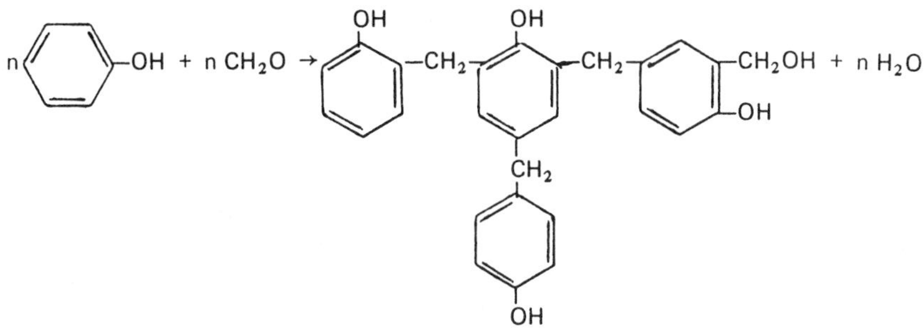

Phenol is not the only reactant used to make phenolic type resins. Also cresols, xylenols, tert-butyl phenol, p-octyl phenol, and cardanol are also used.

Unlike previously discussed polymers such as epoxies and polyurethanes, when phenol formaldehyde resins set, they give off water of condensation which must be ultimately disposed of during adhesive bond formation.

Just as with epoxies, when phenolic resin adhesives set, they form a strong but rigid and brittle bond. In certain select applications such as construction adhesives, it is important to reduce the brittleness and make the adhesive tougher. In order to achieve this, phenolics are commonly compounded with elastomers such as NBR or neoprene. These so called "alloys" are stronger and tougher than adhesives made solely of phenolic resins or just the elastomers. The neoprene/phenolic "alloy" gives high peel strength. The NBR/phenolic combination gives oil resistance as well. In deciding the level of phenolic vs NBR to use, the higher the phenolic, the better the high temperature strength of the compound. While, the higher the NBR level, the better the impact resistance.

Resorcinol-Formaldehyde Resins

Resorcinol is many times more reactive with formaldehyde than phenol. The reason for this is that resorcinol has two hydroxyl groups which are in the meta position to one another. This makes the ortho and para positions extremely active, in fact, much more active and exothermic than the ortho or para positions for phenol. Because of the very reactive nature of resorcinol compared to phenol, resole resins, which are quite stable at ambient temperatures for phenolic resins, are not stable (will gel prematurely) under ambient temperatures if based on resorcinol. Therefore novolac resorcinol formaldehyde resins are used instead along with a part "B" containing a methylene donor.

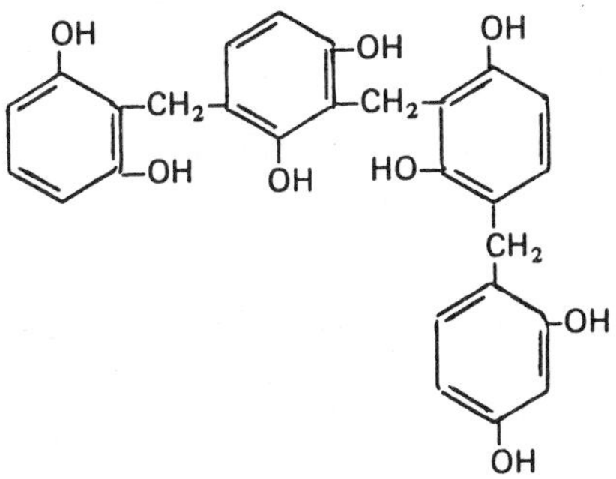

Just as with phenol-formaldehyde condensation, resorcinol-formaldehyde also gives off water of condensation as a by-product of polymerization. Unlike P.F. resins, resorcinol-formaldehyde resins are very water soluble.

Because of resorcinol formaldehyde resin's fast curing properties, in adhesive applications they are used extensively in production of laminated timber and beams. Also, in some applications, resorcinol and phenol are used together with formaldehyde to manufacture resins for use in construction adhesive applications.

Another important application for resorcinol-formaldehyde resins is in tire cord adhesive. The R F resin is mixed with a styrene-butadiene-pyridene latex to achieve good rubber to textile tire cord adhesion. This mixture is called an RFL dip in which the tire cord is pre-dipped before being calendered with the rubber. Upon curing of the rubber product, the adhesive is activated.

Urea-Formaldehyde Resins

This polymer is formed in a similar manner described earlier, ie polycondensation reaction. Here, the subject polymer is formed from the polycondensation of urea and formaldehyde. As with phenol-formaldehyde condensation, water is also given off as a by-product as well.

Unlike phenol-formaldehyde, urea-formaldehyde condensation gives a water soluble product. Usually urea-formaldehyde resins are sold in aqueous solutions for adhesive applications.

Some of the advantages of urea-formaldehyde adhesives over other adhesives are their relatively low cost and clean colorless appearance. However, a major disadvantage of these resins is their poor water resistance relative to a phenolic adhesive. As a result of these qualities, urea-formaldehyde adhesive has found the greatest use by far in the production of particle board (a composite of wood chips and urea-formaldehyde adhesive). Another large area of use for urea-formaldehyde adhesive is in the manufacture of interior plywood. Although plywood made from urea-formaldehyde adhesive may be somewhat water resistant, it is not considered as water resistant as exterior plywood (exterior plywood is commonly made using phenolic resin adhesives.) It is important that a UF adhesive polymer be chosen which was produced from a molar ratio of urea to formaldehyde which will assure that *no* free formaldehyde will be given off later from the product during its uses. (Formaldehyde is toxic and imparts an odor.)

$$NH_2CONH_2 + 2CH_2O \rightarrow HOCH_2NHCONHCH_2OH$$

$$RNHCH_2OH + HOCH_2NHR' \rightarrow RNHCH_2OCH_2NHR' + H_2O$$

These urea-formaldehyde resins are quite reactive and can be used at room temperature. Generally a catalyst is added to cure these resins (usually ammonium chloride or ammonium sulfate). One of the advantages of these polymers is their ease of cure.

Melamine-Formaldehyde Resins

The chemistry involved in the polycondensation of melamine and formaldehyde is very similar to that previously discussed with urea and formaldehyde. One important difference is that melamine is hexafunctional in these condensation reactions while urea is tetrafunctional as seen from a comparison of their structures.

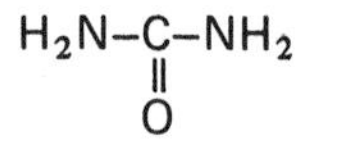

$$H_2N-\underset{\underset{O}{\|}}{C}-NH_2$$

Urea
(Tetrafunctional)

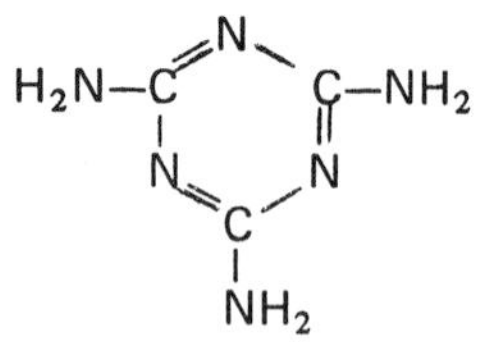

Melamine
(Hexafunctional)

Generally melamine resins are more expensive than urea resins and are therefore not used as widely as urea resins. The end use adhesive applications for urea and melamine resins are very similar; however, melamine resins have some technical advantages over urea. For one thing, melamine, being hexafunctional, can provide a higher crosslink density than ureas. As a result, melamine resins impart a higher degree of water resistance even at elevated temperatures which urea resins cannot provide. Also melamine resins have better heat stability than urea resins.

THERMOPLASTIC TYPE BASE POLYMERS

Unlike the reactive polymers previously discussed, thermoplastic polymers do not chemically react or crosslink on heating. Instead, thermoplastic polymers will melt at a given elevated temperature and can be applied to a substrate as a liquid to enable the adhesive to "wet the surface." Upon cooling, these thermoplastic polymers harden again forming a hard glue line, many times reforming regions of crystallinity in the polymer matrix. For this reason, thermoplastic polymers are widely used in adhesive applications in the so called "*hot melts*".

In addition to the base polymers to be discussed, it should be remembered that most hot melt adhesives contain other additives to aid the polymer in its end use. These additives (to be discussed later) are tackifying resins, waxes, plasticizers, fillers, and antioxidants which render the hot melt more effective as an adhesive.

Polyethylene

This low cost polymer, on a pound-volume basis, is the largest single polymer used in hot melt applications. However, even though large amounts are used, polyethylene homopolymers have rather limited applications being mainly restricted to non-critical porous substrates where high adhesive strengths are not required. These applications consist of the paper products areas (adhesives for cartons, multiwall bags, etc). Unlike other thermoplastic polymers (to be discussed), polyethylenes do not possess functional groups attached to the polymeric backbone which might help enhance adhesive strength and increase compatibility with different adherend surfaces.

$$CH_3 + CH_2 \rightarrow_n CH_3$$

As a result, polyethylene hot melt adhesives display relatively poor strength in application.

One point to note about polyethylenes is that high density polyethylenes (linear structure and high crystallinity) have higher cohesive strengths than low density polyethylenes (branched chains). On the other hand, low density PE has better fluidity and wetting power in the melted state.

Ethylene Vinyl Acetate (EVA)

This thermoplastic polymer is the most versatile and widely used thermoplastic polymer in hot melt adhesive formulations. This is mainly because of the acetate functional group which provides good adhesive strength to polar substrates. Also the polymer is relatively inexpensive.

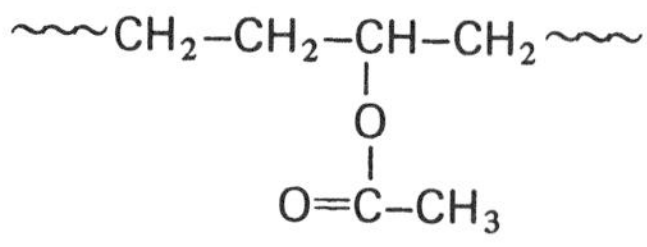

$$\sim\sim CH_2-CH_2-CH-CH_2 \sim\sim$$
$$|$$
$$O$$
$$|$$
$$O=C-CH_3$$

Ethylene Vinyl Acetate

EVA polymers commercially available vary considerably in % vinyl acetate and average molecular weight. The higher the % VA content, the better will be the adhesion with polar adherends. On the other hand, a low % VA content is needed for good adhesion to non-polar substrates such as olefins.

A lower melt index (higher average molecular weight) for an EVA polymer will generally improve low and high temperature adhesive strengths.

One of the true advantages of EVA over other thermoplastic polymers is its wide latitude of compatibility for being compounded with tackifying resins, plasticizers, waxes, etc.

Another quite different acetate polymer called *polyvinyl acetate* homopolymer (PVAc), as well as being used in hot melts, is also used in a latex form to make the very popular household "white glue" which has a large share of the household adhesive market. This aqueous emulsion form of PVAc has gained wide acceptance and is used in bonding porous adherends such as wood, paper, cloth, etc.

Another polymer, called *polyvinyl alcohol*, is also used in aqueous adhesive systems. When polyvinyl acetate is hydrolyzed during manufacture, it is converted to polyvinyl alcohol. Polyvinyl alcohol is water soluble and can be used in water solutions. It competes with other naturally derived water soluble adhesives on the market. Its advantage over natural adhesive is less odor and more uniformity.

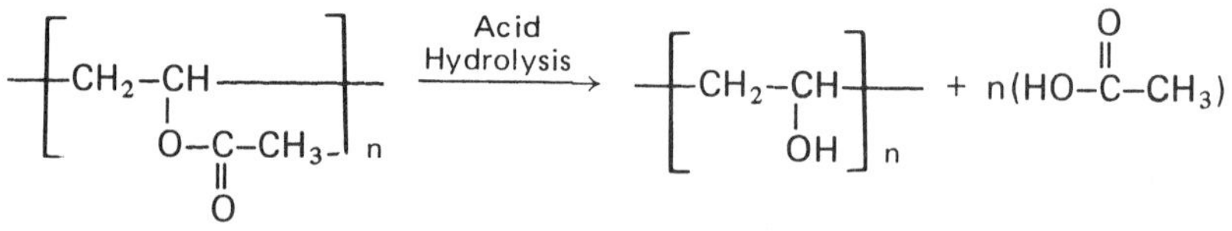

$$\left[CH_2-CH \atop \underset{\underset{O}{\overset{\|}{}}}{O-C-CH_3} \right]_n \xrightarrow[\text{Hydrolysis}]{\text{Acid}} \left[CH_2-CH \atop OH \right]_n + n(HO-\overset{\overset{O}{\|}}{C}-CH_3)$$

Polyvinyl Acetate   Polyvinyl Alcohol

Care should be taken here to not confuse the homopolymer, polyvinyl acetate, with the copolymer, ethylene vinyl acetate.

Ethylene Ethyl Acrylate (EEA)

This is another ethylene copolymer that is commonly used in hot melt applications. Its chemical structure is shown below.

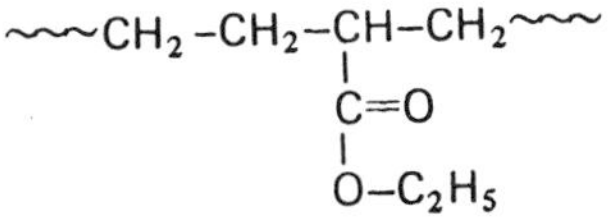

$$\sim\sim\sim CH_2-CH_2-\underset{\underset{\underset{O-C_2H_5}{|}}{\underset{C=O}{|}}}{CH}-CH_2\sim\sim\sim$$

EEA is used in similar end use applications as EVA (discussed previously). However, EEA does have certain advantages over EVA. Some of these advantages are a broader range of compatibility with different substrates, better resistance to heat and a tougher polymer.

Ethylene Acrylic Acid (EAA)

This ethylene copolymer is used as a specialty polymer in hot melt applications where high bonding strength to glass or metallic surfaces is needed. One disadvantage of EAA compared to EVA (besides being more expensive) is its poorer heat stability.

SBS and SIS Blocked Thermoplastic Rubber

These thermoplastic elastomers are unique in that they display high modulus and resilience at room temperature, but behave as thermoplastics at elevated temperatures (these polymers were discussed in detail in the last chapter). They consist of block polymers having a rubbery middleblock (either polyisoprene or polybutadiene) and hard, glasslike polystyrene endblocks. Because each macromolecule contains blocks of units consisting of rubber and plastic units, there are two glass transitions instead of the one glass transition that would result if the units were all randomized (as with say SBR). Thus, these blocked polymers (available under the trade name of Kraton by Shell or Solprene by Phillips) have two separate phases. As discussed in the last chapter, this explains their rubbery and thermoplastic properties. However, the two phases also present special problems for hot melt adhesive applications. For one thing, different hydrocarbon resin loadings affect the polymer in different ways. For example, aromatic coumarone-indene resins solubilize in the endblocks but not in the midblocks. Therefore these resins will improve endblock adhesion, increase the drawing tendency, and stiffen the compound. On the other hand, aliphatic resins (olefin types) tend to solubilize in the midblocks, but not in the endblocks. They improve adhesion of the midblocks to polar surfaces and soften the compound. Also, the aliphatic resins may hurt low temperature properties of hot melts.

Polyamide Resins

These polymers, (formed from a polycondensation reaction between a diacid and a diamine) range from low molecular weight branched polymer liquids to high molecular weight nylons. As mentioned previously, poly-amides are commonly "alloyed" with epoxies in film adhesives for metal-to-metal construction adhesive applications. Also polyamides are com-monly used as thermoplastic polymers in hot melt adhesive applications. One common use for these hot melt adhesives is in shoe manufacture. An advantage of polyamide hot melt compounds is their strong cohesive strength when cooled just a little below the polymers melt point. This high cohesive strength is because of the hydrogen bonding between polar groups of different chains which provide high material strength at rela-tively low molecular weights in some cases. On the other hand, a lower molecular weight high-strength polymer, as just described, can have advantages in having a relatively low viscosity just above the melt point, thus wetting a surface well. In addition, polyamides, because of their functional groups, provide good adhesion to greatly different non-porous substrates.

Polyvinyl Butyral (PVB)

This polymer is commonly used in hot melt adhesive applications, melting at temperatures above 260° F. Its structure is given below.

$$\left[CH_2-CH-CH_2-CH-CH_2-CH\right]_n$$

with substituents OH on the first CH, and the second and third CH groups bridged by O to a CH bearing $CH_2CH_2CH_3$.

Polyvinyl acetals, such as PVB, are not polymerized directly from a monomer, but rather prepared from reacting aldehydes and polyvinyl alcohol. As previously discussed, polyvinyl alcohol is also not obtained directly from a monomer (Vinyl Alcohol does not exist as a monomer per se). Instead, it is derived from the hydrolysis of polyvinyl acetate which in turn is produced directly from its respective monomer. Thus, the syn-thesis of PVB is a rather involved process.

By far, the largest and most well known use for PVB is the production of safety glass such as used in automobiles. The use of a plasticized PVB hot melt for this application is ideal. This is so because the polymer possesses good optical clarity, good tensile and elastic properties, and excellent adhesion to glass surfaces. The adhesive is applied as a clean interlayer between two pieces of glass to help prevent fragments of shattered glass from breaking loose and hitting people.

One interesting point to make, is that PVB by itself has poor moisture resistance. High moisture on glass surfaces gives relatively poor adhesion with PVB. In order to provide good moisture resistance to the PVB hot melt formulation, *polyvinyl ether* is compounded into the adhesive mixture. Polyvinyl ethers are discussed next.

Polyvinyl Ethers

The general molecular structure for these polymers is shown below.

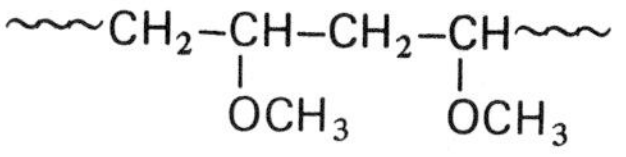

$$\sim\!\sim\!\sim CH_2\!-\!CH\!-\!CH_2\!-\!CH\!\sim\!\sim\!\sim$$
$$\qquad\quad |\qquad\quad |$$
$$\qquad\quad OCH_3\quad\ OCH_3$$

Compared with polyvinyl butyral, polyvinyl ethers have relatively poor cohesive strength. On the other hand, polyvinyl ethers have the special property of being water soluble. Therefore, one of the many uses of polyvinyl ether is in the making of remoistable adhesives which are commonly used in label applications. Polyvinyl ethers are also used in hot melts in bookbinding and label bonding. Also, as discussed earlier, polyvinyl ethers are used to improve adhesion of polyvinyl butyral to moist glass adherends, i.e. safety glass.

ELASTOMERIC BASE POLYMERS

Introduction

Elastomers can be used in adhesive formulations in basically four compound forms.

(1) In a latex compound without any curatives.

(2) In a latex compound with curatives added.

(3) In a compound solvent solution without any curatives.

(4) In a compound solvent solution with curatives added.

The advantage of the rubber latex adhesive form over rubber solvent adhesive is the lower viscosity which gives improved wetting power at a equal total solids level. Also, latex adhesives are safer to work with in that they are generally non-flammable and do not expose the user to organic solvent vapors which might be considered hazardous under certain conditions. On the other hand, solvent rubber adhesives have an advantage over latex rubber adhesives in generally having a faster drying time.

It was also noted that some latex and solvent rubber cements are used without adding curatives while other solvent rubber based cements are commonly used in two part systems in which curatives are added to the cement pot just before application (if the curatives were added too early to the pot, gelation would occur). The addition of a curative to a rubber adhesive will increase its cohesive strength through the initiation of sulfur crosslinking among rubber molecules. On the other hand, rubber adhesives using no curatives, can be used in a one part system and generally have a much longer pot life before application.

One point should be made about the selection of the molecular weight for an elastomer to be used in a solvent based adhesive. The higher the molecular weight of the elastomer, the stronger will be the cohesive strength of the glue line when the adhesive dries. On the other hand, the higher molecular weight elastomers will have a higher solution viscosity when applied and therefore may not wet the substrate as well, thus displaying poorer adhesive strength. Consequently, the selection of the base elastomer's average molecular weight can occasionally get caught in a trade off between good cohesive strength vs good adhesive strength.

Generally, the various base elastomers used do not have enough natural stick and tack to sufficiently adhere to many substrates. Therefore, in order to improve the adhesive strength of rubber adhesives, they are commonly compounded with tackifying resins in order to enhance adhesive strength. (Tackifiers will be discussed later). Also, rubber adhesive formulations contain antioxidants, carbon black, fillers, plasticizers, and colorants. These additives were discussed in the last chapter on the rubber industry.

One adhesive use area that is rather unique to elastomers is pressure-sensitive adhesive applications. Elastomers provide the necessary combination of viscoelastic properties to work for this use. The elastomeric pressure sensitive adhesive must be viscous enough to flow under modest pressure in order to "wet" the substrate surface. On the other hand, the pressure sensitive adhesive must be strong enough to have some degree of cohesive strength.

Not only are elastomers used as the base polymer in a wide variety of adhesive applications, they are also used extensively in *caulks* and *sealants* because of their ability to expand and contract easily with temperature change. Brittle polymers in the same application would crack with temperature changes.

Because elastomers were described in detail in the last chapter, we will only discuss the advantages and disadvantages of using each elastomer in adhesives applications.

Natural Rubber

This elastomer has better tack than other elastomers, but still requires tackifying resins in its application. Natural rubber is used in vulcanizing and non-vulcanizing latex adhesives as well as vulcanizing and non-vulcanizing solvent adhesives. Natural rubber adhesives are used in such applications as carpet adhesives, tile adhesives, and consumer pressure sensitive adhesives.

Styrene-Butadiene Rubber (SBR)

Unlike natural rubber, SBR has poorer tack properties and poorer cohesive strength. This deficiency is compensated for by compounding tackifying resins with SBR. Nevertheless, more SBR is consumed in adhesive applications than natural rubber because SBR has historically been less expensive.

By far the most commonly used process for producing SBR is through the emulsion polymerization process (discussed in the last chapter). There are two types of emulsion SBR: "hot" emulsion SBR (polymerized at 122° F) and "cold" emulsion SBR (polymerized at 41° F with a different catalyst system). "Cold" SBR is by far the more common type used today in rubber compounding because of its superior cured physical properties. However, in relation to adhesive applications, "hot" SBR is preferred because of its broader molecular weight distribution and its lower "trans" content. Hot SBR has a larger low molecular weight fraction which imparts to it superior tack and stick properties than "cold" SBR.

Although SBR is soluble in hydrocarbon solvents, many times, unless it is in a crumb form, it must be "broken down" or masticated on a mill beforehand, in order to shorten solution time in the making of a solvent based cement. On the other hand, SBR made from the "solution" polymerization process, requires less time to dissolve in a given hydrocarbon solvent than emulsion SBR requires. Also, "solution" SBR gives lower solution viscosities than the comparable "emulsion" SBR.

If an even lower viscosity is needed for an SBR solvent cement, then emulsion SBR crosslinked during polymerization with divinylbenzene (DVB) is the proper choice. These crosslinked SBR polymers do not form a solution with a hydrocarbon solvent, but rather a suspension (a sol-gel) Thus crosslinked SBR is used in making sprayable adhesives.

SBR adhesives are used in a wide variety of applications such as carpeting, packaging, pressure sensitive adhesives for consumer use, mastics and caulking compounds.

Butyl Rubber

This elastomer is used in many sealant and caulking applications as well as in making pressure sensitive adhesives such as for electrical tape. It is similar to SBR in that it must be compounded with resins to have sufficient tack properties. Butyl has advantages over SBR and NR adhesives in that it is more resistant to oxidative degradation caused by aging, weathering, or heat. ChloroButyl is a halogenated butyl rubber which is used when a faster cure of better compatibility with other elastomers is needed.

Acrylonitrile-Butadiene Rubber (NBR)

Compared to other elastomers, NBR is very polar in structure. In fact, the higher the bound acrylonitrile content (% VCN), the greater the polarity of the copolymer. This higher polarity gives NBR a unique advantage over other elastomers in that it is compatible at relatively high loadings of a variety of different resins including polyindene, phenol formaldehyde, alkyd resins, natural rosins, epoxies, and PVC. In adhesive applications, many times copolymers with high bound VCN content are selected because the greater polarity allows a higher resin loading. Thus, NBR can be compounded with these resins in adhesive applications to form an adhesive that has very high strength. Not only do NBR adhesives display high cohesive strength, they also display good adhesion to polar substrates such as wood, paper, and textiles. As mentioned earlier, the NBR-resin adhesives are also used in aircraft construction. Another advantage to NBR adhesives is their resistance to oil attack. NBR adhesives are available in solvent-based cements and mastics as well as in latex form. These forms are used in aircraft, automobiles, electronics, and paper industries.

Polychloroprenes (Neoprenes)

This elastomer is more expensive than elastomers such as SBR and natural rubber; however, it is used extensively in adhesives because of the following advantages. First of all, neoprene develops high bond strength quickly because it contains a high trans-1,4, content which provides fast crystallization. In fact, special neoprenes made specifically for adhesives use may contain as high as 90% trans-1,4, (as with Dupont's neoprene AC).

Other advantages to neoprene are superior resistance to attack by solvents, oils, water, acids, chemicals, heat, and weathering. Also some neoprenes have an advantage in displaying good tack properties.

Neoprene adhesives, because of the speed in which they form bond strengths, are used in shoe production (as in cementing the sole), in plastic lamination to wood or steel, and in automobile applications such as adhering vinyl trim, and tops.

Carboxylic Elastomers

These are elastomers available commercially that have carboxyl (-COOH) functional groups in their structure in order to enhance adhesion. Carboxylated forms of polychloroprene, polybutadiene, and acrylonitrile-butadiene rubber are available. It has been found, for example, that cements made from carboxylic Butadiene-acrylonitrile copolymer can give better rubber-to-metal adhesion than when non-carboxylated butadiene-acrylonitile is used. Carboxylic butadiene-acrylonitrile copolymers are used with phenolic resins to achieve strong metal-to-metal adhesion.

One special application of carboxylic polymers is with the use of CRLP (carboxyl reactive liquid polymer). An example of such a liquid polymer is Hycar® CTBN (carboxyl-terminated butadiene-nitrile copolymer) made by B. F. Goodrich Corp. This polymer is added to brittle epoxies in order to add toughness, impact strength, and improve the adhesive performance over a wider temperature range.

Polysulfides

This elastomer is commonly used in adhesive applications in sealants for construction as well as automobiles and marine products. These sealants show good chemical resistance as well as holding up under all kinds of environmental exposures.

NATURAL BASE POLYMERS

The use of naturally occurring polymers for adhesive applications has been steadily declining in the last 30 years. In dollar value of product, these adhesives have slipped from one-fifth of total adhesives production in the early sixties to less than 10 percent. In many cases this switch over from natural polymers (glues) has led to synthetic adhesives that cost more. Some of the reasons for this trend are better uniformity, better performance (such as shorter set times and higher strengths) and faster production outputs in manufacturing processes having an adhesive step. Many times these naturally occurring polymers are referred to as *"glues"* instead of adhesives.

Animal Glues

This natural polymer is derived from the hydrolysis of collagen, a protein. Collagen is either derived from animal hide (hide glues) or from animal bones (bone glues). While the use of animal glue has declined somewhat, it is still used as a packaging adhesive in such applications as tape, paper tubes, and fiber cans. Also, these adhesives are still used in woodworking.

Fish Glue

This is another protein adhesive from fish collagen. This adhesive has a special use in preparing photographic emulsions. This glue is very water soluble with a characteristic odor than can be controlled by the addition of a reodorant. By reacting with alums, fish glue can be made water insoluble. Overall, fish glue has a relatively low consumption.

Casein

This is another proteinaceous adhesive which is a byproduct from production of skim milk. It has declined in usage as an adhesive because of new alternate uses for casein as a food additive. This new use has driven up prices making casein less competitive in the adhesive market. Formerly, casein was used in the production of plywood, but this use has been generally replaced by synthetic adhesives (discussed earlier). Casein, however, is still used in relatively small volume in packaging adhesives.

Dried Animal Blood

This is another proteinaceous glue which today still has a small volume of consumption in adhesives. Historically plywood was made by heat-curing with animal blood glue in order to provide some degree of moisture resistance. Of course, phenol-formaldehyde has displaced this use generally.

Starch

Another naturally occuring polymer, it is based on a polysaccharide structure from plants.

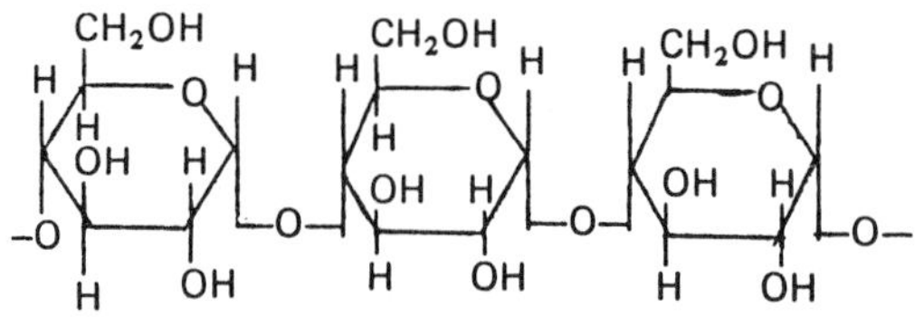

Starch consists of two polysaccharide forms: Amylose and Amylopectin. Amylose is the straight chain water-dispersible form (about 20% of the starch) while amylopectin is the branched form that is less water dispersible (approx. 80% of starch).

Corn starch is the most commonly used starch in America; however other countries may be larger consumers of wheat starch, potato starch, rice starch, and tapioca starch in adhesive applications. The biggest use for starch adhesives is in paper packaging, wall papering, etc.

Dextrin

This polymer is formed by the partial breakdown of the starch polysaccharide chains through controlled hydrolysis to lower molecular weight units. The degree to which hydrolysis is carried determines which form of Dextrin will result. The common dextrins made are British Gum, white Dextrin, and Canary Dextrin.

Dextrin is commonly used to make the adhesive for sealing mailing envelopes and postage stamps as well as many other applications.

Cellulose Derivatives

Cellulose itself is a completely insoluble natural polymer. Wood contains 60% cellulose while cotton contains 90%. The structure of cellulose, a polysaccharide, is similar to starch except that the saccharide units of the macromolecules are connected by beta-linkage instead of the alpha-linkage for starch.

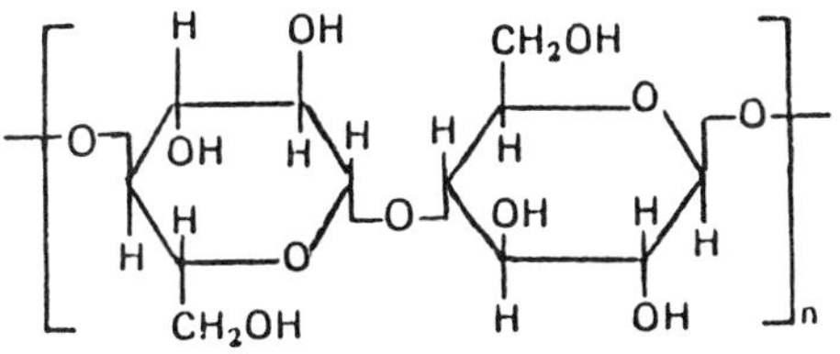

Cellulose

To convert this insoluble polymer into a soluble form, it is reacted with the appropriate acid under the proper conditions to form an ester which is either soluble in an organic solvent or water. Cellulose acetate and cellulose acetate butyrate are soluble in specific organic solvents. On the other hand, methyl cellulose, hydroxyethyl cellulose, and carboxymethyl cellulose are soluble in water. Low molecular weight versions of approx. 25,000–30,000 are commonly used in adhesive applications (> 100,000 for plastic use). These polymers are commonly used for adhesion of leather shoes, plastic, textile and paper.

Sodium Silicate

This material was one of the earliest adhesives used, although technically it is not a naturally occurring adhesive but man-made. Sodium silicate is usually applied as a liquid called "water glass".

This low cost substance is made by fusing silica (sand) with soda ash and dissolving it in water. It is actually an inorganic polymer in solution that forms a glass on drying. It is used still in many paper packaging applications.

TACKIFIERS

Introduction

Tack is the bond strength that is formed immediately when a given material comes in contact with another surface. It has been called "instantaneous adhesion", after which bonding strength may rise to a higher level later in time.

Many elastomers and thermoplastic polymers have very poor inherent tack properties. In order to improve this tack quality, tackifying resins are added to most of these adhesive formulations including evaporation types, pressure sensitive types, and hot melts. Elastomer adhesive formulations almost always require the addition of a tackifying resin to the compound to provide good ultimate adhesion. Many hot melts are greatly dependent on the tackifiers in their formulations in order to work.

The addition of tackifiers can improve performance of a polymeric adhesive in several ways. First of all, the proper selection and loading of a given tackifying resin can improve the compatibility of a given polymer to a selected substrate. For example, a non-polar elastomer such as butyl rubber can be made to have a more polar surface through the use of tackifying resins.

Another way that tackifying resins improve adhesion is through improving the wetting power by changing the viscoelastic properties of the polymer so there is more plastic and less elastic behavior. This allows closer contact of polymer to substrate.

Lastly, tackifying resins improve the tack properties (initial stick) of the adhesive to the substrate which leads to a stronger permanent bond later.

The proper selection of a tackifying resin is critical to the performance of a given adhesive formulation. Above all, the proper selection of a tackifier with the right degree of chemical compatibility with the base polymer is imperative. The variety of different types of resins will be discussed shortly.

Also the selection of the proper molecular weight for the resin (usually determined by softening point) is important. Unlike the tackifying resin used in rubber and plastic compounding, resins used in solvent adhesive compounds tend to be of a lower molecular weight in order to keep the adhesives solution viscosity low (for better wetting). By using lower M.W. resins, higher loadings of resins can be used in solvent adhesives without adversely raising the viscosity.

With elastomeric solvent based adhesives, there may be a critical loading level below which a selected resin does not really improve tack adhesion over those properties already inherent with the base polymer. This critical concentration may be lower for higher M.W. resins than for lower M W resins of the same type. However, the lower MW resin, while requiring higher concentrations to work, generally may impart a higher level of tack improvement than the higher M W resins are capable of displaying. These phenomena are believed to be caused by differences in phase separations.

Rosins

These substances are abietic acid resin mixtures derived from wood.

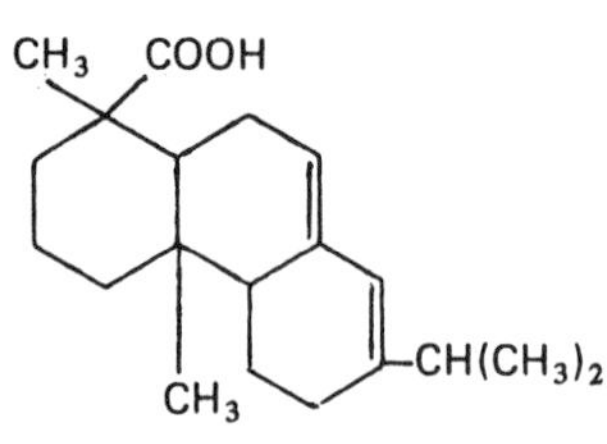

Abietic Acid

If the rosins are derived from the naphtha extraction of old pine stumps, these extracts are called "wood rosins". If the rosins are derived from the destructive distillation of turpentine oil, they are called "gum rosins". Finally, if the rosins are obtained from the destructive steam distillation of tall oil (a byproduct from paper manufacture), they are called "tall oil rosins".

Rosins are one of the earliest materials to be used as a tackifier in adhesive applications. In fact, rosins are sometimes referred to as "naval stores" because they use to be used by sailors to make repairs on wooden ships in the eighteenth century.

Today, rosins are used as tackifiers in all types of adhesive; however, they have a disadvantage in heat applications such as hot melts where they oxidize easily because of their unsaturation. To help solve this problem, they are made available also in hydrogenated forms which display improved thermal stability.

Another class of rosin derivative is the resinate. These rosin products are formed from the reaction of rosin with a metal oxide. Examples of these products are zinc resinate and calcium resinate. These resinates melt at an elevated temperature compared to straight rosins. This enables them to provide better heat resistance in adhesive compounds.

Lastly, ester forms of rosin acids are available which display different solubility properties.

Hydrocarbon Resins

These resins are generally from the polymerization of olefins from petroleum refining operations. Some hydrocarbons are still obtained from coal tar, but this source is much less common.

Basically, the two main types of hydrocarbon resins are the aromatic resins and the aliphatic resins. The aromatic resins are typically the coumarone - indene or polyindene type resin. The aliphatic resins are polymerized mixed olefins which do not have indene as their central structure.

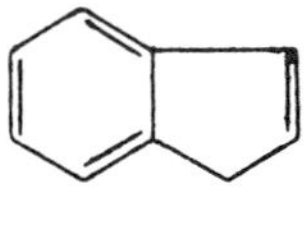

Indene 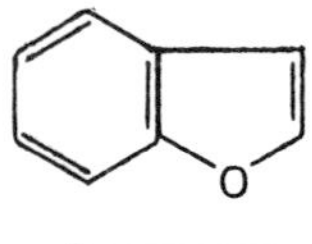   Coumarone

Phenolic Resins

These phenol-formaldehyde tackifying resins are different from the thermosetting type phenol-formaldehyde resins already mentioned. These P.F. resin tackifiers are generally non-heat reactive in that these phenolic resins are alkylated at the para position to prevent further crosslinking three-dimensionally. A typical structure of the non-heat reactive P.F. resins is shown below.

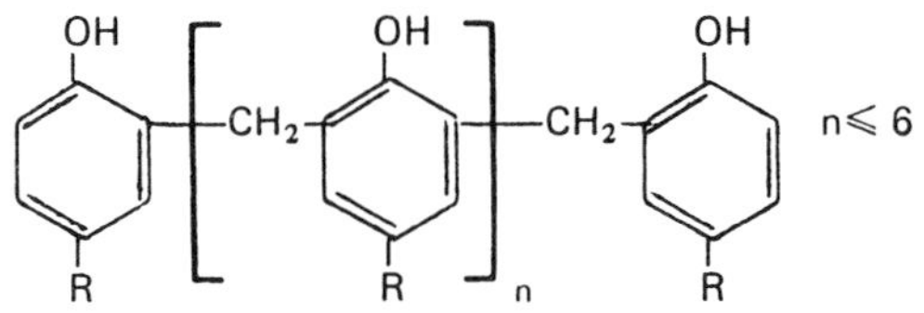

Even though these resins are described as non-heat reactive, still in hot melt adhesive applications, they may be hard to handle in that they have limited thermal stability. Also phenolics can cause odor and color problems in some cases. These resins should only be used in properly ventilated work areas.

Polyterpenes

These resins are polymerized from beta-pinene obtained from wood.

These resins are used extensively in elastomer solvent adhesives, pressure sensitive adhesives, and hot melts. Polyterpene is somewhat resistant to aging effects.

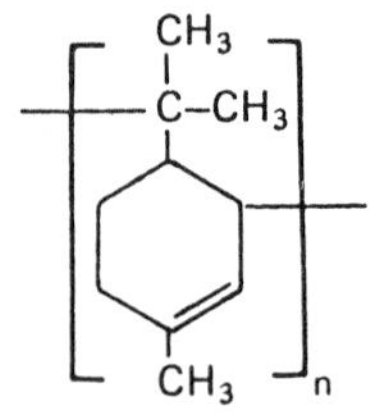

Polyterpene

SOLVENTS

Solvents are used as carriers in evaporation type adhesives in which the base adhesive polymer is dissolved in the solvent to provide a liquid solution that will wet the adherend surface. After application, the solvent escapes from the glue line through evaporation leaving the polymer with good cohesive bonding strenth.

There are six important considerations that should be taken into account in selecting a solvent for adhesive applications.

(1) Solubility with the polymer.

(2) Viscosity stability.

(3) Volatility and drying time.

(4) Flammability

(5) Environmental effects and toxicity.

(6) Cost

As discussed earlier, Hildebrand solubility parameters are used to measure the compatibility of two substances such as a base polymer and a given substrate. In a similar manner, Hildebrand solubility parameters can also be used in predicting the solubility of a given base polymer in a selected solvent.

Generally, it has been found that solvents with similar chemical structures to the solute will dissolve the solute. In other words "likes dissolve likes".

Thus it has been found that if a solvent is within say 1.3 to 2 of the δ values for a given polymer, then that solvent will thermodynamically be a "good" solvent for the polymer assuming that polarity and hydrogen bonding do not distort this comparison. Thus the following comparisons can be made.

Hildebrand Solubility Parameters

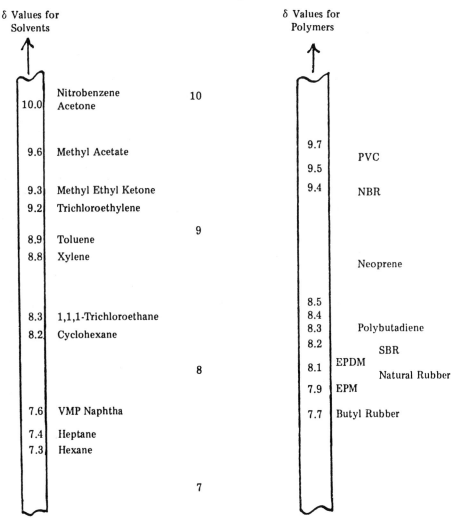

δ Values for Solvents		δ Values for Polymers	
10.0	Nitrobenzene 10 Acetone		
9.6	Methyl Acetate	9.7 9.5	PVC
9.3	Methyl Ethyl Ketone	9.4	NBR
9.2	Trichloroethylene		
8.9	Toluene 9		
8.8	Xylene		Neoprene
		8.5	
		8.4	
8.3	1,1,1-Trichloroethane	8.3	Polybutadiene
8.2	Cyclohexane	8.2	SBR
	8	8.1	EPDM Natural Rubber
		7.9	EPM
7.6	VMP Naphtha	7.7	Butyl Rubber
7.4	Heptane		
7.3	Hexane		
	7		

Thus, when solvent adhesives based on SBR or natural rubber are being prepared, hexane, naphtha, or gasoline may be normally used as solvents (usually with an aromatic solvent such as toluene). On the other hand, when more polar elastomers such as NBR or neoprene are used as the base polymer, solvents such as MEK, ethylene dichloride, and other ketone and ester solvents are used. Polar polymers require polar solvents. The solution time required for a selected solvent (or blend of solvents) to dissolve a polymer is important from a production standpoint. The closer the match, the shorter the solution time.

An adhesive's viscosity and the stability of its viscosity over time are important considerations. For example, MEK will reportedly give lower initial solution viscosities for NBR than nitropropane. However, the NBR adhesive made with nitropropane will not rise in viscosity with aging nearly as fast as the NBR adhesives made with MEK as the solvent. (On the other hand, nitropropane is toxic, and presents a moderate explosion hazard from shock or heat).

The drying time that a solvent will impart to a given adhesive is also important. Obviously, some solvents dry (vaporize) faster than others depending on the average molecular weight, molecular weight distribution, polarity, and degree of hydrogen bonding, if any. Drying time is related to some extent to the boiling point and distillation range of a solvent; however these distillation properties do not provide a perfect prediction of drying time. The best method of determining drying time is by measuring the relative evaporation rates of a particular solvent as it relates to a known standard such as ether or butyl acetate.

Flammability is another important consideration, if a fire risk is involved. Many halogenated solvents (but not all) show good fire resistance.

Environmental considerations in selecting a solvent are also becoming more important today because of Government regulations. Certain solvents such as MIBK have declined considerably in use because they are considered contributors to photochemical smog. Very tight restrictions on certain solvents have been imposed in regions such as Los Angeles County where smog is a common problem. Also there are now tighter restrictions on solvents in general, and halogenated solvents in particular, because of what is believed to be a detrimental effect that these agents may have on the ozone layer in the upper atmosphere. This ozone layer absorbs much of the ultraviolet radiation which might otherwise be harmful to life on Earth. Another problem of concern is worker exposure to solvent vapor. Many solvents have been found to have health risks if workers are exposed to solvent vapor for long periods of time. For example, the use of benzene in solvents has disappeared because benzene has been found to be a carcinogen. Many other solvents are suspected of having health risks as well.

Last, but not least, comparative cost is an important consideraton in selecting a solvent. For example, hydrocarbon solvent blends are less expensive than halogenated or aromatic solvents. Also, a solvent that can dissolve a polymer in a shorter period of time will generate an indirect cost savings through increased production output.

From these six important properties just described, it is not surprising that adhesive formulations will contain blends of different solvents in order to achieve the best balance of these properties. Blends of solvents might be used, for example, to obtain the desired combination of drying time and solvating power. Other blends might be used to achieve a cost advantage.

One last point should be made. Solvents can be used as adhesives by themselves if applied to two polymer adherends bonding together in which both polymers are soluble in the solvent. Thus, all that is necessary is that the solvent be applied to the polymer surfaces and that these surfaces be clamped together while the polymer surfaces are still wet. Then the two surfaces are held stationary till the joint dries completely. Strong bonds are achieved this way.

As can be appreciated, there are literally hundreds of different solvents used in adhesive applications. The following are examples of some of the more common adhesive solvents. Because many of these solvents are considered toxic, adhesives and cements containing them must always be prepared and used in properly ventilated areas. All appropriate safety precautions should be taken in working with these solvents.

Hydrocarbons

Gasoline
Rubber Solvent
VM&P Naphtha
Hexane
Cyclohexane

Ketones

Methyl ethyl ketone
Acetone
Methyl isobutyl ketone

Aromatics

Toluene
Xylene

Esters

Amyl acetate
Ethyl acetate
Butyl acetate

Halogenated

1,1,1-Trichloroethane
Trichloroethylene
Perchloroethylene

Alcohols

Ethanol
Methanol
Butanol

FILLERS

Fillers are important not only because they reduce the material cost of adhesives, but help reinforce the glue line formed, thus improving the cohesive strength of the adhesive. Also fillers help reduce contraction of the glue line in some cases, thus providing a stronger bond.

Fillers are used in virtually all types of the adhesives described. In the previous chapter on the rubber industry, the filler's role as a diluent and reinforcing agent was discussed. Those elastomer property changes from filler loadings examined in the last chapter also apply to elastomer glue lines. Therefore, fillers in elastomeric adhesives improve the bonding strength and reduce raw material cost. Of course, carbon black is by far the most commonly used filler and reinforcing agent used in elastomers. One reason for fillers use in elastomeric adhesives (which was not mentioned previously) is that they sometimes improve tack. Another minor reason, not mentioned, for using filler in rubber is that they can impart some degree of viscosity control to an adhesive. Generally, rubber compounds used in adhesive formulations are not loaded as highly as elastomers compounded for conventional non-adhesive use. Too high a loading may cause dispersion problems when dissolving the compound in a solvent.

With reactive polymer adhesive systems that form a hard rigid bond (such as polyester or epoxies), the role of fillers in reducing shrinkage on hardening is far more important than any improvement in adhesive strength. Reducing shrinkage helps reduce any stress formation along the glue line thus providing for a stronger bond.

The following is a list of fillers commonly used in the adhesive industry.

> Carbon black (excellent reinforcement)
>
> Kaolin Clay (inexpensive)
>
> Titanium Dioxide (for white coloring)
>
> Zinc Oxide (for good thermal conductivity)
>
> Chalk
>
> Ground calcium carbonates (limestone)
>
> Precipitated calcium carbonate (small particles)
>
> Calcium silicate
>
> Barium Sulfate (for acid resistance)
>
> Talc
>
> Iron Oxides
>
> Hydrated precipitated silica
>
> Sand (inexpensive)
>
> Wood Flour (used with phenolics in plywood production)
>
> Glass Fibers (for strength)

Hollow microspheres (to lower specific gravity)

Mica (for arc and heat resistance)

Powdered copper (for electrical conductance)

Powdered silver (for electrical conductance)

Graphite (for electrical conductance)

Walnut shell flour - used in wood adhesives to improve spreading and control penetration

PLASTICIZERS

Plasticizers are used extensively in elastomer adhesives and hot melts. The main purpose of any plasticizer, as the name implies, is to aid in processing or breaking down the elastomer on the mill in order to effectively achieve a higher state of mix on the mill and enable the milled compound to be solubilized more readily by a given solvent. In other words, plasticizers change the rheology of the polymer. This quality was discussed extensively in the last chapter. However, other important effects that placticizers have on the polymer should also be described here. For one thing, plasticizers not only improve processing, but many times have the side advantage of improving tack, just as some tackifiers also improve rubber processing. Plasticizers also affect adhesive use by reducing "drawing", reducing solution viscosities, improving low temperature flexibility of the glue line, and reducing modulus and hardness of the glue line.

With hot melt adhesives, plasticizers serve as non-volatile "solvents" to lower the melt viscosity and thus improve wetting of substrates. In some cases these plasticizers will hurt adhesion somewhat and should not be used excessively. Hot melts can use traditional liquid plasticizers such as those used in elastomers, i.e. dioctyl phthalate, dibutyl phthalate, butylbenzyl phthalate, triaryl phosphate, dioctyl adipate, etc. On the other hand, hot melts also use solid plasticizers (at room temperature) such as microcrystalline waxes, paraffin waxes, as well as diphenyl phthalate and dicyclohexyl phthalate, which are used to impart delayed tack.

HARDENERS

A hardener is used in reactive polymer adhesive systems to initiate the hardening of the glue line by initiating crosslinking or gelling within the polymer adhesive. All appropriate safety precautions should be followed in working with these agents.

With epoxy adhesives, the epoxy ring must be opened by an acid or base to establish a crosslink.

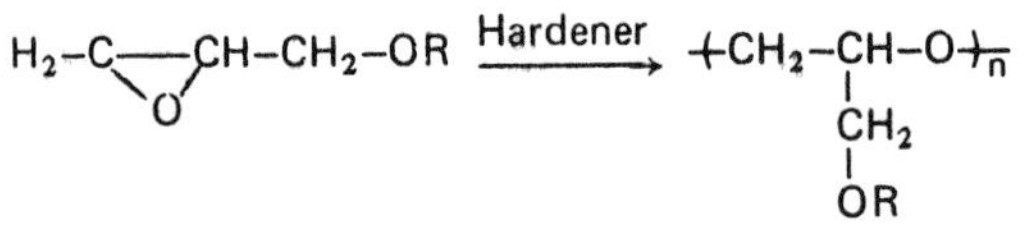

$$\underset{O}{H_2\text{-}C\text{---}CH\text{-}CH_2\text{-}OR} \xrightarrow{\text{Hardener}} \overset{}{\text{+}CH_2\text{-}\underset{\underset{OR}{\overset{\displaystyle |}{CH_2}}}{\overset{\displaystyle |}{CH}}\text{-}O\text{+}_n}$$

This ring opening process is usually initiated by either an acid anhydride or an amine. There are basically four types of epoxy hardeners given below.

(1) Aliphatic polyamines

(2) Aromatic polyamines

(3) Acid anhydrides

(4) Catalytic types

Aliphatic polyamines are very fast and will cure epoxy resins at room temperature. On the other hand, the pot life times are relatively short. Examples of aliphatic polyamines are diethylenetriamine (DETA) triethylenetetramine (TETA), and tetraethylenepentamine (TEPA)

$$H_2N\text{-}CH_2\text{-}CH_2\text{-}NH\text{-}CH_2\text{-}CH_2\text{-}NH\text{-}CH_2\text{-}CH_2\text{-}NH_2$$

TETA

Some disadvantages of using these aliphatic amines are poor peel strength, high volatility, and high toxicity.

Aromatic amine hardeners have an advantage over aliphatic amine hardeners in that they have longer pot life times and better heat resistance; but they may require elevated temperatures to cure. These hardeners are also toxic. An example of an aromatic amine hardener is meta-phenylenediamine (MPDA)

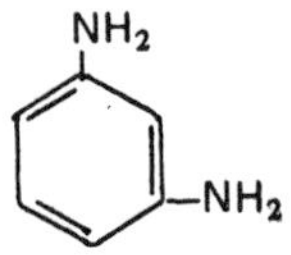

MPDA

Anhydrides are used in epoxy adhesives primarily for high temperature applications. One of the simplest anhydrides, phthalic anhydride is used to cure epoxies, but generally not for adhesive applications because of relatively poor adhesion to metal.

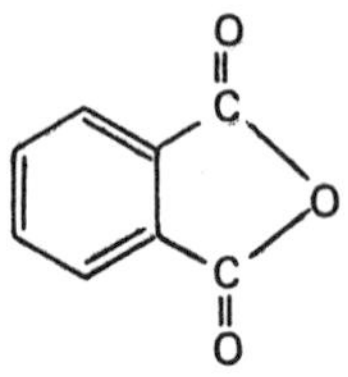

Phthalic Anhydride

On the other hand, anhydrides such as pyromellitic dianhydride (PMDA) give outstanding "hot strength". All anhydrides require a catalyst (e.g., water, hydroxyls, or a Lewis base) to open the anhydride ring.

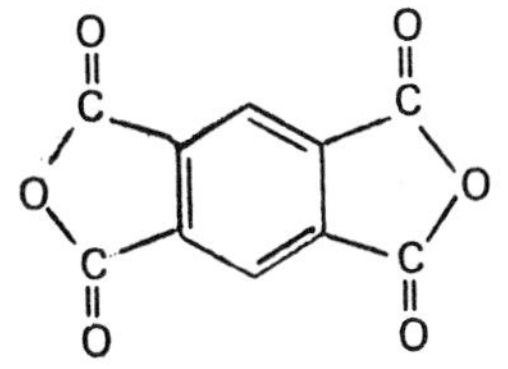

PMDA

Catalysts, such as boron trifluoride (BF_3), are used in hardening epoxies, but usually with another amine hardener. Many times, BF_3 - amine complexes are used as "latent curing agents" in a one-part epoxy system in which the formulation is stable at ambient temperature but sets at elevated temperature. Several amines are used in "latent curing systems". Dicyandiamide (called "dicy") is a very common amine used in latent cure systems for construction adhesives and is known for its superior metal-to-metal adhesion

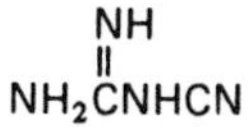

$$NH_2\overset{\overset{\displaystyle NH}{\|}}{C}NHCN$$

Dicyandiamide

A different type of hardener is required for phenol-formaldehyde novolac resins. As you recall, these novolac resins were formed from a polycondensation reaction of phenol and formaldehyde. However, the amount of formaldehyde reactant was held back in order to prevent the resin from reaching gel point and forming a three dimensional matrix. Upon the addition of a methylene donor (added in powder form to the cooled ground up resin), the novolac resin is able to gel when reheated. This type resin is called a two-step thermosetting phenol-formaldehyde resin. The methylene donor commonly used is hexamethylene tetramine.

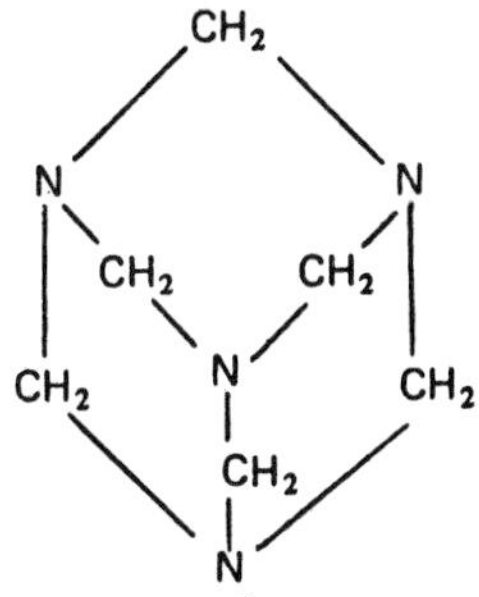

"Hexa"

In a similar chemistry, paraformaldehyde is used as the powder hardening agent (methylene donor) with novolac resorcinol-formaldehyde resins (usually in a liquid form) in order to cure the resins.

$$HO-CH_2-O-CH_2-(O-CH_2)_n-O-CH_2-OH$$
Paraformaldehyde

This paraformaldehyde powder is usually pre-powder-blended with fillers such as pecan or walnut shell flour in order to control the powder consistency. This pre-powder-blend is added to the resin as a curative and hardener.

For urea-formaldehyde resins, an acidic catalyst such as ammonium chloride or ammonium sulfate is added to the resin. The ammonium chloride catalyst effect does not occur till water is added. In the presence of water, hydrogen ions are slowly released from the reaction of ammonium chloride with free formaldehyde in the resin, producing hexamethylene tetramine and hydrochloric acid. This slow formation of HCl will gel the polymer.

Technically, rubber curatives used in a two part system with elastomer adhesives might be considered as hardeners (although this term is generally not applied to elastomer curatives). These rubber curatives were described in detail in the last chapter.

PRIMERS

A primer is a coating which is applied to a given adherend before the adhesive in order to improve the adhesive bond. As mentioned earlier, adhesives are not always compatible with a given substrate. For example, natural rubber adhesives may not be very compatible with certain very polar substrates. In order to bridge this difference in compatibility, a primer is used. A primer may be intermediate in its compatibility between a given adhesive and substrate or the primer may contain a compound with bifunctionality. Primers available on the market are usually proprietary. Depending on what diversity of materials are being bonded, they can consist of epoxides, polyurethanes, polyisocyanates, methacrylates, silanes, or a variety of resins. In applications where primers are applied to metal surfaces to promote good adhesion between metal and rubber, the primer serves an added function of protecting the freshly cleaned metallic surface from corrosion.

THICKENING AGENTS

Some components are added to adhesive formulations in order to increase their viscosity. Thickeners used in latex adhesive can be polysaccharides such as gum arabic or karaya gum. Synthetic polymers can also be added to aqueous systems as thickeners. Soluble synthetic polymers such as polyvinyl alcohol or methyl cellulose can be used for this purpose. Bentonite also is used as an aqueous thickener because of its gel forming tendency. Also colloidal silicas (commonly sold under the trade name of Cab-O-Sil) are also used as thixotropic agents in adhesive systems. The colloidal silica at rest establishes a gel network from bridges of hydrogen bonds formed among particles. On agitation, these bridges are destroyed and the viscosity during agitation decreases.

Also it has been noted with epoxies that chopped glass fibers have an effect on viscosity. In nitrile-MEK cement formulations, a carboxylic vinyl polymer under the trade name of Carbopol® is used as a thickening agent. Lastly, borax (sodium tetraborate) is well known to greatly increase the viscosity of an aqueous starch glue to a usable consistency at a relatively low solids level. Apparently the borax molecules interact with the hydroxyls of the polysaccharides.

Of course, one method of increasing the viscosity of an evaporation type adhesive is to simply raise the total solids level, particularly the base polymer.

NEW DIRECTIONS

As pointed out earlier, there has been a steady shift of adhesive use away from naturally occuring polymers over to synthetic polymers even though in many cases the synthetics were (and are) more expensive. As described before, this is because the synthetics out-perform the natural polymers in adhesive strength, set time, and uniformity. Whether natural polymers will regain any of the lost markets is hard to say, considering the cost of petroleum (from which synthetic polymers are all derived). Most likely natural base adhesives will not regain lost markets to any noticable extent unless petroleum were to fall into extremely tight supply and price differences increased greatly.

One important trend in the future will probably be a continued shift away from solvent base adhesives over to non-solvent adhesives such as hot-melts, reactives, and aqueous base adhesives. The substitution of aqueous base adhesives for organic solvent base adhesives is limited in that aqueous base adhesives have a longer drying time. Therefore, hot melts and reactives are seen to have a larger growth rate in the years ahead. Neither of these types emits hydrocarbons into the atmosphere and hot melts have the added advantage of being fast setting. The driving force bringing about this shift is governmental and public concern over solvent emissions into the environment. As discussed earlier, these concerns have been brought about by the following:

(1) Solvent emissions contribute to photochemical smog conditions in such areas as Los Angeles County.

(2) Concerns over possible long term toxicity effects of certain solvents on workers and consumers exposed to solvent vapors.

(3) Possible long term environmental effects noted in current theoretical beliefs that solvent emissions (particularly of halogenated solvents) may interfere with the formation of ozone at high altitudes in our atmosphere. This layer of ozone protects plant and animal life from excessive exposure to ultraviolet radiation.

In the future, new adhesive systems to meet our growing needs will not necessarily always come from the adhesive industry, per se, but may be developed in other polymer science communities and other polymer industries. Typically, new polymers are developed in the plastics, rubber, or coatings industry, where they later find application in adhesives. Therefore, for the future outlook on technological breakthroughs in the adhesives industry, one must also look not only at this industry, but all other polymer industries as well.

One interesting area of adhesives with much potential growth is the high-temperature adhesives market for applications in aviation, aerospace, and defense. New polymers that are being used in these applications are polybenzimidazoles, polyimides, and polyquinoxalines, to name a few. These new polymers are of the "semiladder" structure (see the Plastics Industry chapter for more details). Some of these adhesives can function at temperatures as high as 530°C for short periods of time (or 230°C for very long periods of time).

SUGGESTED READING

Adhesives Redbook, Adhesives Age, Communication Channels, Atlanta, Ga., annual

Bikales, Norbert M., *Adhesion and Bonding*, John Wiley, N.Y., 1971

Blomquist, Richard F., *Adhesives - Past, Present, and Future*, American Society for Testing and Materials, Baltimore, MD, 1963.

Bruno, E. J., *Adhesives in Modern Manufacturing*, Society of Manufacturing Engineers, Dearborn, MI, 1970.

Cagle, Charles V., *Handbook of Adhesive Bonding*, McGraw-Hill, N.Y., 1982.

Delmonte, John, *The Technology of Adhesives*, Reinhold Publishing Corporation, N.Y., 1965.

DeLollis, Nicholas J., *Adhesives: Adherends, Adhesion*, Krieger, 1980

Flick, Ernest W., *Adhesive and Sealant Compound Formulations*, Noyes Publications, Park Ridge, N.J., 1984

Flick, E. W., *Handbook of Adhesive Raw Materials*, Noyes Publications, Park Ridge, N.J., 1982

Houwink, R. and Salomon, G., *Adhesion and Adhesives, Volume I and II Adhesives*, Elsevier Publishing Co., N.Y., 1965–67

Jackson, B. S., *Industrial Adhesives and Sealants*, Hutchinson, Benham, London, 1976

Landrock, A. H., *Adhesives Technology Handbook*, Noyes Publications, Park Ridge, N.J., 1985

Patrick, Robert L., *Treatise on Adhesion and Adhesives*, Vol. 1, 2, Marcel Dekker Inc., N.Y., 1967–1969

Shields, J., *Adhesives Handbook*, 3rd ed., Newnes - Butterworth, London, 1984

Skeist, Irving, *Handbook of Adhesives*, Van Nostrand Reinhold Co., N.Y., 1977

Wake, William C., *Adhesion and the Formulation of Adhesives*, Applied Science, London, 1982

Wake, Willian C., *Developments in Adhesives*, Applied Science, London, 1977

4

The Coatings Industry

Paints and coating are extremely important to our society. They are not just used for decorative purposes, but to protect wood, metal, and other substrates against deterioration. Without coatings, our cars would quickly rust, bridges would collapse, and houses would rot. Therefore, paints and coatings play a vital role in our modern society.

Before we begin our discussion of this industry, some basic definitions may be in order. First of all, a *paint* is a coating containing a pigment and a polymeric binder. On the other hand, a *varnish* is a clear or translucent coating containing a binder but no pigments. A *stain*, by contrast, is a coating which contains a pigment(s) without a binder. In addition, a *lacquer* is a clear (or colored) coating that forms a film through evaporation of a volatile solvent alone. Lastly, a *primer* is an initial coating which may be applied to a substrate to improve film adhesion to a following coat or to help inhibit corrosion.

As we shall see, paint formulating is quite diverse and complicated. In fact, a paint manufacturing plant may use over 500 different raw materials in formulating a wide variety of paint products.

MARKETS

The coatings industry has sales in excess of 5 billion dollars annually (in 1980 dollars). This industry is larger than the Adhesives Industry, but smaller than the Rubber Industry. It is a very diverse industry, meeting the needs of many different markets.

Broadly speaking, the coatings markets can be divided into two major groups, i.e., the *Trade Sales Markets* and the *Industrial Markets*.

The trade sales markets are characterized as standard commodity-like paints that are shelf items generally distributed through wholesaler and retailer outlets. Examples of paint markets of this group are interior and exterior house paints, masonry paints, trim paints, varnishes, and other architectural coatings. These markets represent about 2 billion dollars in annual sales.

The industrial markets are the second major group of markets. Paints and coatings in this group are generally specially formulated to meet the specific needs of an industrial customer. This group can be divided into two principle subgroups, i.e., *Original Equipment Market* (OEM), and *Industrial Maintenance*. The OEM represent a sales value of approximately 2 billion dollars annually. Examples in this group include original automobile and truck paints, container coatings, aircraft coating, furniture coating, appliance enamels, and original machinery finishes. On the other hand, the industrial maintenance paints include refinishes of automobiles and trucks, machinery, boats, etc. In some refinish applications, a standard stocktype paint may be used. These industrial maintenance markets are approximately 1 billion dollars in sales volume.

As can be seen, the market structure for the paint industry is very diverse. There are over 35 separate markets for coating products. Because of the diverse markets and their unique technical requirements, it is very difficult for any one paint company to serve every market. Generally, paint companies will tend to specialize in a limited number of markets.

The structure of the paint industry itself is quite diverse, consisting of large and small firms. While there are estimated to be approximately 1200 paint producers in the industry, the top 10 firms represent approximately one-half of the total sales. The top ten paint producers in approximate descending order according to paint sales are given below.

(1) Sherwin Williams

(2) PPG

(3) Dupont

(4) Glidden

(5) Mobil

(6) DeSoto

(7) Inmont

(8) Benjamin Moore

(9) Grow Group

(10) Reliance

While somewhat concentrated, the paint industry is very competitive. Many firms specialize in a select group of markets in order to achieve a competitive advantage.

The coatings and paint industry can be accurately described as a mature industry. The production of paints, in "dry gallons", is projected to grow at about the same rate as the general economy (the actual output in liquid gallons is declining because of concern to reduce solvent emissions into the environment). However, while many traditional markets will have slow growth, other specialty markets will have high growth potentials. Examples of these growth areas might be powder coatings, or high-solids paints.

HISTORY

Paints date back to the cave paintings in Lascaux, France, over 15,000 years ago. While these historic paintings are mostly pigments, they represent man's first efforts.

In these early days, water was the only solvent available. Binders were rather limited. Such materials as egg, milk, tree sap, and berries were commonly used. Also, many of the pigments were ground minerals. The Egyptians, 4000 years ago, used gum arabic, starches, egg white, and casein to make water based paints to decorate interiors.

Over a thousand years ago, linseed oil was used as a drying oil with lead oxide and lime as driers. Varnishes became quite common by the 16th Century. Whitewash (slaked lime in water) was used in America during the colonial period. By 1900, tung oil and rosin were used in making faster drying oleoresinous varnishes. In 1923, fast drying nitro-cellulose lacquers were introduced to the automobile industry in order to increase production rates. Also in the 1920s, alkyd paints were introduced. Alkyds are unique to the paint industry. They were developed exclusively through the R&D efforts of the paint industry for use in coating applications (commonly new polymers used in the paint industry originate from other industries). Alkyds have been found to be one of the most versatile binders to be used in coatings. These polymers have gained wide acceptance in many markets and continue to be perfected even today.

Urea-formaldehyde binders were introduced to the coatings industry in 1929, followed by the introduction of melamine-formaldehyde in 1939. Also an important coating improvement was noted with the use of poly-urethanes in 1940. Epoxies started to be used in coatings around 1950.

A silent revolution began in the paint industry with the introduction of styrene-butadiene latices in 1948. At first there were some technical problems with these paints, but these problems were corrected and latex paints have gained wide acceptance especially in architectural markets because of their ease in application, ease of clean up, their non-flammability, and their lack of hydrocarbon emissions. Latex paints were improved greatly with the introduction of polyvinyl acetate and acrylic based latices in the 1950's.

Acrylics have greatly changed the coatings industry. In the late 1950s, both acrylic lacquers and acrylic thermoset resins were adopted by the coatings industry. These resins have found wide acceptance in such markets as automotive paints.

The following is a chronology of innovations in the coating industry.

Innovation	Approximate Date
First paints	14,000 BC
Use of casein, gum arabic, etc. as binders	2,000 BC
Use of linseed oil with lead driers	1000 AD
Common use of varnishes	1500 AD
Common use of whitewash in America	1700
Use of tung oil with rosin for faster drying varnishes	1900
Fast drying nitrocellulose lacquers for cars	1923
Introduction of titanium dioxide	1924
Introduction of alkyds	1928
Introduction of urea-formaldehyde	1928
Molybdate orange as pigment	1930
Phthalocyanine blue introduced	1937
Introduction of melamine-formaldehyde	1939
Introduction of polyurethanes	1939
Introduction of silicones	1944
Introduction of S-B latex paints	1948
PVA$_c$ latex paint	1950
Acrylic latex paint	1950
Epoxy coatings	1950
Acrylic lacquers and thermosets	1958
Powder coatings, UV curing, high solids coatings, water dispersible coatings	1970s
Applications by robotics	1980s

MANUFACTURING PROCESS

The polymer (binder) producers and producers for many other raw materials, are few in number because of the economies of scale required to maintain an efficient operation. On the other hand, just as with the plastics and adhesives industries, the paint industry also consists of many paint factories, both large and small. In fact, small paint plants can be quite efficient because they generally require capital equipment on a modest scale. Essentially, a paint factory operation consists of grinding and dispersing pigments in the paint vehicle (medium)-called a "mill paste"-in a disperser or grinder (to be discussed), followed by a "let down" in which solvent, oils, or other liquids are added to thin out the paint and reduce its viscosity. Of course, this description is a broad generalization; specific operations may vary greatly.

Most pigments consist of aggregates and agglomerates of particles.

These clusters of particles must be broken up during the grinding process. There are three stages of dispersion during this grinding process. The first phase is called *wetting* and involves the substitution of the vehicle for air at the pigment particle's interface.

The second phase is called *deagglomeration* and consists of the separation or breaking up of particle clusters into primary particles (or at least into small aggregates).

The third phase of dispersion is called *stabilization* and consists of the formation of a homogeneous dispersion of particles with minimal tendencies toward flocculation (unlike agglomerates and aggregates, flocculates will break up easily with shear).

Three-Roll Mill

This is a very effective type of dispersion equipment (disperser) used in the paint industry. It works by compressing and shearing high viscosity paint pastes between heavy rolls rotating at different speeds (to give more shear) as shown below.

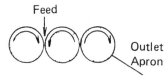

While this method is very effective in grinding paints that might normally be quite difficult by other methods, it does have limitations. First of all, this equipment is very dependent on operator supervision. It also requires a high viscosity paste to be effective. Lastly, it has a relatively low level of output compared to some other methods.

Ball Mill

This method consists of a rotating hollow cylindrical container. Inside this cylinder are steel or porcelain balls which rotate and tumble against each other as the cylinder rotates and grind the paint base through shear and compression as shown below.

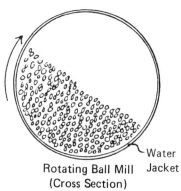

Rotating Ball Mill   Jacket
(Cross Section)

The advantage of this type of disperser is its simplicity. You simply charge it and leave it. It requires very little operator attention and can be left to run overnight.

One problem with the use of ball mills is the need to thoroughly clean them when switching to another color. Many times light colored paints are avoided for this reason. Also it is important that the charge, to be ground, have the correct range of viscosity. If the viscosity is too high, poor cascading of the balls will result. If the viscosity is too low, balls will be thrown against the sides of the container during rotation, resulting in a loss in efficiency. Also there is an optimal fill factor and rate of rotation for the best grind.

Sand Mill

This disperser is the "workhorse" of the industry because of its relatively high output at a low operating cost and its versatility.

A sand mill consists of a hollow cylindrical shell which contains a rotating shaft with discs that move sand particles against one another in the paint medium. This very fast movement of the sand causes a great deal of shear action to occur in a relatively short period of time. This greater dispersion efficiency, in comparison to the ball mill, for example, is due to the fine particle size of the sand which imparts greater surface area for shearing (more chances for sand particle collisions with the pigment particles). While sand mills are more efficient than ball mills in deagglomerating a pigment in a shorter time period, sand mills do not impart the higher level of shear that the ball mill is capable of achieving. The ball mill can produce a higher level of shear necessary to reduce primary particle size of the pigments. Nevertheless, this primary reduction in particle size is seldom needed. Therefore the shear level imparted by the sand mill is sufficient for most paint applications.

It is important to note that unlike the other dispersers described, which are batch operations, the sand mill is part of a continuous process. For this reason, the sand mill requires a premix holding tank and at least one positive displacement pump to work. The premix is usually pumped into the sand mill at the bottom and the dispersed product discharged near the top of the sand mill by way of a screened outlet trough as shown below.

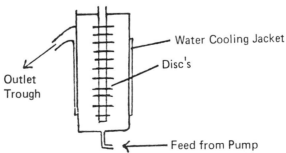

During the operation of the sand mill, heat is generated which must be removed through the walls by means of a cooling water jacket. Depending on the dwell time (through put) in the sand mill, a second pass may be needed to achieve sufficient dispersion.

Sand mills are most popular because they can give a relatively high production output at low cost. Usually they require only minimal amounts of an operator's time and one operator can service several sand mills running simultaneously. Also when changing colors, clean out is not a big problem.

High Speed Impeller

An impeller is a disperser using a rotor blade at very high speed to disperse a pigment in the paint medium. Motors of 75 horse power or more are sometimes used. Pigment agglomerates are dispersed from the shear forces of the blade rotation, impact of particles against the blade, and impact with other pigment particles.

It is important that the formulation have the correct viscosity range. If its viscosity is too low, no shear will develop. On the other hand, if the viscosity is too thick, the particles will not move fast enough. Dilatant formulations disperse much better than thixotropic or pseudo-plastic formulations.

Impellers are mostly batch operations. These mixers are sometimes preferred for light colored paints because of the relatively easy clean up needed for a color change. Also this disperser is relatively simple and cheap to operate. On the other hand, this disperser can not impart the high shear levels of most of the other dispersers discussed. Therefore, it is mainly used on less difficult pigments.

TESTING

As in other polymeric industries discussed, there are numerous standard test methods which are used in the paint industry to assure that processing and product requirements are met.

Fineness Gauge

During the grinding process, just mentioned, it is important to be able to measure the fineness of grind (the degree of breakdown of agglomerates and aggregate particles) resulting from the disperser run. If "the grind" is not fine enough, poor color uniformity, gloss or hiding power can result. One fast method to measure the "fineness of grind" is to use the *Hegman Gauge*. This gauge is a steel plate having grooves of declining depth from zero to 100 microns. Samples from the disperser are drawn over this gauge with a straight edge. The depth of the groove filled by the sample which starts to show a disruption or rippling of the film is taken and read as the Hegman reading for particle size of the pigment. This end point where the reading is taken is due to the depth of the pigment particles, causing them to disrupt the film surface at that point.

This gauge is a very fast method to determine the approximate state of grind. Other similar gauges are also available to quickly measure this property in the paint plant.

Viscosity

Viscosity measurements are performed as both process tests as well as end product tests. Many plant viscosity tests (such as the Ford Cup and Zahn Cup tests) simply measure the time required for a given volume of sample to pass through a standard size orifice. The Brookfield viscometer is also commonly used to measure viscosity from shear resistance placed on a standard spindle rotating at a constant speed. A Stormer viscometer, on the other hand, measures the time required for a standard paddle to revolve 100 revolutions from a constant force (weight and pulley). The Stormer is sometimes preferred when measuring the viscosity of a heterogeneous mixture. Lastly, for homogeneous varnishes, the Gardner-Holdt bubble viscometer is used. It quickly indicates varnish viscosity by measuring the time required for a bubble to rise to the top of a standard tube after being inverted under standard conditions.

The problem with most of these viscometers is that they measure viscosity at relatively low shear. They can be used as good production control tests and may relate to degree of disperion; however, paints are usually applied at high shear by brush, roller, or spray. As will be discussed later, paint and coatings are non-Newtonian, meaning that their viscosity will change with changes in shear rates. (See "Thixotropes and Thickeners" under Additives section). Since many paints display pseudoplastic behavior (viscosity drops with increased shear rate), the viscosity tests described may not relate to the paint's end use application.

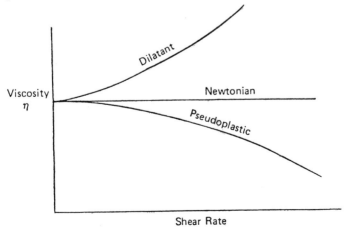

Solids Content

The "solids" of a paint include those non-volatile components which remain part of the paint film. These "solids" can include not only the binders, pigments, and additives, but also liquid plasticizers of very low volatility which also remain in the film after drying. Testing the solids content of a paint is very important to assure good control of the manufacturing process. Formulations that are off specification in their solids content can cause processing and end product performance problems. Paints off on solids may have variations in color and film durability. Also the solids content is important in calculating the amount of coverage a paint will impart.

Specific Gravity

This is an important, easy-to-measure property which indicates the weight of the paint for a given volume. While many of the raw materials used to formulate a paint may be purchased on a weight basis, paints are usually sold on a cost per volume basis (dollars per gallon). Therefore, the higher the specific gravity of a given raw material, the greater its cost contribution for a given volume.

Specific gravity is not only a simple quality control test for raw materials; it is also used as an in-process quality control test as well as an end-product test. Running a specific gravity is an easy way to determine if all the major components were added in the correct amounts.

Specific gravity can be measured very easily by filling a tared container of known volume with the sample and weighing. For exact measurements to a high level of precision, special laboratory glass containers called pycnometers are used. Also, hydrometers can be used to quickly measure specific gravity.

Color

Obviously color is an important property to control during paint manufacture. Poor dispersion can affect the color appearance of the paint. Many times color is controlled by a simple visual comparison of samples against a standard. Also several instrumental methods are available to measure color.

Developmental Tests

Many of these tests are used in the development of a new paint formulation. These tests, such as weatherability, are not usually performed as routine quality control tests. Instead, many of these tests may only be used to determine if a new or modified paint formulation will meet the required end use performance requirements. Also these tests may be used to audit a regular paint product from time to time.

A complete discussion concerning these tests is beyond the scope of this work. Instead listed are some of the most prominent examples of these tests. For more information regarding these tests, consult ASTM standards under the jurisdiction of D1, committee on coatings.

Development Tests

Film drying time

Sag resistance

Spray properties

Freeze-thaw resistance

Brush properties

Roller properties

Mechanical stability (latex)

Dry film hiding power

Film hardness (Sward Rocker, Pencil Hardness, etc.)

Dry film thickness

Detergent resistance

Chemical resistance

Fire resistance

Dry film adhesion (tab pull, peel tape)

Film impact resistance

Film elongation and flexibility (over mandrel)

Film tensile strength

Film gloss (light reflectance)

Film abrasion resistance

Scrubbability (number of cycles)

Weatherability (chaulking, flaking, etc.)

Salt spray resistance

Light stability

BASIC PRINCIPLES

The two main functions of a coating are to protect the substrate and/ or to be decorative. Depending on the end use application, one function may have a far greater degree of importance than another. For example, protecting a multimillion dollar steel bridge from corrosion would be far more important than decorative features.

Convertible and Nonconvertible Coatings

Coatings, by their nature, are usually applied in a liquid form (to coat) after which they form a solid film which possesses a degree of adhesion to the substrate. There are two general categories of coatings- convertible and non-convertible. A convertible coating is one that undergoes some sort of chemical change during the film forming process. This chemical change results in the conversion of the polymeric binder into a three dimensional crosslinked network or matrix. This film is permanently altered and can not be reliquified by heat or contact with most solvents. This crosslinking of the polymeric binder may be achieved through an oxidative mechanism (as with oil and alkyd paints), catalysts (as with urea-formaldehyde coatings), amine curing agents (as with epoxy systems), or simply U.V. radiation. All of these means of film hardening will be discussed later.

Latex Film Formation

Non-convertible coatings, on the other hand, do not form a film through any chemical reaction or change. Instead, non-convertible coatings form films usually through the evaporation of the solvent or liquid dispersion medium, leaving the polymeric binder to form a deposit film on the substrate. Generally these films formed are thermoplastic in nature. An example of this type of coating can be seen historically with nitrocellulose lacquers which formed film through solvent evaporation. As the solvent evaporated, the resin concentration increased. As this happened, it was important that the solvent remaining be compatible with the resin in order to keep the film homogeneous in nature (the resin should not precipitate out of solution). As the solvent continued to evaporate, a tacky film formed. Depending on how tenaciously a solvent held to a film determined how long it took the resulting film to reach final hardness. It can be seen that to assure the formation of a homogeneous film with a reasonable drying time will require careful selection of the solvent(s). As we shall learn later, different solvents impart different solubility properties and evaporation rates.

Another special means of film formation can be seen with latex paints. The mechanism through which these emulsion coatings form a film is quite different from that discussed above for solvent systems. A latex emulsion consists of colloidal polymer particles, (usually smaller than 0.5 micron in size), which are coated with an emulsifier to prevent these particles from flocculating and coalescing together. In other words, the paint's latex base is stabilized. Also particles in this range of size display Brownian motion. After the latex paint is applied to the substrate, water begins to be lost from the wet film through evaporation. As a result, these polymer colloidal particles become more and more concentrated. Next the particles touch and pack together, still leaving voids of water which must eventually be lost to evaporation or porous absorption of the substrate.

Finally the particles completely coalesce and fuse into a continuous film without voids.

| Wet latex film | Intermediate latex film | Dry latex film |

While these processes may at first seem simple, there are many factors which can prevent the coalescing of particles into a continuous protective film. First of all, the polymer itself must be soft enough and have a sufficiently low glass transition temperature to allow this fusion. For this reason, many of the base latex polymers used are copolymers which are usually softer than homopolymers (this will be discussed later). Also coalescing agents (temporary plasticizers) may be included in the paint formulation. As the water evaporates, these coalescing agents have an increasing softening effect on the polymer particles, helping them to coalesce more easily.

Another important consideration is the temperature at which the latex coating is being applied. If the temperature is too low, the colloidal particles may be too hard to coalesce. Also if some of the latex polymer particles are too large, a problem in film formation can occur.

Lastly, the specific paint formula itself is important. If the paint formulation contains too much pigment, this can interfere with the smooth formation of a continuous film. In fact, any imbalance or excess of paint additives can potentially affect latex film formation.

Film Property Requirements

As we shall learn, different coating applications demand diversity in paint properties. So in the case of exterior house paint, it must protect the wood substrate principally by reducing the rate of moisture pickup and loss by the wood. This helps prevent wood swelling and shrinkage and prevents buckling, etc. A good exterior house paint must not only have good original film flexibility, it must retain that flexibility through severe outdoor weathering. In other words, it must resist losing plasticizer or having the film become brittle through oxidation. An exterior house paint film must be able to accommodate normal expansions and contractions of the substrate.

On the other hand, paints used to coat automobiles have other requirements. These coatings should be very impermeable to help protect a ferrous metal surface from oxygen and water (two conditions promoting corrosion). Also the film should electrically insulate the surface to reduce galvanic action and the flow of electrical current needed to induce corrosion. Surface preparation is very important. If a car is being repainted, the surface should be cleaned down to the bare metal. Automobile coatings, of course, must protect the ferrous substrate against corrosion. It must resist exposure to salt on the road in the winter. The protective film should also have high impact resistance to protect against "flying" gravel when driving at high speed. Also appearance is very important. The paint must be able to retain a high gloss appearance and have good color stability. For the original equipment market, the speed with which the finish can "dry" is important in order to maintain production rates.

Overall, most paint films have to display a degree of hardness and flexibility to impart good service life. These two differing properties may not be obtained well if the binder used is a single homopolymer. Therefore, in many paint applications, the binder system will consist of binder blends, copolymers, or terpolymers to provide the best trade-offs between film hardness and flexibility.

Types of Technologies

The technologies and techniques used today in the paint industry are diverse.

In the following discussion there will be references to seven general types of technologies used in the coating sciences today.

(1) **Conventional Solvent Systems.** This technology includes both convertible oil based paints and non-convertible (lacquer type) paints. This technology is somewhat mature.

(2) **High Solids Solvent Systems.** This is a newer technology which utilizes coatings with less than 30% solvents and tries to overcome the inherent problem of high viscosity, etc., resulting from the high solids levels. This technology has evolved because of EPA requirements to reduce hydrocarbon air emissions.

(3) **Aqueous Emulsion Systems.** The markets for these latex based paints have grown greatly in the last thirty years because of their ease of handling and applying, ease of clean up, non-flammability, and lack of hydrocarbon solvent exposure. These latex paints have been improved over the years in stability, film formation, and durability.

(4) **Water Soluble Binder Systems.** These coatings are based on binders that are soluble in water. Upon drying, the binder remains to be crosslinked, usually through oxygen conversion. Convertible resins are modified through the addition of carboxylic, hydroxyl, or amino groups to render them water soluble.

(5) **Two Part Catalyzed Systems.** These are two part coatings in which one component contains a reactive polymer (such as an epoxy) while the other component contains the cross-linking agent. Once the two parts are mixed, the resulting mixture has a limited "pot life" before which it must be applied. Usually crosslinking can occur at room temperature after the film is applied. By using such systems, it is possible to apply relatively thick films at 100 percent solids and minimal hydrocarbon emissions.

(6) **Powder Coating.** This is a system in which a polymeric powder is

applied to a substrate usually in a fluidized bed or as a spray. Sometimes the powder particles may be electrostatically charged. With the application of heat, a thick continuous film is formed. A major advantage in powder coating is that a film can be formed quickly and a thick film can be applied. A possible disadvantage is that some techniques require the substrate to be heated. In general, these systems are very efficient with practically no hydrocarbon emissions into the environment.

(7) **Radiation Curing.** This is another emerging technology which was developed to help reduce the overall solvent hydrocarbon emissions from applied coatings. To date, this technology has only been applied to very limited specialty applications. These special coatings can be cured quickly from exposure to photons from ultraviolet, infrared, or electron beams. Also microwaves have been used for some films. One problem is that the surface being coated must be relatively flat (line of sight) for equal exposure. These coatings in general cannot be highly pigmented.

Methods of Application

The methods used in applying coatings to a substrate are changing. Many years ago, paint was only applied by brush. Then rollers were invented for quick applications to flat surfaces. Now the spraying techniques that are used are even faster. Also dipping is used in factory settings. For coating quickly flat tinplate, steel plate, or other flat sheets, a capital intensive method called "roller coating" can be used which applies the coating by using a feed tank and a series of rollers after which the sheet is oven dried before rerolling.

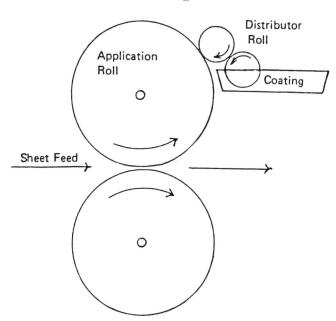

Roller coating techniques are commonly used to coat metal sheeting which is used to make beverage cans, etc.

Newer techniques for improved efficiency include electrostatic paint spraying and electrodeposition (for immersion applications). These techniques can produce a very uniform film, greatly improved efficiency and reduce spray fog (in the case of electrostatic spraying). In fact, electrostatic techniques can be used in applying powder coatings through spraying. Also heated articles can be powder coated through dipping into a fluidized bed (discussed earlier).

It should be noted that brushing, rolling, or spraying will generally apply a paint under high shear conditions as shown below.

Application Method	Rate of Shear (sec^{-1})
Brushing	5,000–20,000
Rolling	3,000–40,000
Spraying	1,000–40,000
Dipping	10–100

As discussed earlier, most viscosity tests measure viscosity under low shear conditions. As seen from the data above, viscosity at high shear should also be taken into account because most all coatings are non-Newtonian (change in viscosity with changes in rate of shear).

FORMULATING

Paint formulating may be more of an art than a science. A paint formulator must choose from hundreds of different materials in selecting a combination which will have the various film forming and service properties desired as well as being cost effective.

Components

A paint formulator chooses raw materials from basically four broad categories as shown below.

(1) Binders

(2) Solvents

(3) Pigments

(4) Additives

The binder is the "backbone" of film formation. The polymeric binder is key to the formation of a continuous film. It literally binds and holds the pigment particles together.

Solvents are important because they are needed to solubilize the binder. The coating must be in the fluid state to be applied. Afterward, the solvent is lost to evaporation during film formation. Latex emulsions do not use solvents; instead, they utilize a fluid medium-water-in which colloidal binder particles are suspended.

Pigments are usually added to impart a particular decorative color. They also help protect the binder against U.V. degradation from outdoor exposure. Clear coatings do not contain pigments.

Additives are a very broad category including literally hundreds of different chemicals. These ingredients may be used at only a fraction of a percent and usually are not used at levels greater than 5 percent. Even with their relatively low concentration, they have a very large effect on the paint properties.

Examples are shown below of types of additives used for both solvent based and latex paint systems.

Solvent Base Additives	Latex Additives
Driers	
Antiskinning agents	
Thixotropes/thickeners	Thixotropes/thickeners
Biocides	Biocides
Corrosion inhibitors	Corrosion inhibitors
	Surfactants
	Defoamers
	Coalescing agents
	Freeze-thaw stabilizers
	Anti-rust agents
	pH buffers

Formulation Example

The paint formulation is a key part of any paint factory operation. It indicates which ingredients are required and the quantities necessary for the formulation. A single paint formulation can call for as many as 20 different ingredients. Paint formulations will usually denote number of pounds and gallons required for each ingredient per batch. Formulations can be expressed as percent by weight or pounds per 100 gallons. Also these formulations will indicate which ingredients are added initially in the disperser and which are to be added in the "let down" stage afterward.

The following shows an example of a latex paint.

Exterior Latex House Paint

Ingredients	Pounds	Gallons
Water	115.0	13.81
Ethylene glycol	25.00	2.69
Thickener	4.20	0.36
Dispersant A	10.00	1.01
Dispersant B	2.50	0.12
Surfactant	2.00	0.22
Defoamer A	1.00	0.14
Titanium dioxide	235.00	7.08
Zinc oxide	50.00	1.07
Talc	175.00	7.38
Let Down		
Water	142.31	17.09
Propylene glycol	35.00	4.06
Biocide (mildewcide)	7.00	0.87
Acrylic latex (50%)	375.00	42.37
Defoamer B	2.00	0.27
Coalescent Agent	10.00	1.26
Ammonium hydroxide	1.5	0.20
Total	1,192.51	100.00

Besides this formula listing ingredient pounds and gallons per batch, it can also show percent non-volatiles (NV), "dry pounds", "dry gallons", or dollar costs. Customarily, "formulation constants" are provided at the bottom of the formulation to enable the formulator to check that all the ingredients and their proper quantities are included. These "constants" may include specific gravity, solids, viscosity, or perhaps pH. Also calculated constants may be provided such as pigment/binder ratio, pigment volume concentration, non-volatile content by volume, or square feet of coverage at 1 mil dry film. Commonly a computer is employed to calculate these constants and recalculate them when formula adjustments are made. These calculated constants and their importance are discussed next.

Pounds vs. Gallons

As shown before, paint formulations are commonly given in both pounds and gallons. Gallons of a given ingredient can be converted to pounds as follows:

$$\text{wt. (lb)} = 8.33 \times \text{sp. gr.} \times \text{vol. (gal)}$$

Here 8.33 is the number of pounds that one gallon of water weighs at

77°F. If the ingredient in question were only water, you would only have to multiply the number of gallons by 8.33 (at 77°F) to determine its weight in pounds (the specific gravity of water at 77°F is very close to 1.0). On the other hand, it can be seen from the equation that materials having a specific gravity greater than 1.0 will also weigh more than 8.33 lb./gal. Therefore, when comparing two different materials on an equal weight basis, it is important to note that the material with a higher specific gravity will occupy less volume. This fact is important to remember because generally paint is sold on a volume basis (in gallons). However, raw materials many times are purchased on a price/pound basis. Therefore, true paint cost reduction should be calculated in cost savings per gallon.

Non-volatiles

From a customer's perspective, the percent non-volatiles is very important in that the volatile components are lost after application. While these volatile components are necessary vehicles for applying a coating, they do not have value to the customer in contributing to dry film thickness or total coverage. Therefore, knowing percent non-volatiles can be important to the customer in estimating paint coverage at a given dry film thickness.

Some paint formulations will give the percent NV or "dry pounds" for each ingredient. The weight percent non-volatiles can be calculated as follows:

$$\text{wt. \% NV} = \frac{\text{wt. of non-volatiles}}{\text{wt. of formula}} \times 100$$

The volume percent non-volatiles is calculated as follows:

$$\text{vol. \% NV} = \frac{\text{vol. of non-volatiles}}{\text{vol. of formula}} \times 100$$

Film Coverage

The wet film thickness can be calculated as follows:

$$\text{wet film thickness} = \frac{\text{vol. of applied paint}}{\text{area covered by paint}}$$

To calculate the dry film thickness, one must use the calculated volume percent non-volatile content (discussed above) to make this correction.

The above equation can also be used to calculate area that can be covered by a given volume of paint for a given film thickness.

Pigment to Binder Ratio

In addition to the non-volatile content of a paint, there are other calculations to address the quality or nature of these "solids" that remain after film formation. One of these properties is called the pigment-to-binder ratio (P/B).

$$P/B = \frac{\text{wt. of total pigment and extender}}{\text{wt. of total binder solids}}$$

Coatings with P/B values that are above say 4:1 have a relatively small quantity of binder to hold the high quantities of pigment together in the film. Thus paints with high P/B values are flat in appearance and can not provide a continuous matrix for effective protection against the elements. On the other hand, paints with very low P/B ratios (say, less than 1) are characterized as more durable and possessing a higher gloss (provided that a flattening agent has not been applied.)

Pigment Volume Concentration

While the weight ratio of pigment/binder roughly denotes the quality characteristics of a paint, the volume concentration of the pigments present is a more accepted scientific method of predicting how a paint film will perform.

Pigment volume concentration or %PVC is calculated as follows:

$$\% PVC = \frac{\text{vol. of pigments}}{\text{vol. of total non-volatiles}} \times 100$$

Here "volume of pigment" includes the volume of pigments and extenders while "volume of total non-volatiles" includes binders, pigments, extenders, and other non-volatiles.

Another important term used is called "critical pigment volume concentration" or CPVC. CPVC is the minimum level of binder required to completely wet the pigment particles and fill the voids. From a paint formulator's point of view, it might be considered a point of discontinuity with increasing PVC in that many film properties change rather abruptly. This is because at pigment volume concentrations above CPVC, there is not enough binder to completely fill the voids between the pigment particles. As a result, at PVCs above CPVC, the film becomes more porous, permeable, and weak. CPVC usually occurs between 40 to 55% PVC. The following graph illustrates the changes in film properties that occur at CPVC.

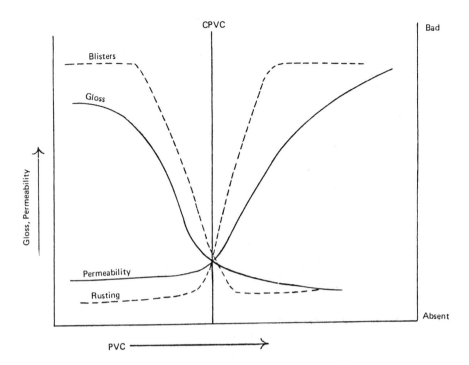

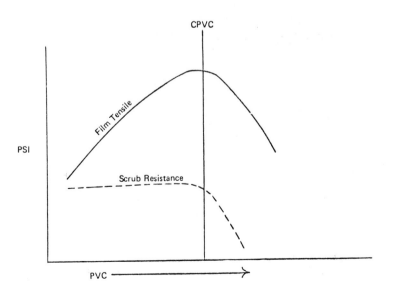

As explained previously, at PVCs above CPVC, the film porosity is greatly increased. Since the film above the CPVC can no longer protect a ferrous substrate, rusting occurs. Also because of the increasing porosity and permeability, blistering is reduced. Another effect above CPVC is that the pigment particles begin to disrupt the smoothness of the film surface and cause light to be scattered and film gloss to be reduced. So at high PVCs above the CPVC, the film surface becomes very rough with pigment particles, and the film will have a very flat appearance.

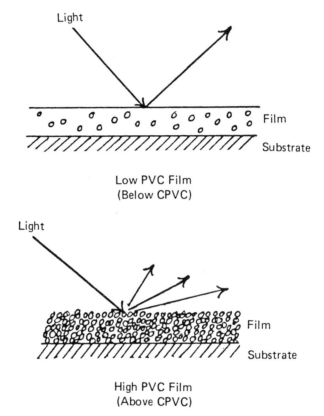

Low PVC Film
(Below CPVC)

High PVC Film
(Above CPVC)

Thus, paints that impart very high gloss appearances, such as automobile finishes, must have PVC values well below the CPVC. Conversely, paints that impart a flat appearance, such as some interior house paints, must have relatively high PVC values.

Another consideration regarding PVC is cost. Sometimes, depending on price fluctuations, pigments and extenders can be less expensive than a binder. If this is the case, higher PVC paints may be less costly to formulate than lower PVC paints. However, as seen from previous illustrations, there are limits to how high one can extend a paint film before certain customer property requirements are lost. Part of the challenge to a paint formulator is to select a balance between cost and product performance.

Important Properties

The remainder of this chapter will involve a description of the wide scope of raw materials from which a paint formulator may select and the comparative advantages of one material vs. another. Below is given a listing of properties which a paint formulator may have to consider when developing a new paint formula.

Cost per gallon

Ease of application

Sprayability

Hiding power

Brightness

Hue

Sag resistance

Freeze-thaw resistance

Storage stability

Drying time

Dry film hardness

Dry film flexibility

Dry film durability

Dry film weather resistance

Chalking properties

Flaking resistance

Detergent resistance

Chemical resistance

Lightfastness

Fire Resistance

Scrubbability

Dry film gloss

Dry film adhesion

Dry film impact resistance

EPA hydrocarbon emission standards

Salt resistance

Corrosion protection

Blister resistance

Ease of manufacture

Odor

Toxicity

Any number of these properties may be considered in establishing the requirements for a specific product. For a paint formulator to meet these requirements in a cost effective manner requires years of experience, many times learned through trial and error. There is a great deal of art in paint formulating. The purpose of the following sections is not to make the reader a paint formulator, but rather to serve as an overview of the many different types of functional raw materials used in paint formulating today.

BINDERS

Binders are the polymeric components in a coating formulation which form the protective film. The binder literally binds the pigment particles together in the film. If the coating contains no pigments or extenders, then the binder will usually form a glossy and maybe transparent coating. As discussed earlier, binders may be either convertible (analogous to thermosetting) or non-convertible (analogous to thermoplastics).

In 1982, over 2 billion pounds of binders (resins) were consumed by the coatings industry. A breakdown concerning a percentage by types of binders consumed is shown below.

Type Resin	Weight % of Total Consumption
Alkyds and polyesters	33%
Acrylics	20%
Vinyls	20%
Epoxies	5%
Urethanes	5%
Amines, phenolics, and miscellaneous	17%
Total	100%

The following will cover each type of resin in more detail.

Drying Oils

These natural oils have been used for hundreds of years as film formers and binders in paint manufacture. Drying oils are vegetable or marine (fish) oils and chemically are triglycerides of the general structure shown below.

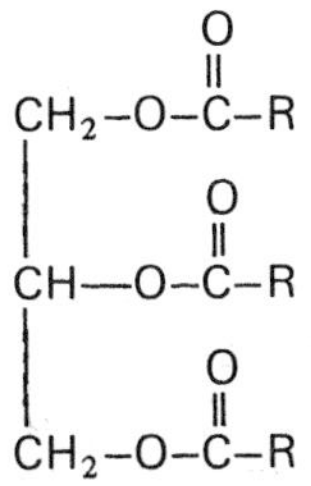

$$CH_2-O-\overset{\overset{\displaystyle O}{\|}}{C}-R$$

$$CH-O-\overset{\overset{\displaystyle O}{\|}}{C}-R$$

$$CH_2-O-\overset{\overset{\displaystyle O}{\|}}{C}-R$$

These triglyderides are based on a blend of saturated, unsaturated, and polyunsaturated fatty acids. Some common examples are shown below.

Caprylic acid	$CH_3(CH_2)_6-COOH$
Capric acid	$CH_3(CH_2)_8-COOH$
Lauric acid	$CH_3(CH_2)_{10}-COOH$
Myristic acid	$CH_3(CH_2)_{12}-COOH$
Palmitic acid	$CH_3(CH_2)_{14}-COOH$
Stearic acid	$CH_3(CH_2)_{16}-COOH$
Oleic acid	$CH_3(CH_2)_7CH=CH(CH_2)_7COOH$
Linoleic acid	$CH_3(CH_2)_4CH=CHCH_2CH=CH(CH_2)_7COOH$
Linolenic acid	$CH_3CH_2CH=CHCH_2CH=CHCH_2CH=CH(CH_2)_7COOH$
Eleostearic acid	$CH_3(CH_2)_3CH=CHCH=CHCH=CH(CH_2)_7COOH$
Licanic acid	$CH_3(CH_2)_3CH=CHCH=CHCH=CH(CH_2)_4\underset{\underset{\displaystyle O}{\|}}{CH}(CH_2)_2COOH$

Triglyceride oils containing mostly saturated fatty acids will not form films after exposure to the air and are called non-drying oils. They are only used in paints when blended with drying oils. It is the unsaturated fatty acid constituent, particularly the polyunsaturated fatty acid component, that promotes oil "drying" and film formation. (A polyunsaturated fatty acid is one that possesses more than one unsaturated site.)

Drying oils form flexible films from the formation of oxidative crosslinks between the polyunsaturated fatty acid constituents of different triglyceride molecules. The chemical process of forming an oxidative crosslinked matrix after air exposure of a gylceride based partially on linoleic acid is illustrated below.

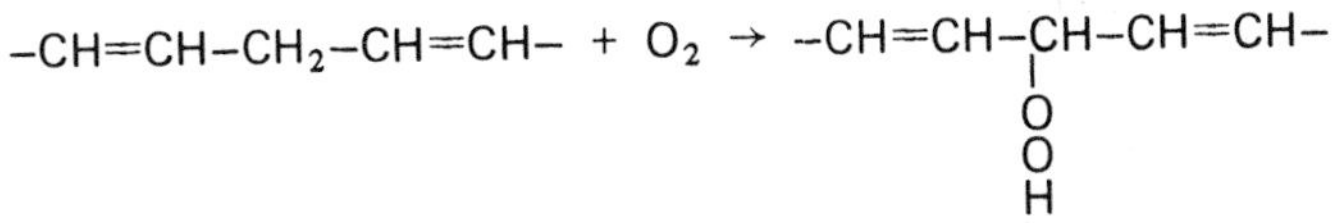

$$-CH=CH-CH_2-CH=CH- + O_2 \rightarrow -CH=CH-\underset{\underset{\underset{\underset{H}{O}}{O}}{|}}{CH}-CH=CH-$$

This formation of the hydroperoxide after air exposure, decomposes in the presence of driers (to be discussed later). Driers are usually soaps of cobalt, lead, or manganese.

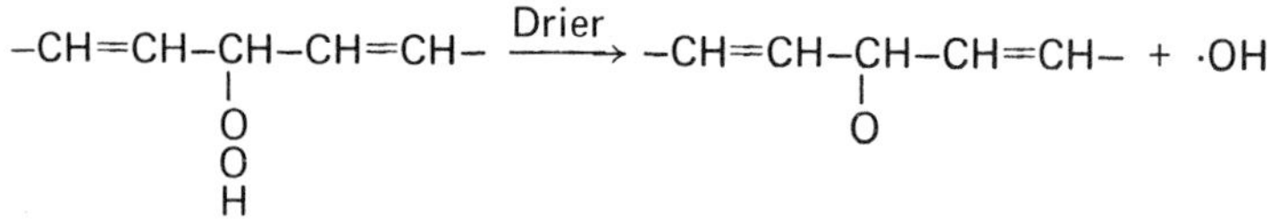

$$-CH=CH-CH-CH=CH- \xrightarrow{\text{Drier}} -CH=CH-CH-CH=CH- + \cdot OH$$

with $-CH=CH-CH-CH=CH-$ bearing $\overset{|}{O}-\overset{}{O}-H$ and product bearing $\overset{|}{O}$

$$\cdot OH + -CH=CH-CH_2-CH=CH- \rightarrow -CH=CH-CH-CH=CH + HOH$$

Thus free radicals, $R\cdot$ and $RO\cdot$, combine to form ether, peroxy, and carbon-carbon crosslinks between oil molecules.

$$
\begin{aligned}
RO\cdot \ \ &+\ R'O\cdot \ \ &\rightarrow\ &ROOR' \\
ROO\cdot \ &+\ R'\cdot \ \ &\rightarrow\ &ROOR' \\
ROO\cdot \ &+\ R'OO\cdot \ &\rightarrow\ &ROOR' + O_2 \\
RO\cdot \ \ &+\ R'\cdot \ \ &\rightarrow\ &ROR' \text{ (Ether Crosslinks)} \\
R\cdot \ \ \ &+\ R'\cdot \ \ &\rightarrow\ &R-R' \text{ (Carbon-Carbon Crosslinks)}
\end{aligned}
$$

(Peroxy Crosslinks) — bracketing the first three reactions

These crosslinks form between the fatty acid branches of the triglyceride molecules to form a three dimensional matrix as shown below.

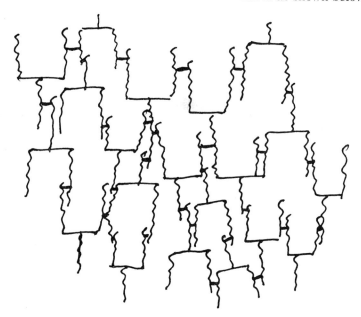

Triunsaturated fatty acids, such as linolenic acid, are much more rapid in forming oxidative crosslinks than diunsaturated fatty acids, such as linoleic acid. On the other hand, conjugated, triunsaturated fatty acids, such as eleostearic acid, are faster yet.

Drying oils differ in the film forming properties they impart to a paint. These differences in performance are attributed to compositional differences between oils. Linseed oil, tung oil, oiticica oil, and fish oil are examples of commonly used drying oils. When a drier (usually a cobalt, manganese, or lead soap to be discussed later) is added to these oils and they are exposed to air, they form a film quickly. These oils are described as follows:

Linseed oil. This oil is the most popular oil used by the paint industry. Over fifty percent of its fatty acid component is linolenic acid. It is obtained from flaxseed. The high linolenic acid content tends to cause some yellowing; however in highly pigmented paints, this is not usually a problem.

Tung oil. This oil is a very fast drying oil because the fatty acid component is 80% eleostearic acid. Because of eleostearic acid's conjugated, triple unsaturation, this oil is faster drying than linseed oil, giving good film hardness and flexibility. Tung oil (also called China wood oil) comes from nuts of the Tung tree indigenous to China.

Fish oil. This oil contains both high levels of fast drying acids and non-drying acids (saturated and mono-unsaturated acids). As a result it dries more slowly than linseed oil and gives a softer film because of the saturated fatty acids present. Fish oils can impart an undesirable odor.

Oiticica oil. This oil is from nuts of a Brazilian tree and almost 80% of its fatty acid composition is based on licanic acid which is similar to eleostearic acid in reactivity. This explains its fast drying speed. It is still slightly slower drying than tung oil, however. Also when it is used alone in a paint, it gives lower film strength.

Paints based solely on drying oils as the only binder have been declining steadily in the last fifty years. Alkyds, acrylics, and other synthetic polymers have been replacing these oils as the sole binding agent. Drying oils alone, impart a softer film than other synthetic binders. These oils alone have a low gloss and are only used in highly pigmented paints. Even though these paints have relatively poor durability and chemical resistance, they are still used in small volumes for limited applications.

Oleoresinous Binders

These binders are used in making varnishes. They are produced by "cooking" a given quantity of a natural drying oil with a predetermined amount of resin in a large reactor vessel. This "cooking" process increases the molecular weight and renders the oil and resin compatible with each other and soluble in a thinning solvent. A kind of "copolymer" between the oil and resin is formed. After the so-called "cooking" process, the oleoresinous product is thinned with a solvent.

Oil Length. The ratio of oil to resin determines many of the final varnish properties. A higher ratio of oil to resin means a slower drying speed, greater softness, increased flexibility, and reduced gloss. With a lower ratio of oil to resin, there is an increase in drying speed, high film hardness, and increased gloss. The ratio of oil to resin is commonly referred to in the varnish industry as *oil length*. Oil length is defined as the number of gallons of oil to 100 pounds of resin. Thus the following terms are defined below.

Type Varnish	Oil Length
Short oil	5-15
Medium oil	16-30
Long oil	>30

Of course, not just the oil length, but also the types of oil or resin also affect a varnish's physical properties. The oils discussed previously (linseed, tung, fish, and oiticica oils) are also used to make varnishes. Linseed oil, however, is not as commonly used in varnish making because it is slower drying and yellows badly. On the other hand, tung oil is considered the best oil for varnish manufacture.

Types of Resins. Many types of resins are used in making oleoresinous varnishes. These resins include the following:

Polyterpene resins

Petroleum resins

Coumarone indene resins

Gum rosins

Wood rosins

Tall oil rosins

Rosin soaps

Rosin esters

Novolac phenolic resins

Rosin modified phenolic resins

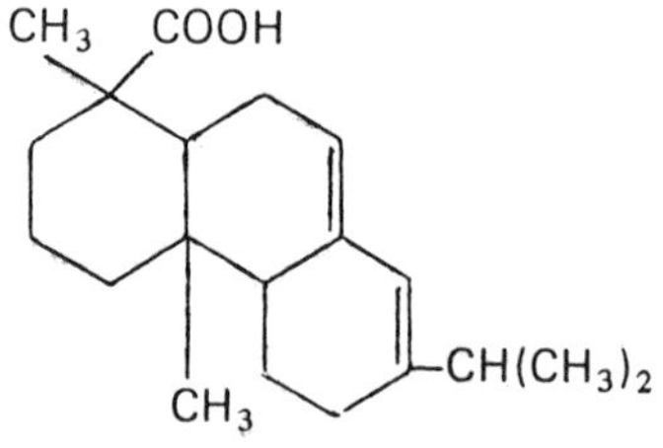

Rosin (Abietic Acid)

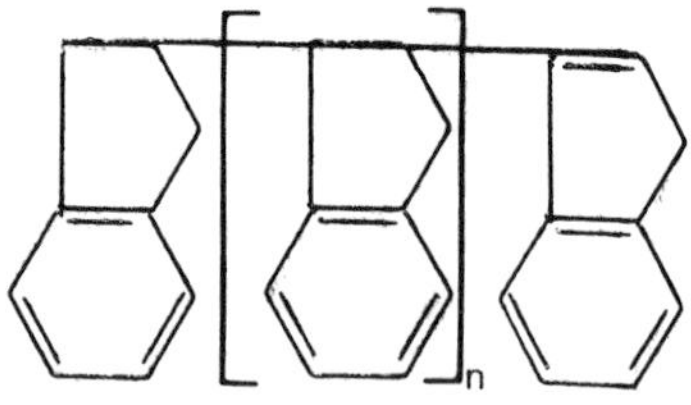

Polyindene Resin

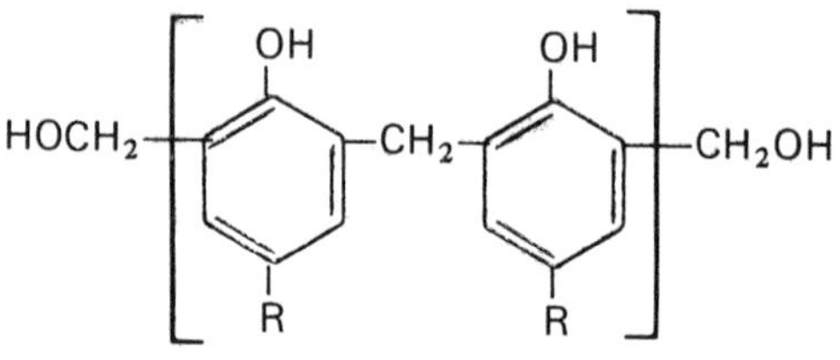

Phenolic Resin

Usually these resins will react with the oils during the "cooking" process in a manner shown below (as an example).

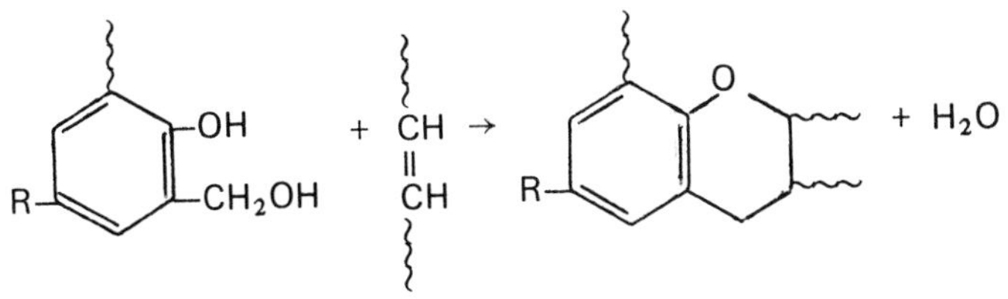

These reactions do not occur with every resin however. For example, the heating process may help solubilize terpene or coumarone-indene resins in the oil; however these resins do not chemically react with the unsaturated oil itself.

Most oleoresinous varnishes are made from rosins and phenolic resins. The gum, wood, and tall oil rosins are usually interchangable in this coating application. Usually the straight rosin is not used per se because of its high acidity, brittleness, and weak water resistance. Instead, the rosin is used as rosin ester or soap. Rosin esters are obtained by reacting rosin with alcohol such as glycerin. These products have low acid numbers, but are soft (low softening point). The softening point can be increased by using polyols such as pentaerythritol. Also these resins can be hydrogenated or polymerized to increase hardness. On the other hand, rosin soaps are formed by reacting rosin with an alkali such as lime. These products have higher softening points and are harder and more brittle. Use of these materials improves the gloss of a varnish.

The synthetic phenolic resins impart superior water resistance and weather resistance to a varnish. On the other hand, phenolic resins can cause color problems with varnish as well as an oil compatibility problem. Modifying phenolic resins with rosin esters may help color retention. Longer alkyl groups substituted at the para position will also improve a phenolic resin's solubility in oils.

As a last note, it should be mentioned that these "soft" oleoresinous varnishes have been declining in use. The new "hard" polyurethane varnishes for the most part have replaced them. Also the polyurethane varnishes possess superior chemical resistance. Remember how a glass of ice-cold water might leave a ring on an old soft varnished tabletop. Polyurethane varnishes are water resistant.

Alkyd Resins

These resins are produced by the reaction of a polybasic acid with a polyol and a monobasic fatty acid (or oil). They are called "alkyds" from "al" which means alcohol and "kyd" which stands for the "cid" in acid.

In the early part of this century, experiments were conducted to react dicarboxylic acids with a polyol such as glycerin, to form a polyester coating. However, these resins were insoluble in common solvents and could not be air dried. Then in the 1920s, fatty acids were used to modify these resins to render them soluble in aliphatic solvents and capable of being air dried. Since then, alkyd resins have grown to be one of the largest classes of binders used in paints today.

The most common example of an alkyd resin is from the polycondensation reaction of glycerin, phthalic anhydride, and fatty acid. Glycerin, with three hydroxyl groups, is considered trifunctional. Phthalic anhydride is difunctional. Therefore, a polycondensation reaction between these two reactants could gel before the polycondensation was complete. But with the addition of a monocarboxylic fatty acid, the formation of a somewhat linear polymer is possible. This polymer is soluble in aliphatic solvents if the content of fatty acid is high enough.

Although fatty acids can be used directly, usually less expensive vegetable or marine oils are used in producing these resins in a two stage process. The first stage involves heating the oil and part (or all) of the glycerin with a suitable catalyst, such as calcium hydroxide, to around 230°C. This results in alcoholysis of the oil as shown below.

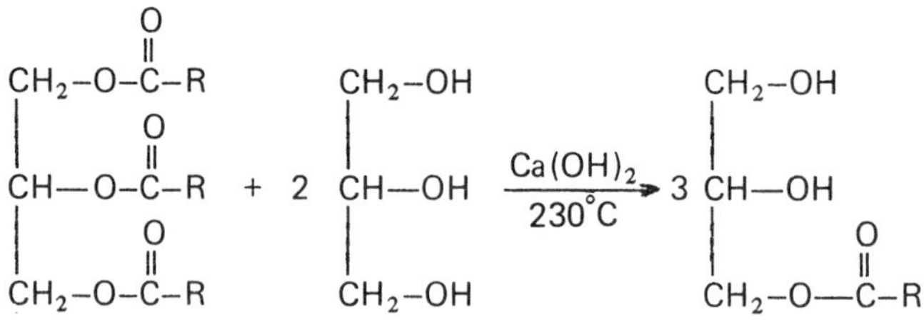

Triglyceride   Glycerin   Monoglyceride

In a second stage of the process, the phthalic anhydride is added to the reactor. The phthalic anhydride reacts with the monoglycerides to produce the alkyd resin. It is important to note that during these reactions, the reactor vessel should be purged of oxygen by using an inert gas such as nitrogen since oils are attacked by oxygen at elevated temperatures and "body".

The variety of alkyd resins used today in the paint industry is seemingly endless because of the wide variety of reactants and differences in compositions that are possible. Although glycerin is the most common polyol used, the following other polyols may also be used as well.

Ethylene glycol

Pentaerythritol

Sorbitol

Trimethylol ethane

Trimethylol propane

Diethylene glycol

Propylene glycol

Next to glycerin, pentaerythritol (PE) is the most important polyol used. Unlike glycerin, it is tetrafunctional.

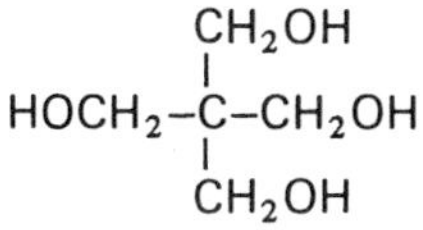

$$CH_2OH$$
$$|$$
$$HOCH_2-C-CH_2OH$$
$$|$$
$$CH_2OH$$

Pentaerythritol

Because it is so reactive, it is commonly used with other polyols, such as glycerin or ethylene glycol. PE is used to improve water resistance, raise viscosity, increase film hardness, and accelerate drying of an alkyd resin.

The other alcohols are less often used. Of the diols, ethylene glycol is most commonly used. Also diethylene glycol, and propylene glycol are used as diols. Of course, diols are difunctional. Also other trifunctional polyols besides glycerin include trimethylol propane and trimethylol ethane. Although sorbitol, with six hydroxyl groups, may be thought to be more reactive than PE, in reality it reportedly behaves more like glycerin in reactivity and functionality.

Although phthalic anhydride is the most common polybasic acid used in producing alkyd resins, also the following are used.

Isophthalic acid

Terephthalic acid

Trimellitic anhydride

Maleic anhydride

In the last twenty years, isophthalic acid and terephthalic acid have been used on occasion as substitutes for phthalic anhydride. Maleic anhydride is used (with other polybasic acids) as a modifier to increase viscosity and film hardness. Trimellitic anhydride has gained use in making alkyds because it reportedly renders these resins water soluble for special applications.

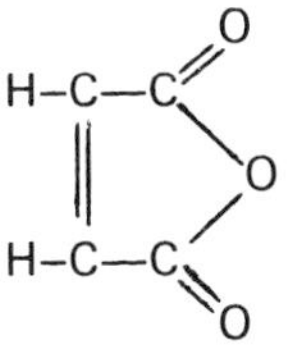

Maleic Anhydride 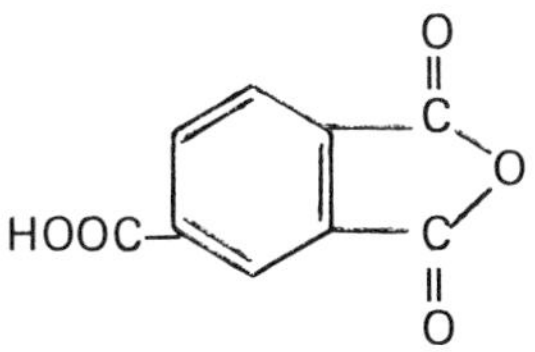   Trimellitic Anhydride

The selection of an oil can also affect the properties of an alkyd resin. The most common drying oil used in alkyd production is linseed oil. Also, soybean oil is used when yellowing is to be avoided. Other drying oils that are used occasionally include fish oil, tung oil, and oiticica oil. Just as was the case with oil paints, it is the use of these vegetable drying oils which enables these alkyds to oxidatively air dry (crosslink) through a similar mechanism discussed above under "Drying Oils". As mentioned before, cobalt, manganese, and lead salts are used to promote oxidative crosslinks. Sometimes a non-drying oil, such as cottonseed oil or coconut oil, is used when the alkyd resin is to serve as a plasticizer, not as a film former.

In addition to a variety of different types of polyols, polybasic acids, and oils, other ingredients are sometimes added as well. Non-fatty mono-carboxylic acids, such as benzoic acid, are sometimes used in place of a portion of the fatty acid to terminate the alkyd chains. These additives impart special properties related to gloss, flexibility, and drying time. Also other modifying agents such as styrenics, acrylics, silicone, phenolic resins, and natural rosins are occasionally used in modification to impart special properties to the alkyd.

Not only do differences in reactive components affect alkyd proper-ties, but also oil length. Here oil length may be defined as the number of grams of oil used in making 100 grams of alkyd resin. Short oil alkyds are heated (baked) to form a film. Their solubility is limited to aromatic solvents and other non-aliphatic solvents; they are generally restricted to industrial coatings. On the other hand, long oil alkyds are easily air dried and are readily soluble in aliphatic solvents. They can easily be applied by brush and are used extensively in architectural applications. Generally as oil length increases, viscosity will decrease, hardness decreases, flex-ibility increases, and water resistance is reduced.

As mentioned earlier, unlike other polymers, alkyds were developed in the paint industry for use in paints. Ninety-five percent of alkyd use is in coating applications. The ability of alkyds to be modified in so many ways helps explain their wide use in the coatings industry. They are used in more different paint applications than any other binder. Their ver-satility is their main advantage.

A noted disadvantage in using alkyds is the lack of the film's chemical resistance. Alkyds, being modified polysters, do not possess high alkali-resistance (subject to attack by hydrolysis). Conventional alkyds are also dependent on the use of organic solvents which may have environmental considerations.

Acrylics

This class of binder has gained much popularity since the early 1950's. One reason for its wide use is its inherent clarity, color stability, and resistance to yellowing. Acrylics have gained wide acceptance in both solvent based coatings and latex paints. Acrylic polymers have one of the largest volume usages in the coatings industry.

Acrylics represent a large number of different polymers of which polymethyl methacrylate is by far the largest volume used. These polymers can be represented by the two general structures given below.

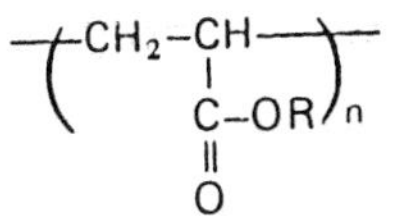

$$\left(\!\!CH_2\!-\!\underset{\underset{\parallel}{C-OR}}{\overset{CH}{|}}\!\!\right)_n \qquad \left(\!\!CH_2\!-\!\underset{\underset{O}{\overset{|}{C-OR}}}{\overset{CH_3}{\underset{|}{C}}}\!\!\right)_n$$

Polyacrylates 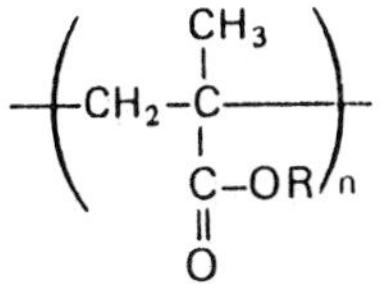   Polymethacrylates

(where R = alkyl group)

Alkyl groups can include methyl, butyl, isobutyl, 2-ethylhexyl, isodecyl, lauryl, etc. As the number of carbon atoms in the alkyl group is increased, the glass transition temperature and brittle point for both types of polymers decreases till about C8, for polyalkyl acrylates, and C12, for polyalkyl methacrylates, then they rise again. It should be noted that the polyalkyl methacrylate has a considerably higher glass transition temperature and brittle point than the corresponding polyalkyl acrylate. As a result, polymethyl methacrylate is a hard, tough polymer at room temperature while polymethyl acrylate is fairly soft. Moreover, polyethyl acrylate is soft and tacky while polybutyl acrylate is *very* soft and tacky.

In paint applications, sometimes a combination of methacrylates and acrylates is used to achieve needed film properties. While polymethyl methacrylate is commonly needed for strength and hardness of a film, a polymer such as polymethyl acrylate will improve film flexibility. As a result, a "plasticizing" monomer, such as ethyl acrylate or ethylhexyl acrylate is sometimes added with the methyl methacrylate monomer by the polymer manufacturer in an addition-polymerization process to produce a softer copolymer for use in coating formulations. In paint film formation, a softer and more flexible polymer is usually needed, especially for latex film formation. The latex particles must be able to coalesce together on drying to produce a continuous film. A softer copolymer enables this to happen.

Acrylic based coatings can be either thermoplastic or thermosetting. The polymers just discussed, which are not modified, are thermoplastics and do not crosslink at elevated temperatures. An acrylic polymer can be made thermosetting by the polymer manufacturer by adding a functional monomer during the free radical polymerization process. This additional monomer provides functional groups off the backbone to give crosslinking sites. These functional groups can include carboxyl, hydroxyl, and amide groups. Two examples of "functional monomers" that can impart these functional groups to an acrylic polymer are shown below.

$$CH_2\!=\!CHCONH_2 \qquad\qquad CH_2\!=\!CHCOOH$$

Acrylamide   Acrylic Acid

These thermosetting acrylics are baked with other reactive resins, such as epoxies, to form a crosslinked film with greater hardness, toughness, and chemical resistance that would not be attainable from a thermoplastic acrylic. Lastly, it should be noted that polymer functional groups imparted from monomers, such as acrylic acid, can also improve film adhesion.

Acrylic and methacrylic monomers are polymerized by using the emulsion, suspension, and solution processes.

Acrylic thermoplastic organic-solvent based paints are commonly used in such high volume applications as automotive finishes. Organic soluble, thermosetting, acrylics are used to bake on a hard enamel for appliances such as washers or dryers. Acrylics also offer the added advantage of possessing good color stability, resistance to yellowing, and good gloss.

On the other hand, acrylic emulsions, used to make latex paints, have revolutionized the exterior house paint market. Although more expensive, acrylic emulsion paints have greater color stability than styrene-butadiene latex paints and better masonry alkali resistance than polyvinyl acetate water based paints. Also they are ideal for outdoor use because of their durability and flexibility. Their film adhesion is also quite excellent. These films resist blistering and chalking as well. In addition, acrylic latex paints have also gained wide acceptance as an interior house paint.

Vinyl Resins

In the coating industry, vinyl resins usually refer to either polyvinyl acetate or polyvinyl chloride.

Today polyvinyl acetate has a large usage in the interior and exterior latex paint markets. Mostly produced by the emulsion polymerization process, a polyvinyl acetate homopolymer by itself is too hard to allow its colloidal latex particles to coalesce well into a continuous film at ambient temperatures. Therefore, most polyvinyl acetate emulsions used in the paint industry are copolymers with another "plasticizing" monomer such as dibutyl maleate, 2-ethylhexyl acrylate, n-butyl acrylate, dibutyl fumarate, isodecyl acrylate, or ethyl acrylate. Also by polymerizing under pressure, copolymers of vinyl acetate and ethylene have also been used in paints. Lastly, another technique for lowering the glass transition temperature of polyvinyl acetate is by using an external plasticizer such as dibutyl phthalate. All these methods not only soften the polymer to allow the latex particles to coalesce into a continuous film, but also impart sufficient film flexibility which is especially needed in exterior house paints.

PVAc emulsions are less expensive than acrylic emulsions. In addition, PVAc imparts very good color stability and durability. These properties have allowed PVAc to penetrate a large portion of the interior and exterior house paint markets. One advantage of PVAc copolymers over acrylics is their greater moisture permeability which allows houses to "breathe". One disadvantage of PVAc based latex paints is their relatively poor alkali-resistance compared to acrylics on masonry substrates which may lessen durability.

The other vinyl polymer, polyvinyl chloride, is manufactured by one of four processes, i.e., bulk, solution, suspension, or emulsion polymerization. Basically, the main application that the PVC homopolymer has in the coatings industry is in powder coatings applications at elevated temperatures. This is because the homopolymer displays too poor solubility in most solvents to be used in solution coatings and is too hard at room temperature to allow the formation of a continuous film in latex paint applications. To overcome these problems, PVC resins are used as copolymers, usually with vinyl acetate. The bulkier acetate group prevents the polymer macromolecules from packing as tightly (as is the case with the PVC homopolymer), thus allowing a much greater range of solvent solubility. Also the acetate contributes to greater polarity which aids in solvation with some solvents. In latex applications, a "soft" monomer, such as vinyl acetate, can reduce the resin hardness and allow latex particles to coalesce into a continuous film on drying. As a result, vinyl chloride/vinyl acetate copolymers (as well as other vinyl chloride copolymers) are used extensively in solution coating applications (such as container and can linings) and in latex coating applications (such as paper coatings). Other forms in which these resins can be used in the coatings industry are plastisols and organosols. (These forms were covered in detail in the chapter on the Plastics Industry and need not be reviewed again here).

Non-reactive vinyl chloride copolymers applied at room temperature to smooth metallic surfaces may not display high film adhesion. This can be overcome, however, by utilizing vinyl chloride copolymer with reactive groups to enhance adhesion. Also vinyl chloride resins are susceptible to discoloration at baking temperatures if the proper pigments and heat stabilizers are not used in the paint formulation.

Styrene-Butadiene Binders

These polymers are used in such applications as interior latex house paints. The main monomer, styrene, can be polymerized by solution, emulsion, or suspension processes. Just as with other addition polymers already discussed (PMMA, PVAc, PVC), the homopolymers of styrene are too hard to enable a continuous film to form from a latex. Therefore, a "plasticizing" monomer is copolymerized with styrene to produce a softer polymer. This "soft" monomer is butadiene. A ratio of 2 parts styrene to one part butadiene is commonly used to produce the styrene-butadiene latexes used in coatings. (This is almost the inverse combination used to produce SBR, discussed in the chapter on the Rubber Industry).

S-B latexes were the first practical latexes to be used in interior architectural applications in the 1950s. S-B latex paints still are used in home interiors. One advantage of S-B emulsion based paints is its superior alkali and chemical resistance. S-B has greater alkali resistance than either PVAc or acrylic emulsion paints. One large disadvantage of S-B emulsion is its susceptibility to oxidation because of the unsaturation contained in the butadiene units of the polymer backbone. This oxidation can lead to yellowing and ultimately film embrittlement; however, countermeasures can be taken in formulating these paints to minimize these effects.

Not only are S-B polymers used to make latex paints, but they are also used in solvent based coatings.

Polyurethane Resins

Many finishes are based on these resins because they are characteristically tough, yet flexible, very resistant to abrasion, impart good film adhesion, and can be made to harden at ambient temperatures. In the coating industry, any polymer containing the urethane linkage in the backbone, even if that same polymer may also contain many other non-urethane linkages as well, is termed a polyurethane coating.

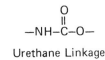

Urethane Linkage

Generally the polyurethane finishes involve the reaction of a diisocyanate with a hydroxyl compound.

Polyurethane coatings are used as a single-pack system or a two-pack system. There are three basic types of the single-pack systems and two basic types of the two-pack system. We shall describe each of these five types in the following discussion.

Oil Modified Type. These finishes are based on a polyurethane modified alkyd. These systems are also called "polyurethane-alkyds" or simply "uralkyds". They are prepared in a similar manner to alkyds (see previous discussion on alkyds). As the reader will recall, alkyds are usually made in a two stage process. In the first stage, an alcoholysis reaction between a drying oil and a polyol, such as glycerin results in a preponderance of monoglycerides. But in the second stage, instead of reacting these monoglycerides with a polycarboxylic acid, they are reacted instead with a diisocyanate, usually TDI (tolylene diisocyanate). The ratio of isocyanate to available hydroxyl groups is carefully controlled in a closed system to assure that there is no unreacted free isocyanate which is toxic. Thus, the reaction product possesses urethane linkages, as shown below, instead of the ester linkages associated with traditional alkyds.

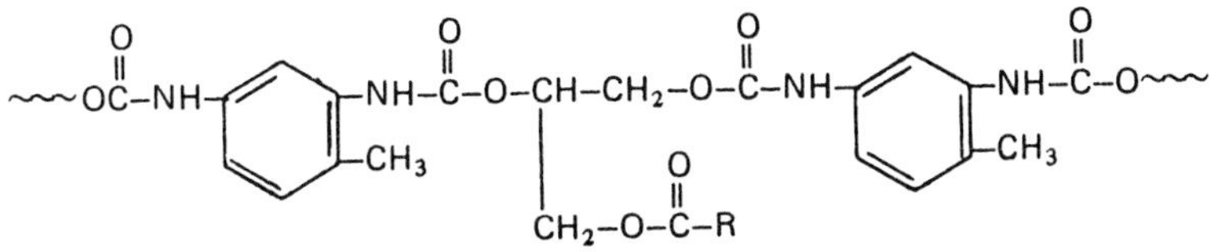

The percent isocyanate or conversely "oil length", determines the properties the resulting film will possess. The higher the isocyanate content (the shorter the oil length), the more reduced film flexibility and impact resistance, increased film hardness, increased solvent resistance, and shorter drying times. With some commercial coatings, some of the isocyanate may be replaced by phthalic anhydride to make a phthalic modified uralkyd.

Uralkyds are cured in the same manner as regular alkyds, i.e., through oxidative drying. As with alkyds, metallic catalyst combinations can be used to accelerate drying.

Uralkyds are one of the most commonly used forms of polyurethane resins finding applications in making varnishes for floors, boats, and general use. While these resins do not impart the best gloss, they do give excellent abrasion resistance and reportedly better water and alkali resistance than an alkyd.

Moisture Curing Prepolymers. These are single-pack polyurethane systems that cure from exposure to moisture present in the air. Initially a diisocyanate, such as TDI (tolylene diisocyanate), is reacted in a closed system with a high molecular weight polyol, such as castor oil. The TDI reacts with the hydroxyl groups as shown below.

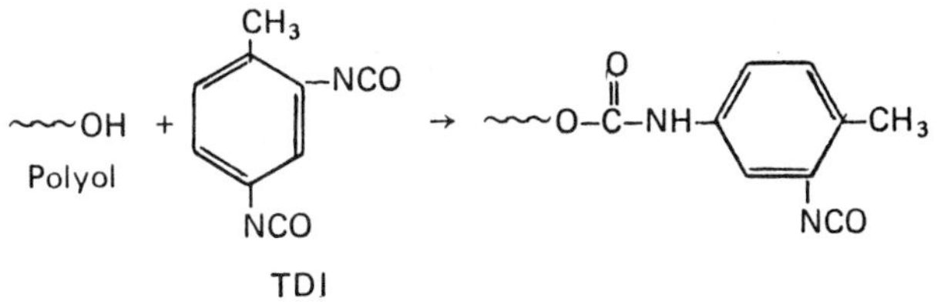

This reaction should occur with a minimum of three hydroxyl groups per molecule in order to render this prepolymer trifunctional. This prepolymer is then cured into a hard film during application from exposure to moisture in the air. Two isocyanate groups, in the presence of water, bond to form a urea linkage and carbon dioxide gas.

$$\sim\sim NCO + OCN\sim\sim + H_2O \rightarrow \sim\sim NH-\underset{\underset{O}{\|}}{C}-NH\sim\sim + CO_2\uparrow$$

During film formation, the carbon dioxide diffuses slowly from the film without leaving bubbles. Cure rate is greatly dependent on the relative humidity present. With these polyurethane systems, care must be taken that all pigments are free of moisture. This system results in a tough film and a common application is in floor varnish.

Blocked Isocyanates. In this system, a polyol is reacted in a closed system with a diisocyanate, such as TDI, to form a structure with isocyanate adducts. These adducts are then blocked by reacting the intermediate with a blocking agent, usually phenol, again in a closed system.

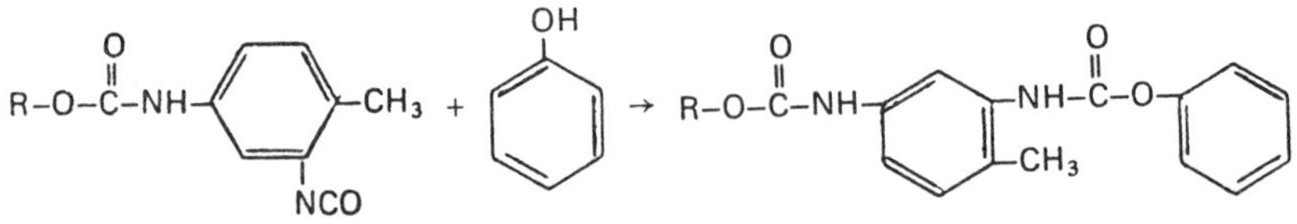

The bond between the phenol and isocyanate is stable up to 150°C. Therefore, at room temperature, no reaction will take place while the coating is being stored. However, after a coating is applied containing this resin, a film will form upon stoving at a temperature in excess of 150°C. At this temperature, the phenol splits off to allow the resin to crosslink. Since volatilized phenol is dangerous, the vapors must be safely exhausted from the oven. One common application for this type of resin is in coating electrical wires and parts.

Catalyzed Moisture Cures. This system is the same as the one package moisture cure system discussed earlier, except here a tertiary amine catalyst, such as N,N-dimethylethanolamine, is added just before application in order to shorten cure time as well as increase the crosslink density of the film for better chemical resistance. Of course, as soon as the catalyst is added, you have a limited pot-life before gelation occurs.

Two Component Systems. These are two package polyurethane systems. One pack contains a low volatility isocyanate. This may be a polyisocyanurate (a polymerized TDI).

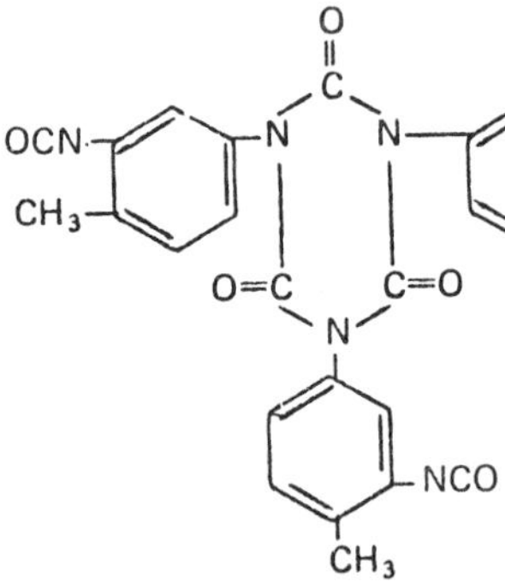

Isocyanurate Ring

Another way to prepare a low volatility isocyanate component is to react approximately two moles of TDI with one mole of a low molecular weight polyol such as trimethylol propane in a closed system. Low volatility of this package is important to minimize worker exposure to isocyanates during application. Isocyanates are considered toxic. Also workers can become sensitized to isocyanates.

The other package consists of a polyol component. This polyol may be either a polyether or saturated polyester resin. Polyethers may be less expensive but do not reportedly impart as good color retention as polyesters. The distance between the hydroxyl groups contained on a polyol molecule relates to the resultant film flexibility and hardness. If some film flexibility is desired for greater exterior durability, sufficient space between these hydroxyl groups is needed. On the other hand, if a high crosslink density is desired for greater film hardness, the distance between these reactive hydroxyl groups should be less.

When the two packs are mixed, the isocyanate functional groups react with the hydroxyl groups contained on the polyol as shown.

$$2\left(\underset{\text{OH}\quad\text{OH}}{}\right) + 2\ (\text{OCN–R–NCO}) \rightarrow$$

$$
\begin{array}{cc}
\text{O} & \text{O} \\
| & | \\
\text{C=O} & \text{C=O} \\
| & | \\
\text{NH} & \text{NH} \\
| & | \\
\text{R} & \text{R} \\
| & | \\
\text{NH} & \text{NH} \\
| & | \\
\text{C=O} & \text{C=O} \\
| & | \\
\text{O} & \text{O}
\end{array}
$$

It is not uncommon for water vapor present to also enter into the reaction resulting in urea crosslinks as well. This is illustrated below.

$$2\left(\underset{\text{OH}\quad\text{OH}}{}\right) + 2\ (\text{OCN–R–NCO}) \xrightarrow{+\text{H}_2\text{O}}$$

$$
\begin{array}{l}
\text{—OH} \qquad\qquad\qquad\qquad\qquad\qquad\qquad \text{HO—} \\[4pt]
\qquad\ \ \overset{\text{O}}{\overset{\|}{}} \qquad\qquad\ \ \overset{\text{O}}{\overset{\|}{}} \qquad\qquad\ \ \overset{\text{O}}{\overset{\|}{}} \\
\text{—O–C–NH–R–NH–C–NH–R–NH–C–O—} \quad + \ \text{CO}_2\uparrow
\end{array}
$$

Because of the presence of moisture, usually a slight excess in the stoichiometric ratio of NCO/OH is needed. When using this system with an appropriate solvent, care should be taken to select a "moisture free" solvent. Only use solvents which will not react with isocyanates. For example, never use alcohols.

Although this polyurethane two-component system has disadvantages in having a finite pot life, it has a short drying time which has resulted in numerous applications such as use in coating floors, furniture, boats, metals, and concrete.

Overall, polyurethane coatings of all five types have found a wide variety of applications; however, one problem that remains in some limited cases is the risk of exposure to toxic isocyanates during application. In many cases, however, this risk can be minimized or eliminated. However, only work with these coating systems under adequate ventilation.

Another point worth noting is the use of aromatic isocyanates which can result in yellowing from prolonged exterior exposures to light. One way to overcome this problem is to use aliphatic isocyanates, even though they are less reactive than the aromatics. Aliphatic isocyanates, however, can be extremely toxic. Some are more dangerous to handle due to higher volatility.

Epoxy Resins

Approximately half of all the epoxy resins produced in the United States are used in the coatings industry. Some reasons for this high usage of epoxy resins include the polymer's superior chemical resistance, its high film adhesion strength, and its ability to be made tough but yet flexible. On the other hand, epoxies are relatively expensive, slow drying (hardening), and do not render as high a gloss retention as many other binders.

An epoxy resin is one containing an epoxy group.

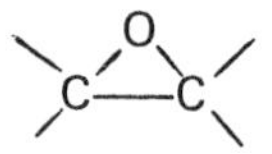

Many different types of epoxy resins have been used in the coatings industry. The most common epoxy resin used is one produced from the reaction of bisphenol A with epichlorohydrin.

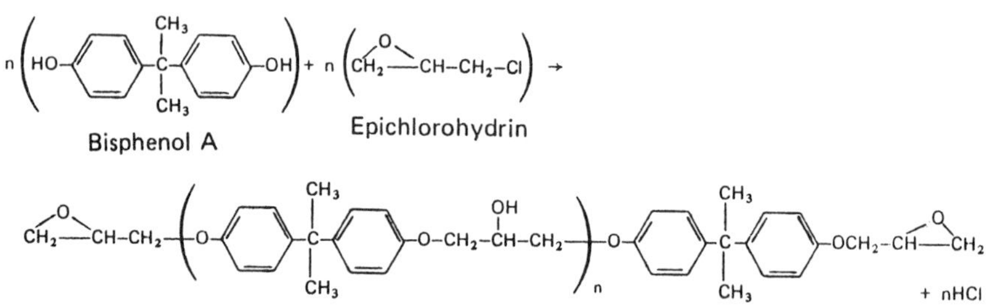

These epoxy resins can be cured in a two pack system using amines, polyamines, polyamides, or acid anhydrides. Aliphatic amines are faster curing at room temperature than aromatic amines. Of course, these amines and amides are toxic and all the appropriate precautions in their use must be taken. The higher molecular weight polyamines and poly-amides might be considered a little less hazardous because of their lower volatility. Nevertheless, all safety precautions must be followed when working with any amines. Only work with these materials in a well ventilated area.

An amine combines with an epoxy group in a ring opening reaction as shown below.

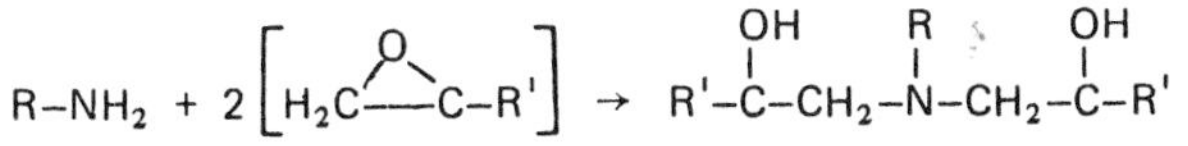

$$R-NH_2 + 2\left[H_2C\overset{O}{\overbrace{}}C-R'\right] \rightarrow \underset{\underset{R'-\overset{OH}{\overset{|}{C}}-CH_2-\overset{R}{\overset{|}{N}}-CH_2-\overset{OH}{\overset{|}{C}}-R'}{}}{}$$

It can be seen that through this type of reaction with a trifunctional amine and diepoxides, a three dimensional crosslinked matrix will form. These types of systems find application in industrial maintenance coatings and marine coatings to name a few. These coatings have superb resistance to solvent and chemical attack. Some systems can be slow drying however.

Although the systems just described have noted resistance to solvent and chemical attack, these properties can be improved even further by blending the epoxy with a phenolic resin and baking the applied coating. At elevated temperatures, the phenolic resin will react with the epoxy groups, through a ring opening mechanism to form a strong bond as shown below.

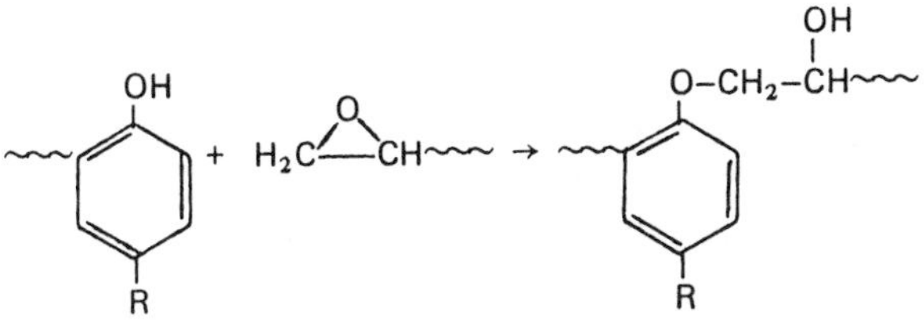

In addition, the secondary hydroxyl groups react further to form a tight, crosslinked network which imparts the ultimate in solvent and chemical resistance. On the other hand, flexibility of the film is reduced compared to other epoxies and discoloration can be a problem. If a lighter color is needed, urea-formaldehyde resins can be used with the epoxy in place of the phenolic resin with little loss in chemical and solvent resistance. These types of coatings are applied to the interior of cans as well as serving as lining for drums, tanks, and piping.

A third class of epoxy used in coatings is called the epoxy resin esters. These resins are formed from reacting an epoxy resin with fatty acids from drying oils in the same type of equipment used to make alkyds. The fatty acids react with both the epoxide and hydroxyl groups of the epoxy resin to form the epoxy ester as shown below.

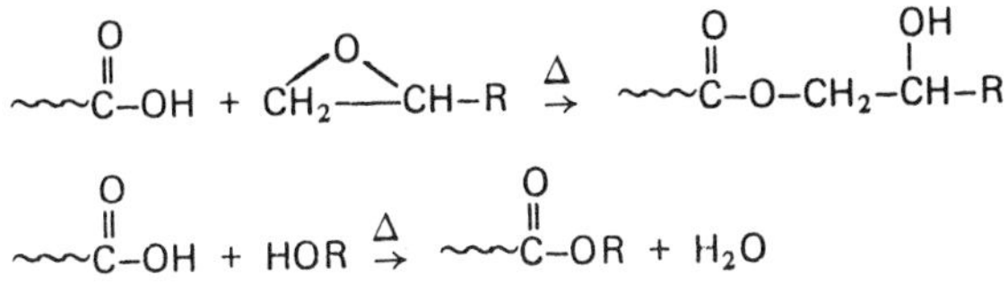

These epoxy resin esters are less expensive than other epoxy resins because they are partially composed of lower cost fatty acid. They also have superior flexibility compared to other epoxies. Their oil length determines the degree of film flexibility as well as cure conditions (air drying vs. stoving). On the other hand, these epoxy esters have poorer solvent and chemical resistance when compared to the other epoxy types previously discussed. However, these epoxy esters may display better chemical and alkali resistance as well as better adhesion when compared to conventional alkyds. Also, just as with alkyds, these epoxy esters are cured by autooxidative air drying with the aid of metallic driers such as cobalt or manganese soaps. These coatings have found applications as industrial and maintenance paints, automotive primers, and varnishes. However, these resins may not compare as well to alkyds in outdoor applications because of their reported greater tendency to chalk.

Other types of epoxy binders are also used in the coatings industry. For example, phthalic anhydride is sometimes used to partially replace fatty acids in producing epoxy resin alkyds. Also short oil length epoxy esters (sometimes based on a non-drying oil), are occasionally modified with an amino resin to yield a coating binder capable of forming a very hard, chemically resistant film. Lastly, there are epoxy systems based on modifications of the epoxy resin with acrylic resins or polyisocyanate curing agents, to name a few.

As can be seen, the variety of chemical modifications for epoxy resins used in the coatings industry are quite numerous.

Amino Resins

This term is generally applied to urea-formaldehyde resins (UF) and melamine-formaldehyde resins (MF) that are used in the coatings industry as binders. One common use historically for these coatings is in baking finishes that are used to coat appliances, metal cabinets, metal parts, etc.

In preparation of urea-formaldehyde resins by the manufacturers, usually urea, formalin, and an alcohol (such as butanol) are charged into a reactor vessel. In the first stage, under alkaline conditions, urea reacts with formaldehyde to form dimethylol urea.

$$NH_2-\overset{\overset{\displaystyle O}{\|}}{C}-NH_2 \ + \ 2\,CH_2O \ \xrightarrow[\text{Solution}]{\text{Basic}} \ \begin{array}{l} NH-CH_2OH \\ | \\ C{=}O \\ | \\ NH-CH_2OH \end{array}$$

DMU

This water soluble product (DMU) is rendered soluble in organic solvents by a reaction in the second stage under acidic conditions with an alcohol, such as butanol. In this stage, two competing reactions occur, i.e., polymerization and etherification.

Polymerization

$$-NH-CH_2OH + -NH-CH_2- \xrightarrow[acid]{} -NH-CH_2-N-CH_2- + H_2O$$

Etherification

$$-NH-CH_2OH + HOR \xrightarrow[acid]{} -NH-CH_2-OR + H_2O$$

The reactions are short stopped before the gel point is reached by cooling and neutralizing the acid.

The higher the C_n of the alcohol used and the greater the degree of alkylation, the greater the organic solubility of the resin and the better the compatibility of this resin with other resins. On the other hand, higher alkylation results in imparting a slower resin cure rate.

Melamine-formaldehyde resins (MF) are somewhat analogous to the urea-formaldehyde resins just discussed. Unlike urea, which is tetrafunctional, melamine has a structure which is hexafunctional as shown below.

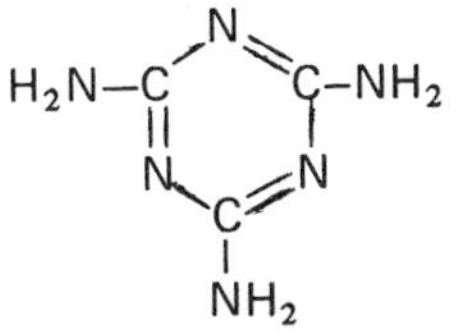

In a similar manner to UF, MF resins are manufactured by reacting melamine with both formaldehyde and an alcohol, such as butanol, under acidic conditions (an initial stage to prepare a methylol melamine is not necessary). In this one step reaction, available active hydrogens and methylol groups react with one another in a polymerization reaction as well as hydroxyl groups reacting in an etherification reaction. This process is short stopped in a similar manner discussed with UF before the gel point is reached.

MF resins are more expensive than UF resins but are sometimes used in place of UF resins because of their greater color stability and superior outdoor durability. MF resins are less sensitive to moisture attack because they do not contain carbonyl groups. Also cured MF films can have higher crosslink densities than UF films.

In many coating applications, the use of UF or MF as the sole binder makes the resulting film very hard and brittle with perhaps only fair film adhesion. Therefore, it is common for alkyds to be used with these amino resins as a plasticizer. The alkyds are acidic, which helps accelerate the cure. The excess hydroxyl groups contained on the alkyd react with methylol groups as shown below.

$$-CH_2OH + HO- \rightarrow -CH_2O- + H_2O$$

The alkyds used may be based on drying or non-drying oils. Also the amino resins are commonly used with epoxies or acrylics in coating applications to overcome brittleness and enhance other properties.

As can be seen, the properties of the coatings based on these amino resins depend on many variables. The following is a list of important variables that affect these coating properties.

Ratio of urea or melamine to formaldehyde

Degree of alkylation (from alcohol)

Chain length of alcohol

Type of catalyst

Level of catalyst

Cure temperature

Type of modifying resins

These amino coatings can be applied as either one package (stoving type) or two package type (room temperature or warm cures). As noted earlier, an amino coating can remain stable if it is stored at cool temperatures and contains low acidity. Many stoving amino coatings also have an alkyd which helps accelerate the cure at elevated temperatures. These stoving amino coatings are used in baking operations to protect metal surfaces of appliances, etc.

Also, the amino coatings can be used as a two pack system. The second package contains an acid, which serves as a catalyst for curing at room temperature. Also, once the second package is added to the first, the coating has a finite pot life. The room temperature cure properties allow these coatings to be applied to wood articles such as furniture, etc.

Cellulose Binders

These binders are used mainly in lacquers. Because they are used in this way, they display all the advantages of a lacquer such as quick room temperature drying and no limitations on pot life.

Cellulose itself is quite insoluble; however, through a nitration reaction, it can be rendered soluble in organic solvents through its conversion to nitrocellulose (a misnomer for cellulose nitrate). The degree of nitration determines the polymer's solubility characteristics. Thus, if the polymer only contains 11% nitrogen, its solubility is limited to alcohol. On the other hand, if its nitrogen content is 12%, it is soluble in esters and ketones. There are generally three different grades of nitrocellulose recognized in the coatings industry based on degree of nitration.

Also the molecular weight of this polymer is important. The MW is influenced by the degree of hydrolysis occurring under conditions of heat and pressure during the polymer's manufacture. Greater hydrolysis results in a lower molecular weight and a lower solution viscosity. The cellulose nitrate polymers themselves are extremely flammable and are classified as explosive.

Nitrocellulose itself is generally too brittle to be used alone in coating applications. Therefore, coating formulations will also contain a plasticizer to impart some flexibility to the film. An example of a plasticizer used is dibutyl phthalate.

Fast drying nitrocellulose lacquers historically were used in large volumes in coating automobiles; however other coatings have displaced this application. These lacquers are still used in coating furniture which can not tolerate baking.

Other types of cellulose binders used in coatings are *ethyl cellulose* and *ethyl hydroxyethyl cellulose*. These binders are commonly used in making printing inks.

SOLVENTS

Introduction

Solvents represent another important group of components which are used in paint and coating formulations. The solvent is the fluid vehicle in which the binder is dissolved and allows the binder to be applied to the substrate as a wet film after which the solvent volatilizes, leaving a solid protective film. In the case of non-convertible coatings, film formation is solely dependent on volatilization (loss) of the solvent carrier. Of course, in the case of convertible coatings, solvents are also used to enable the binder to be applied to the surface; however solvent volatilization here is only one of the necessary conditions for the formation of a hard, useful protective film.

Organic solvents are not the only fluid vehicles used in paints and coatings. In the case of aqueous emulsion paints, the water represents the continuous medium in which colloidal particles of a polymeric binder are suspended. Since these particles are not actually dissolved in the aqueous medium, water here is not considered a true solvent. Also the term diluent is used to describe a volatile organic liquid which is used to extend another organic solvent but by itself will not dissolve the polymer. Diluents are generally used to reduce a coating's cost; however they are not considered true solvents. Lastly, other liquids such as liquid binders (before cure) or plasticizers are also commonly included in the paint formulation; however these fluids become part of the dried film after applications.

Important Properties

Six important properties are listed below which should be considered when selecting a solvent.

(1) Solvency

(2) Volatility

(3) Odor

(4) Toxicity

(5) Flammability

(6) Cost

The chemical structure of a solvent determines its solvation power concerning a selected resins(s). If the attractive forces among solvent molecules are greater than the intermolecular forces between the solvent and resin, then the resin will be insoluble. Thermodynamically, for a given solvent to be able to solubilize a selected resin, the "heat of mixing" (ΔH), must be less than the product of temperature (T) and entropy change (ΔS), i.e., change in degree of disorder. In other words, Gibbs free energy (ΔG) must be negative in order for a solution to occur.

$$\Delta G = \Delta H - T\Delta S$$

Heat of mixing is reduced when the cohesive energy densities of the solvent and resin are similar. The square root of the cohesive energy density (CED) is called the Hildebrand solubility parameter (δ).

$$\delta = \sqrt{CED}$$

Solubility parameters were discussed in more detail in the last chapter on adhesives. Solvents and resins having the same or similar δ values should be compatible provided that they have the same degree of hydrogen bonding. If these parameters are to have value, they should be used to compare only solvents and resins with the same approximate degree of hydrogen bonding, i.e., resins of weak hydrogen bonding with solvents of weak hydrogen bonding or resins of strong hydrogen bonding with solvents of strong hydrogen bonding.

The solvation power of a solvent is not only important for dissolving a resin binder, but it also determines what the coating's solution viscosity will be at a specific solids level. Also this property can determine the maximum concentration possible for a specific resin without surpassing a viscosity limit needed for effective application. Sometimes a blend of solvents is used to achieve the desired solubility parameters.

Another important property to consider when selecting a solvent is volatility. This property relates to evaporation rate. The evaporation rate of a simple solvent can be directly measured easily and is usually expressed relative to n-butyl acetate which is assigned a value of one. The boiling point of a simple solvent can crudely relate to its evaporation rate; however there are solvents, such as alcohols with strong hydrogen bonding, where boiling point is not a good indicator of volatility. In any event, while evaporation rates can be measured for simple solvents, one can not mathematically predict from simple calculations what the evaporation rate of a blend of two or more solvents will be because of interaction among the solvent components. Furthermore, when these solvent blends are added to a coating formulation, the evaporation rate is reduced on application because of the attractive forces between the solvent components and the resin.

Volatility properties of solvents are very important concerning coating applications. If a solvent based paint is to be sprayed, it should contain fast drying solvents with boiling points below 100°C, in order to prevent "running". On the other hand, if a solvent based paint is to be applied by brush, it should contain slower drying solvents to allow good flow on application. The evaporation rate of a solvent used in a paint can also affect sagging, wet edge, and gloss properties.

The odor of a particular solvent can be a problem during paint applications especially for indoor household use. Aliphatic mineral spirits have less noticeable odor than solvents such as esters or ketones. The "odor problem" of organic solvent based paints is one of the principle reasons for the increased popularity of latex paints which have less odor.

The toxicity of different solvents is another important consideration when selecting a solvent. Solvents such as benzene are now considered far too toxic to be used in paints. Many other solvents, such as MIBK, are considered photochemically active and contributors to smog and air pollution. Government regulations regarding the use of organic solvents are constantly changing and the latest rules should be read before making any decision in the selection of a solvent. In general, these new regulations also help explain the trend toward latex paints or toward new solvent based paints with higher total solids (less solvent content). Adequate ventilation is required when working with any solvents or coatings containing these solvents.

Flammability is another important consideration in selecting a solvent. The lower the flash point temperature of a solvent (or blend of solvents), the greater the fire hazard potential.

Cost is another consideration. Aliphatic solvents can be less expensive than some esters or ketones. Some "solvents", which are not true solvents, are sometimes used as diluents to reduce total cost.

The following discussion will concern specific classes of solvents used in the paint industry.

Hydrocarbon Solvents

These solvents represent the most commonly used solvents in paint applications. Hydrocarbon solvents consist of three classes of components, i.e., paraffins, naphthenes, and aromatics. Most of the H.C. solvents now used in the paint industry are aliphatic in nature, containing mostly normal paraffins and isoparaffins. Some H.C. solvents will also contain fractions of naphthenes (cycloparaffins) and aromatics, which improve the solvation power of these solvents, but increase their odor.

Probably the most commonly used organic solvent is called "mineral spirits". The so-called "odorless" grades are relatively low in odor because they contain mostly isoparaffins. On the other hand, the lack of naphthenes or aromatics can hurt the solvency of this solvent with many resins. These grades of mineral spirits vary in their boiling ranges. The choice of the proper volatility range is important for smooth brush applications and controlling "wet edge" times.

VM&P naphtha is chemically very similar to mineral spirits except this naphtha cut has a lower boiling point. Because of its higher volatility, this solvent can be used in paints for spraying applications.

Aromatic Solvents

These solvents consist of mainly aromatic components. Examples are toluene and xylene. Usually these aromatics are used with other solvents in a blend. Many times they are needed in, say, aliphatic solvents to help solubilize alkyds of shorter oil length.

Ester Solvents

These solvents usually have a sweet, pleasant odor. Examples are methyl acetate, ethyl acetate, and butyl acetate. These solvents are commonly used to help dissolve such binders as cellulose acetate.

Ketones

These solvents have strong solvency power. Examples of solvents in this group are methyl ethyl ketone (MEK), methyl isobutyl ketone (MIBK), and acetone. These solvents are used to dissolve binders such as vinyl resins, polyurethanes, epoxy, acrylics, and cellulose acetate butyrate. MEK has the greatest volume of usage in this group. While not as high a usage, acetone is one of the fastest solvents in evaporation rate.

Alcohols

These solvents are usually used as a cosolvent in a solvent blend. Methanol (wood alcohol) is the most volatile of the alcohols (too volatile and toxic for most paint applications). It possesses good solvency power when used with nitrocellulose.

Denatured ethanol, on the other hand, is slower in evaporation rate. It is used to solubilize shellac and some synthetic resins such as polyvinyl acetate. Ethanol is commonly used as a cosolvent with other solvents in flexographic inks.

Other alcohol solvents include isopropanol, normal propanol, and various butanol isomers. n-propanol and n-butanol have the slowest evaporation rates within their respective isomer series. A small level of n-butanol is sometimes used to help in a downward adjustment of viscosity for alkyd resin solutions even though alkyds are insoluble in 100% n-butanol.

Lastly, ethylene glycol monethyl ether (cellosolve) is a good solvent that is added to a coating because of its slow evaporation rate.

Terpenes

These solvents have declined in use because of their higher cost. They are mainly obtained from pine wood. Turpentine and dipentene are two of the most common terpene solvents used in coatings. These solvents consist of varying proportions of alpha-pinene, beta-pinene, and dipentene.

α-Pinene   β-Pinene   Dipentene

PIGMENTS

Introduction

Pigments are used to impart the desired color to a paint. White pigments achieve a white appearance by reflecting all wavelengths of light equally. Black pigments absorb all wavelengths equally. On the other hand, a specific colored pigment imparts a desired color (or *hue*) by selectively absorbing and reflecting different wavelengths of the visible light spectrum.

Pigments differ not only in the type of color, called hue, but also in such qualities as *tinting strength* (ability to color a white base) and brightness. Different pigments can even appear to have exactly the same shade of color under indoor lighting, but have differences in color in outdoor light. This phenomenon, called *metamerism*, is due to differing spectrophotometric absorption characteristics between the pigments.

There are other important pigment characteristics which should be discussed as well. For example, *lightfastness* is an important property to consider when selecting a pigment. This property relates to the pigment's resistance to fading or changing color from prolonged light exposure, especially after ultraviolet exposure from outdoors. Also some pigments are sensitive to prolonged exposure to heat.

Another important property to consider when choosing a pigment is a property known as *bleeding*. This property is important when a top coat is applied over a base coat. Some pigments in the base coat may "bleed" into the applied coat if they are soluble in the top coat's solvents. This can cause discoloration of the top coat.

Another important property of a pigment is its *opacity or hiding power.* This term is particularly applied to white pigments such as titanium dioxide. This quality relates to the ability of the pigment particles in the film to prevent light from penetrating through the film to the substrate and back through the film to an observer. In other words, it's the pigment's ability in a dried film to obliterate any color from the substrate. Hiding power can be measured as the number of square feet covered from one pound of a pigment. If a given paint contains a pigment with good opacity, it can mean that only one coat may be needed. If opacity is poor, two or more coats may be needed.

There are two important characteristics of a pigment which help determine its degree of opacity. First, particle size is important. The smaller the particles, the greater the number of interfaces between the dispersed pigment and binder. This increases the opportunity for light scattering till the particle diameters are about one-half the wavelength of light (0.2 to 0.4 micron). With particles smaller than this range, the scattering power diminishes. This is because light is a wave and will go around smaller particles.

The other characteristic which relates to opacity is the refractive index of the pigment. The greater the difference between the pigment particle's refractive index and the binder's refractive index, the greater the degree of light scattering and the higher the opacity. On the other hand, if the refractive indexes of the pigment and binder are very similar, little hiding power will result. It is interesting to note that "flat" paints may have greater hiding power than high gloss paints because these flat paints are so highly loaded with pigment that the pigment particles partially interface with the air. Since air has a very low refractive index, a greater difference occurs between RIs and a higher opacity results.

Chemical reactivity can also be an important property to consider in pigment selection. For example, some pigments can react with air pollutants causing discoloration.

There are two broad classes of pigments used in paints, i.e., *organic* and *inorganic.* In most cases, there will be several organic and inorganic pigments for the same color group. In general, both organic and inorganic pigments each have their respective comparative advantages. For one thing, organic pigments are generally more brilliant than corresponding inorganic pigments. On the other hand, inorganic pigments, in general, have better lightfastness, heat stability, and bleed resistance than their organic counterparts.

Many organic structures may absorb electromagnetic radiation in the ultraviolet portion of the spectrum, but not in the lower energy visible portion. Thus the vast majority of organic compounds are clear white powders. In order to achieve a selective absorption in a portion of the visible spectrum, conjugated organic structures consisting of fused aromatic rings, chromophore groups (such as azo groups), and/or organometallic complexes are used. Generally speaking, absorption in the visible spectrum is achieved from interaction of the electrons of the aromatic structures with those of the so-called *chromophore* groups, such as azo linkages (-N = N-). Also the presence of electron donor groups, such as amino or hydroxyl groups, can enhance color appearance. These groups are called *auxochromes*. Of course, this discussion applies only to organic pigments. Inorganic pigments selectively absorb in the visible region through a different mechanism.

Lastly, the color that is seen reflected from a given pigment is the complement of the color actually absorbed by the pigment. For example, if a pigment selectively absorbs in the violet wavelength, the color produced will be yellow.

Many of these pigments (particularly those containing lead and chromium) are very toxic. Breathing pigment dust can be extremely hazardous to workers. Every effort must be undertaken to prevent pigment dust exposure to workers in the workplace. Dustless forms should be used where possible.

The following describes some of the pigments, both organic and inorganic, that are used in the paint industry.

White

These pigments do not have color. In other words, they do not absorb light at any wavelength. In fact, if the particles of many of these pigments were fused together to form a solid, homogeneous mass, it would be clear, not white. The whiteness comes from the light scattering and diffusion of light resulting from the many particle interfaces with the polymeric medium. The smaller the particles and the greater the differences in refractive indexes between the pigment particle and the medium, the greater the light scattering which imparts the "white" appearance.

Many years ago the chief white pigment used in the paint industry was basic lead carbonate. Because of well known toxicity problems associated with the use of lead compounds, this pigment's use has been almost completely replaced by other white pigments, chiefly titanium dioxide. TiO_2 is ideal as a white pigment because it is usually sold with an average particle size close to one-half the wavelength of light (the optimal for light scattering). Also it has a relatively high refractive index (2.5 for anatase and 2.7 for the rutile crystallographic form) and is chemically inert. Titanium dioxide is the best overall general purpose white pigment for providing the best hiding power. Rutile is used when the greatest hiding power is needed, while anatase is used when chalking is desired.

Other white pigments are used as well, but they all have shortcomings compared to titanium dioxide. Zinc oxide is used as a white pigment in paint; however, it can be chemically reactive within the resin and is not as cost effective in hiding power. Lithopone (71% barium sulfate and 29% zinc sulfide) is generally too costly to be used in high volumes in the United States. It is mostly used in Europe and elsewhere. Other white pigments, such as talcs, clays, calcium carbonate, and mica are not effective compared to titanium dioxide because they do not have a sufficiently high refractive index as well as other disadvantages. These materials are generally used as extenders in paint to help reduce the cost per gallon of the paint.

Black

These pigments absorb all wavelengths of light. Carbon black is the most common black pigment used; however, it is one of the most difficult pigments to disperse well. Carbon blacks are insoluble in all solvents but do reportedly have some affinity for aromatic solvents. They have very high tinting strength and opacity. There are three types of carbon blacks available for use in the paint industry, i.e., furnace blacks, thermal blacks, and channel blacks. Furnace blacks are produced from thermal decomposition of petroleum oil feedstocks and are the most commonly used blacks. They have a medium jetness (black intensity). On the other hand, thermal blacks, obtained from thermal decomposition of natural gas, have larger particle size, are more expensive, and do not display the same degree of jetness. Lastly, channel blacks provide the highest degree of jetness and are still used in paints even though they must be imported from abroad. Channel blacks are obtained from a rather inefficient, dirty process using natural gas as a feedstock.

Other black pigments are also still used in paints. Some of these include lampblack, mineral black (from coal), and bone black from charred bones.

Red (Wavelength 610–700 nm)

Red iron oxide (Fe_2O_3) is perhaps one of the most common inorganic pigments used in paints. Because it is inorganic, this pigment displays excellent heat stability and lightfastness. However, because of its relatively high specific gravity, it can impart settling problems.

Also, organic red pigments are used as well. Unfortunately many of these pigments are known to display bleeding problems because of their solubility characteristics. Some of these red organic pigments are shown below.

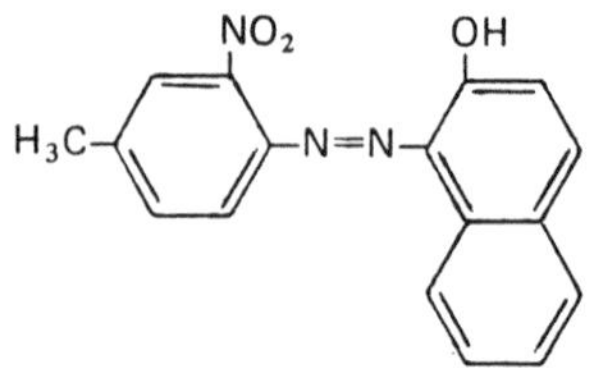

Toluidine Red 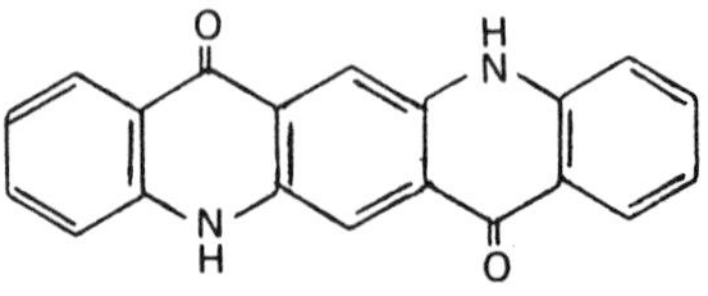   Quinacridones

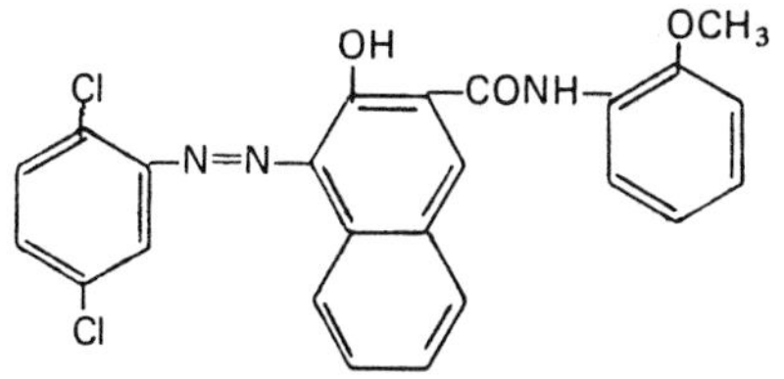

Arylamide Red

Orange (Wavelength 590–610 nm)

Just as with the reds, there are inorganic oranges that can be used for better lightfastness. These inorganics include chrome orange ($PbCrO_4$ and PbO) and molybdate orange (a mixture of lead chromate, lead molybdate, and lead sulfate); however, their chief disadvantage is their high toxicity which severely limits their use.

Organic orange pigments that are used are shown below.

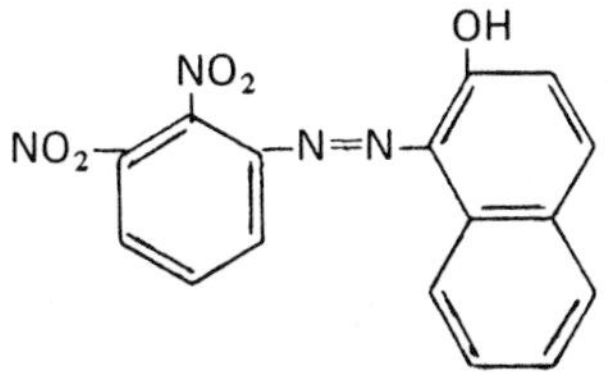

Dinitroaniline Orange

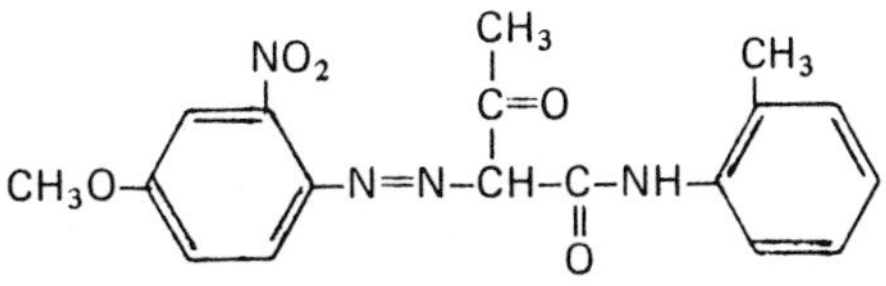

Hansa Orange

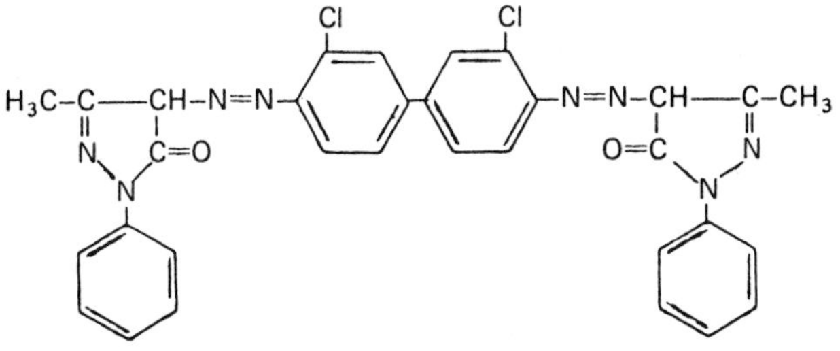

Benzidine Orange

Yellow (Wavelength 570–590 nm)

For this color, hydrated iron oxide ($Fe_2O_3 \cdot H_2O$) is a good low cost pigment which is not considered highly toxic like other alternatives. On the other hand, its use may present some settling problems. Other inorganic yellow pigments used are the zinc chromates, such as zinc yellows ($4ZnO\ K_2O\ 4CrO_3 \cdot 3H_2O$), lead chromate ($PbCrO_4$), and cadmium yellow ($CdS$). All these pigments have toxicity problems.

Some organic yellow pigments that are commonly used are shown below.

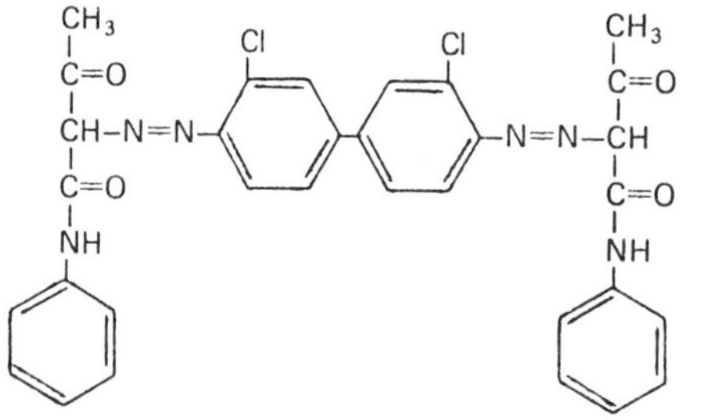

Benzidine Yellow

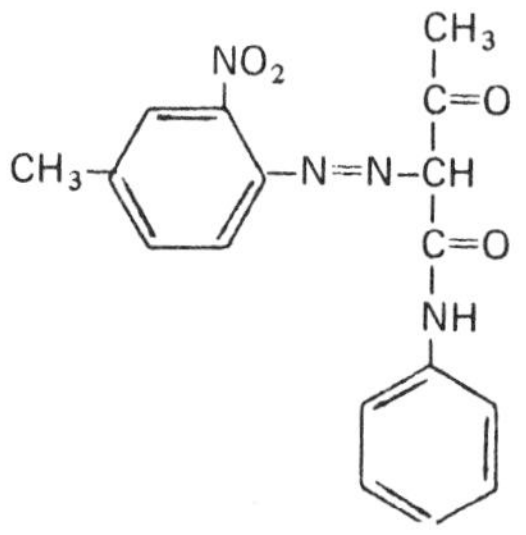

Hansa Yellow

Green (Wavelength 500–570 nm)

The commonly used inorganic green pigments are mostly based on chromium compounds. The so-called "lead chrome green" is a mixture of lead chromate ($PbCrO_4$) and iron blue $KFe (Fe(CN)_6)$ at different proportions to give different shades. Also chromium oxide (Cr_2O_3) and hydrated chromium oxide ($Cr_2O_3 \cdot 2H_2O$) are used as green colorants. In fact, chromium oxide has been used to impart the olive shade used in the Army for camouflage. Of course, toxicity considerations should be taken into account when using any of these pigments.

One commonly used green organic pigment is copper polychlorophthalocyanine green. It is similar to the structure shown for copper phthalocyanine blue (see "Blue" section) except the green structure is chlorinated with 14 to 16 chlorine atoms to shift the absorption band in order to impart a green appearance. Another commonly used organic green is Pigment Green B shown below.

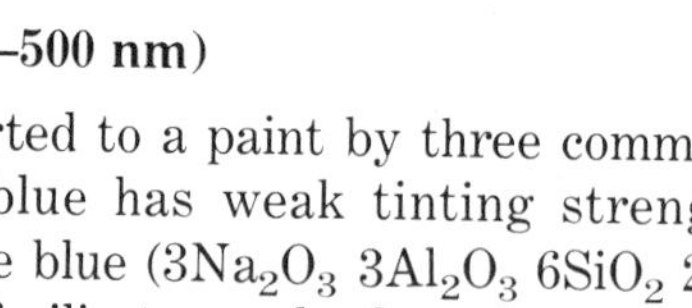

Blue (Wavelength 450–500 nm)

This color is imparted to a paint by three commonly used blue pigments. Ultramarine blue has weak tinting strength but good heat resistance. Ultramarine blue ($3Na_2O_3 \ 3Al_2O_3 \ 6SiO_2 \ 2Na_2S$) is a complex sodium aluminum sulfosilicate made from heating china clay, sodium carbonate, silica, and sulfur together as a mixture.

Two commonly used organic blue pigments in paints are shown below with rather complex structures necessary to absorb light at the longer wavelength (lower energy) portion of the visible spectrum.

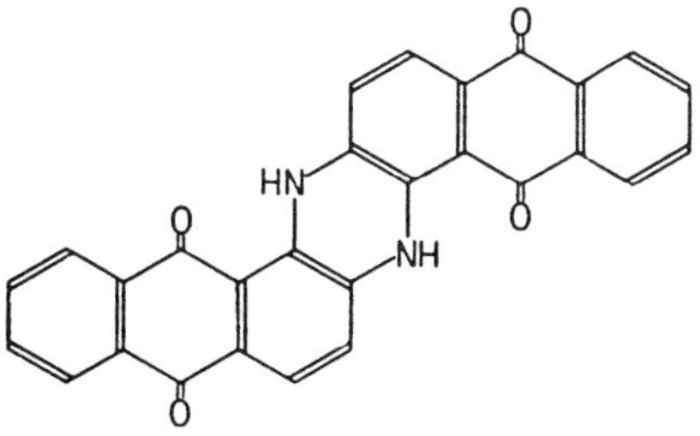

Indanthrone Blue

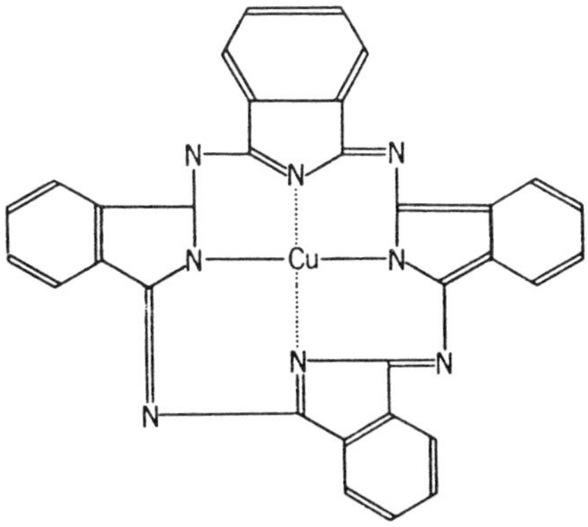

Copper Phthalocyanine Blue

Extenders

These are mineral fillers of low hiding power (low refractive index) which are used primarily to reduce the paint formulation's cost, reduce gloss, and sometimes adjust viscosity. Common extenders used are as follows.

Calcium carbonate (whiting)

Clay

Calcined clay

Mica

Talc

Calcium sulfate

Silicas

Selected grades of calcium carbonates can be very white in appearance and low in cost. Calcium carbonates are available in ground or precipitated forms. Some grades are used to impart a flattening effect to the film. Many times, calcium carbonates are used with talcs.

Clays are also being used in the paint industry as extenders. The plate-like shape of the clay particles reportedly helps toughen the film. The more expensive calcined clays are used because loss of the water of hydration results in a higher refractive index and better hiding power.

Micas are used because their plate-like particles line up and help to reduce moisture penetration through a film as well as reportedly improving the film's durability.

Synthetic silicas are used as thixotropes (to be discussed) as well as flatteners for lacquers or varnishes.

Diatomaceous silicas (which consist of the highly irregular-shaped skeletal remains of diatoms) are highly effective flattening agents. For this reason they have been used in many flat home paints. However, diatomaceous silicas are now believed to pose an environmental risk to workers in the factory from the dust they impart to the work place. All the appropriate government safety regulations must be followed in the factory when handling diatomaceous silicas or any other extender or pigment used.

ADDITIVES

As discussed earlier, additives are those chemicals which are added to a paint formulation at relatively low concentrations to impart some specific properties to the coating. The following will discuss each of these types of additives in more detail.

Driers

Driers are additives used with alkyd or oil based coatings to accelerate or promote their oxidative film hardening. These so-called "driers" are usually soaps of transition metals which have more than one oxidation state. These metals are usually used as *naphthenates*, *octoates* or *neodecanoates* to enable their use in an oil soluble form. Such soap forms as *rosinates* have also been used occasionally.

Primary driers (driers that alone can promote oxidative film curing), are considered true catalysts. These driers promote oxidative crosslinking by (1) catalyzing oxygen absorption into the film and (2) serving as a means to accelerate the decomposition of the formed peroxides into free radicals which propagate the crosslinking mechanism (discussed earlier). The mechanism through which driers accelerate oxygen uptake is not clearly understood; however the destruction of formed hydroperoxides is believed to be from repeated shifting of the transition metal back and forth from one oxidation state to the other as shown below.

$$M^{++} + ROOH \rightarrow M^{+++} + RO\cdot + OH^-$$
$$M^{++} + ROOH \rightarrow M^{+++} + RO^- + \cdot OH$$
$$M^{+++} + ROOH \rightarrow M^{++} + ROO\cdot + H^+$$
$$M^{+++} + RO^- \rightarrow M^{++} + RO\cdot$$
$$M^{+++} + OH^- \rightarrow M^{++} + \cdot OH$$

Here M could be represented by a transition metal such as cobalt which has two common oxidation numbers of $+2$ and $+3$.

Historically three commonly used driers have been based on cobalt, manganese, and lead. Cobalt driers are the fastest and operate through both the mechanisms described above. Manganese driers also function through both mechanisms, but are slower than cobalt. Lead driers only function by catalyzing oxygen uptake and are not nearly as reactive as the cobalt or manganese. Because lead driers must be used with other driers to be effective, they are considered secondary driers. On the other hand, a single drier is not normally used alone but in combination with other driers for reasons to be explained.

Cobalt driers are so fast that they are seldom used alone. This is because they quickly dry at the outermost layer of the film, establishing a barrier against further "through" drying of the film. If only cobalt is used, wrinkling of the film can result. Because cobalt is so effective for "surface" drying, usually not more than 0.05% cobalt per vehicle solids is used.

Lead driers, as discussed earlier, are slower and not effective for surface drying. They had traditionally been used at levels between 0.5 to 1.0% metal per vehicle solids with other driers to achieve "through" drying further inside the film itself. Because of more recent concerns over lead's toxicity, lead drier concentrations have been greatly reduced or eliminated through the use of other non-lead driers such as Zirconium. Besides the toxicity problem, lead driers can also contribute to discoloration from exposure to sulfide pollution in the atmosphere.

Manganese driers are similar to cobalt driers in some ways. However manganese driers are slower than cobalt in surface drying which may reduce the tendency for wrinkling on the surface. Also manganese driers, because of their slower rate, provide some through drying as well; however still they are usually used with other driers to achieve the best balance of drying properties. Manganese driers are generally used at concentration ranges similar to those used for cobalt.

Other metallic soaps are also being used as driers as well. Calcium driers are used as auxiliary driers mainly with lead driers.

The calcium driers reportedly help provide uniform solubility for the lead drier. Sometimes, in non-toxic applications where no lead can be used for "through" drying, calcium has been used as a much less effective substitute. Calcium driers are very sensitive to changes in humidity. High humidity can adversely affect drying rates.

Another drier that has been used more recently as a substitute for lead is zirconium. The zirconium soaps have been reported to impart effective "through" drying similar to that imparted by lead. Also the zirconyl compounds do not discolor like lead soaps.

Iron driers are used occasionally as catalysts for heat curing a film. However, cure temperatures must be above 100°C before catalyzed activity is seen. These high temperature cures impart high hardness to the film.

Zinc compounds are also occasionally used with other driers to dry a film. Zinc itself is not a true drier. It is used with cobalt to retard its activity and slow down its surface drying in order to prevent wrinkling.

As can be seen, single driers are not normally used by themselves in paints but are utilized in combinations to achieve the best balance of film properties.

Antiskinning Agents

These are chemical additives used in paints to prevent the formation of a solid "skin" on the surface of paint usually while it is still in the can. This skin is formed from oxidative crosslinking at the paint's surface. If this skin formation is not effectively removed before stirring, the skin may form permanent lumps in the paint.

The antiskinning agents that are added to prevent skinning are essentially volatile antioxidants which prevent the oxidative crosslinking process by absorbing free radicals as well as complexing with the metallic driers in the paint. When the paint is applied to a surface, however, these antiskinning agents are lost through volatilization leaving the free radical mechanism unimpaired (see previous discussions). Care must be taken not to use too high a level of antiskinning agent or the paint's drying time may be lengthened. Only a very low level of antiskinning agent is usually needed.

Oximes are by far the most commonly used antiskinning agents. Some examples of these oximes are shown below.

$$C_3H_7CH=NOH$$

Butyraldoxime

$$C_2H_5-\overset{\overset{\displaystyle CH_3}{|}}{C}=NOH$$

Methylethylketoxime

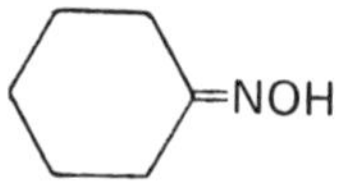

=NOH

Cyclohexanoneoxime

Low molecular weight substituted phenols have also been used as antiskinning agents.

Thickeners and Thixotropes

For most paints to be effective, they must have "body", or "thickness" which well prevent the pigments from settling during paint storage or the paint from "sagging" after application. On the other hand, the paint can not be so thick that it is not easily applied to a surface by brush or spray. These flow properties (rheology) are dependent on the binders, solvents, and pigments used; however, thickener or thixotrope additives are also used by paint formulators to help adjust to the flow properties needed.

Viscosity (η) is a coefficient which is equal to shear stress over rate of shear.

$$\eta = \frac{\text{Shear Stress}}{\text{Rate of Shear}}$$

If viscosity always stays constant with increasing or decreasing rate of shear, then the liquid is *Newtonian*. Paints are usually mixtures and are distinctly *non-Newtonian*. If the viscosity increases with increasing shear rate, the fluid is termed *dilatant*. This is generally undesirable in paints and occurs only at very high pigment volumes. On the other hand, if viscosity decreases with increasing shear rate, the fluid is considered *pseudoplastic*. This is a desirable condition for paints because when the paint is brushed or sprayed on to a surface, the high shear rate reduces the viscosity for better application. Conversely, when the paint is at rest (little or no shear), the viscosity is greater to resist settling of pigments in the container or sagging after application. Even more desirable, in some cases, is when the viscosity, at very low or zero shear rates, will increase with time from the formation of a temporary gel structure. A liquid displaying this property is said to be *thixotropic*. The formation of this "gel" when the paint is at rest, helps resist pigment settling before application and sagging afterwards. Under the shear force of brushing or spraying, the gel structure is destroyed and viscosity reduced. Additives which enhance this formation of a "gel" under rest conditions are called *thixotropes*.

Some additives that are used in organic solvent based paints to alter flow properties are shown below.

Modified clays

Aluminum soaps

Hydrogenated castor oil

Polyamide-alkyd resins

Modified clays (called Bentones) are bentonite clays (Al_2O_3 $2SiO_2 \cdot 2H_2O$) which have been pretreated with quaternary ammonium ions (R_4N+) where R is representative of an alkyl group. Because of these alkyl groups, organic molecules are attracted to the surfaces of the bentonite platelets and cause swelling. Sometimes a pretreatment with an alcohol may be needed to start swelling. These modified clays establish a gel structure at rest, thus imparting thixotropic properties. One advantage of these "Bentones" is that they can be used in aromatic or polar solvents. Some of the other agents to be discussed will not work in these solvents.

Alumium soaps, such as aluminum stearate, have also been used in the paint industry to impart gel structures. The higher the percent aluminum content of the soap, the higher the gel strength. These soaps work well in mineral spirits. If gel strength is too high, "leveling" will be impaired. Also, it may be difficult to maintain product uniformity using this type of "bodying" agent.

Another additive used in solvent paint systems for flow control is hydrogenated castor oil. This additive has a molecular structure based on the trigylcerides. Hydroxyl groups are present on the fatty acid chains and account for this material's ability to form a colloidal network at rest from the polar bonding among these functional groups. Thus when this additive is present at sufficient concentration, a thixotropic property is imparted to the paint. The use of a polar solvent reportedly can destroy the colloidal network because the polar solvent molecules will be attracted to the functional groups of the additive destroying the network links.

In a similar manner just described, polyamide-alkyd resins also impart thixotropic flow properties through polar attraction among functional groups. The polyamide-alkyd resin may here be used as the binder or as an additive. Through manufacturing modification, these resins can be made to impart a wide range of thixotropy to a paint. These resins are in common use; however, one possible disadvantage in their use is a tendency to yellow under certain conditions.

In aqueous based latex paint systems, a whole different set of thickening agents are used to adjust flow properties. Some of these thickeners for aqueous systems are given below.

Bentonite clay

Methylcellulose

Hydroxyethylcellulose

Sodium carboxymethylcellulose

Polysaccharides

Proteins

Fuming silicas

Bentonite clay particles swell when added to water due to absorption of water between the particle's plate-like layers. This swelling contributes to the formation of a gel in the water-clay system. This property can be used to alter flow properties in an aqueous coating.

Cellulosic thickeners are commonly used in aqueous based paints, especially PVAc latex paints. As discussed in previous chapters, straight cellulose is quite insoluble in water due to the strong attractive forces of the OH groups present on the polysaccharide macromolecules (the cellulose structure is shown below).

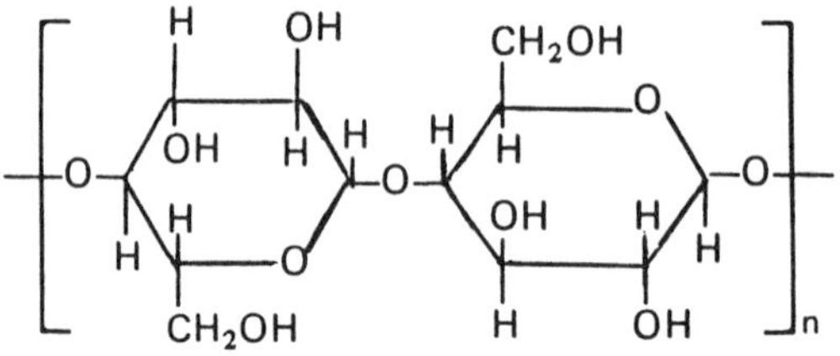

By first converting the cellulose to an alkali cellulose (from reaction with caustic soda) and then conducting a substitution reaction with an alkyl chloride (or other reactant), a water-soluble, viscous polymer such as methylcellulose, hydroxyethylcellulose, or carboxymethylcellulose is formed.

As can be seen, there are three OH groups available per anhydroglucose ring to participate in the substitution reaction, discussed above. Not all three hydroxyl groups are substituted to achieve the desired water soluble thickening properties. The extent of substitution does relate directly to the performance of these thickeners. The three most common cellulosic thickeners and their corresponding R group substitutions for OH groups are shown below.

Thickener	R Group Substitution for Hydroxyl
Methylcellulose	$-OCH_3$
Hydroxyethylcellulose	$-OCH_2CH_2OH$
Sodium salt of carboxymethylcellulose (CMC)	$-OCH_2\overset{\displaystyle O}{\overset{\displaystyle \|}{C}}-ONa$

Generally, the addition of a cellulosic thickener at about 1 to 2% of paint solids is sufficient to thicken a latex paint.

Water soluble acrylic polymers are also commonly used as thickening agents for latex paints, especially in acrylic latex paints. These thickeners are water soluble polymers based on polyacrylic acid or its sodium or ammonium salts. The higher the polymer's molecular weight, the greater the thickening effect it imparts to the aqueous system. On the other hand, these acrylic thickeners are reportedly not as effective as the cellulosic thickeners but may give better leveling properties.

Naturally derived thickeners such as polysaccharides or proteins are still used in aqueous coating systems. For example, alkali-proteins have sometimes been used in styrene-butadiene latex coatings. The proteins may be obtained from soybeans or milk (called casein). These proteins are made soluble by reacting them with an alkali such as ammonia at elevated temperatures. The use of these thickeners has declined in use somewhat. One disadvantage in their use is a finite shelf life after which putrefaction (protein degradation) and/or discoloration can occur.

Fuming silicas (derived from hydrolysis of silicon tetrachloride) have been used for many years as thickeners and thixotropes in both solvent and aqueous paint systems. They are also used as flattening agents. Their unique thixotropic properties are attributed to their very small particle size, ability to form chainlike structures, and their formation of hydrogen bonds.

Biocides

These additives are used in both solvent and aqueous based paints as *fungicides* and /or *mildewcides* to resist the growth of microorganisms on dried film surfaces. Also these additives are used in water based latex paint as *preservatives* to prevent the biodegradation of thickeners and other components of the coating. Some biocide additives may be used exclusively as preservatives, some are used exclusively as fungicides and/ or mildewcides, and some are used for both purposes.

Biocides serving as fungicides or mildewcides function in the dried film coating by poisoning the microorganisms trying to grow on the film surface.

On the other hand, biocides can also function as preservatives in aqueous based coatings. Preservatives, as such, are not needed in organic solvent based paints in that bacteria and other microorganisms can not survive without water present. However, preservatives may not work in aqueous coatings if enough time has passed before their addition to allow the bacteria or fungi to produce enzymes—biological "catalysts". These enzymes will attack cellulosic thickeners causing a viscosity drop in the paint even after the paint has been sterilized by the preservative additives. In fact, if the enzyme contaminated batch is used for blending with other uncontaminated paint, viscosity reduction with the blended paint may also occur. Many times enzyme contamination can even occur from raw materials used, such as latexes or thickeners. Obviously sterilizing with preservatives in these situations is ineffective.

As can be seen from the examples above, preservatives are not effective if the enzyme is already present. On the other hand, the early use of preservatives can prevent the formation of enzymes in the first place.

The following are examples of biocide additives used as either fungicides/mildewcides, and/or preservatives in paints.

> Zinc oxide
>
> Calcium carbonate
>
> Barium metaborate
>
> Mercury compounds
>
> Organotins
>
> Copper compounds
>
> Dithiocarbamates
>
> Chlorinated phenols
>
> Phthalimide types
>
> Benzamidazole

For years zinc oxide has not only been used in paints as an extender, but also as a mildewstat as well. It prevents mildew formation from occurring on the surface of the paint's film. To be effective, a relatively high concentration of zinc oxide is required, perhaps as high as 30% of total paint solids.

Calcium carbonates are commonly used also as extenders in paints. While these extenders are not true mildewcides for a dry film, they can impart an alkaline environment which may retard some mildew growth. On the other hand, calcium carbonate can hurt the water resistance of a film.

Just as with zinc oxide, barium metaborate is sometimes selected as an extender in a paint formulation because of its fungicidal quality in dry film. Just as with zinc oxide, to be effective, this barium compound reportedly must be used at concentrations greater than, say, 15% of total paint solids. Because of this material's slight water solubility, it may have restricted use in exterior paint applications.

Mercury compounds have been one of the most effective biocides used in the paint industry as both a preservative in latex paints and as a fungicide/mildewcide in latex and solvent based paints for dry film protection. *Phenylmercury propionate* and *phenylmercury oleate* are examples of mercury compounds used by the paint industry. These mercury biocides are very effective requiring only very low concentrations to work. On the other hand, because of the high toxicity of these substances, the government has restricted the use of these biocides and established a maximum concentration level. Also, some of these mercury compounds have poor resistance to leaching from the film. Because of the high toxicity, handling and environmental concerns regarding the use of mercury biocides, other alternate biocides are starting to be utilized as substitutes when possible.

Organotin compounds, such as tributyl tin oxide (TBTO), have been used as mercury substitutes in latex paints as a preservative as well as a fungicide. These compounds are generally not as toxic as the mercury biocides just discussed. On the other hand, a higher concentration is needed for organotin compounds to be effective. Therefore organotins cost more to use than mercury compounds. Organotin compounds are also used as a fungicide for dry film protection. Levels up to 1% of paint total solids may be needed to impart protection. Organotin compounds are toxic to humans; therefore, all appropriate precautions should be followed.

Copper compounds are also used as fungicides and occasionally as anti-fouling agents to retard the growth of marine organisms such as barnacles on submerged surfaces. Copper-8-quinolinolate is a commonly used copper biocide which has a mild degree of toxicity. It imparts a yellowish green color to a paint. Cuprous oxide has also been used occasionally as a fungicide. It too can alter a paint's color and is toxic.

Dithiocarbamates, such as zinc dimethyldithiocarbamate, are also used as fungicides in paints. Many may be used up to 5% of paint's total solids. The use of these fungicides is avoided for paint systems which are air cured because they may lengthen drying time.

Overall, great care should be taken in selecting a biocide for use in coatings. Toxicity, proper concentrations, and proper handling procedures should be taken into account when selecting a biocide and all government regulations on the use of these agents should be followed.

Corrosion Inhibitors

As discussed previously, corrosion of a metal substrate is an oxidative process which occurs through an electrochemical mechanism. The corrosion occurs on the metal surface at "local cells" having both the cathode and anode sites at the points of corrosion. The corrosion always occurs at the anodes, not the cathodes. This is an important principle which is used to advantage to inhibit corrosion of a metal surface. Corrosion inhibitors are ordinarily used in pretreatments, conversion coatings, and primers of coating systems in order to protect a metallic substrate through mechanisms to be discussed. Although ferrous alloys are the most commonly coated metal surfaces, also other nonferrous metals such as aluminum, magnesium, tin, copper, brass, zinc, etc., are protected by organic coating systems.

Coatings basically prevent corrosion of the metal substrate by the following means.

(1) Providing a barrier (increasing electrical resistance)

(2) Increasing the polarization at the anode or cathode

(3) Artificially keeping the metal surface cathodic by containing a less noble metal (usually zinc) which serves as an anode and is sacrificially corroded.

The "protective barrier" function works through a principle that may not be immediately obvious. It is true that for corrosion to occur, there must be the presence of moisture and oxygen at the metallic surface. While many protective films do reduce the permeability of moisture and oxygen, they can not stop it. In fact, such extenders as *mica* and *stainless steel flakes* have been used to form a protective barrier in dried films by the alignment and packing of the plate-like particles. This may help reduce some water permeation; however, the true protection of a "barrier coating" is from its ability to electrically insulate the metal surface. If the electrical resistance of the coating is maintained at a very high level, the electrochemical corrosion process can not occur. This is the basic principle applied when using a "barrier" coating to protect against corrosion.

Another way to prevent corrosion of a metal substrate is to use passivator inhibitors. These are inorganic inhibitors which are slightly

soluble in water. In the film, they slowly dissolve from moisture permeability. The inhibitive ions that are formed are carried to the film-metal
interface by moisture diffusion through the coating. These ions at the
interface cause the corrosion-electrical-potential to be elevated to the
"passive potential" where the corrosion rate is dramatically inhibited
through polarization at the anode. It is important to select a passivating
inhibitor with the optimal solubility for the coating application being
considered. If the passivating salt is too water soluble, it will be lost with
time and impart only short term protection. On the other hand, if the
passivating salt is too insoluble, then insufficient concentrations of ions
will reach the metal-coating interface from moisture diffusion. Ideally, the
chosen passivating inhibitors should have only enough water solubility to
provide sufficient ions to inhibit at the metal-coating interface, thus
providing the optimal time release.

The classes of passivating corrosion inhibitors commonly used in
coatings include the following:

> chromates
>
> molybdates
>
> phosphates
>
> lead compounds

The chromates are one of the most commonly used corrosion inhibitors in coating applications. They have been used not only to protect
ferrous alloys, but also aluminum, tin, magnesium, and other metals.
Common examples of chromate salts used in coatings are given below.

> zinc chromate
>
> lead silicochromate
>
> strontium chromate
>
> zinc yellow ($4ZnO \cdot K_2O \cdot 4CrO_3 \cdot 3H_2O$)

These chromate compounds are soluble enough to allow the permeation of moisture through a film to "leach out" enough chromate ions to the
film-metal interface to inhibit corrosion. Salts such as sodium chromate
are *too* soluble while salts such as lead chromate are not soluble enough to
be effective. Chromates can protect ferrous alloys not only by passivating,
but also by absorbing onto the steel substrate forming a complex. One
limiting factor in the use of chromates is the high toxicity that they
possess. Chromates have been connected with lung cancer and only safe
handling procedures should be used with these materials. All government
regulations must be followed in using chromates.

With growing concerns regarding the toxicity of chromates and lead
compounds, more attention has been directed to using phosphates as

passivating corrosion inhibitors with only limited success. Phosphates, such as zinc phosphate, have been used in latex paints with some reported success.

Just as with phosphates, molybdates have also gained new interest because of the high toxicity of chromates and lead compounds. Examples of molybdates which have reportedly shown some success are basic zinc molybdate and basic calcium zinc molybdate. One problem reported in the use of molybdate is the occurrence of osmotic blistering. Also molybdates can be toxic.

The passivating inhibitors mentioned so far have been soluble enough to allow moisture to carry sufficient ions to the interface. On the other hand, there are other "insoluble" basic passivating inhibitors which work through a different mechanism. A prime example of this is red lead oxide (plumbous orthoplumbate, Pb_2PbO_4). "Red lead" historically has been used as a very effective corrosion inhibitor in primers to protect ferrous substrates. Red lead was found to be most effective when used in primers based on linseed oil. The red lead apparently reacts with fatty acid decomposition products of the linseed oil to form soaps which inhibit corrosion. Red lead may not be as effective with some other binders however. Other lead compounds, such as lead chromates, have also been used as corrosion inhibitors. Because of the recent environmental considerations (discussed earlier), the use of lead compounds has declined because of their high toxicity. In fact, new government regulations either prohibit or severely limit the use of lead compounds in coatings.

Corrosion many times is less likely to occur in an alkaline environment than an acidic one. Therefore basic extenders, such as *calcium carbonate*, have been reported to help reduce corrosion rates. *Zinc oxide* extenders have also been reported to slow corrosion as well. *Red iron oxide* is also used as a corrosion inhibitor. Occasionally *barium metaborate* has been used in primers as a corrosion inhibiting pigment.

In addition to the protective barrier and the passivating inhibitors, there is a third way in which coatings can be used to prevent corrosion of metallic surfaces, i.e., by cathodic protection. This cathodic protection works by loading the protective film with a "less noble" metallic powder which eletrochemically becomes the anode and sacrificially corrodes instead of the metallic substrate (which is driven to a protected cathodic state). The most common method of applying this type of protection is through the zinc rich primer. The concentration of zinc powder in a dried primer must be sufficiently high enough to allow the particles to make contact and allow an electrical current to be established between these particles in the primer and the metallic substrate. Since zinc is less noble than iron, it becomes anodic and is sacrificially corroded. When the zinc is completely oxidized, there is no further protection. Therefore, there is a finite protective period; however, zinc rich primers have gained wide acceptance as very effective corrosion inhibitors. Care should always be taken, however, when working with zinc powder to keep it free of moisture in that hydrogen can be generated. Free hydrogen poses an explosion risk. Also zinc dust can form explosive mixtures with air posing a dangerous risk.

Flame Retardants

Occasionally these additives are used in coatings to reduce the risk of fire. For a description of these agents, refer to the flame retardant discussion in the previous chapter on the "Plastics Industry".

Plasticizers

These are non-volatile liquids which are added to a paint in order to reduce the binder's glass transition temperature and render the film more flexible (less brittle). Plasticizers achieve this by disrupting inter-molecular forces among the molecular chains of the binder, thus allowing them to move more freely. Some polymeric binders may be copolymerized to improve their flexibility (this is called *internal* plasticization). However, when a plasticizer is added to a paint formulation to achieve this effect, it is called *external* plasticization. If a plasticizer is highly compatible with a given binder (has a similar Hildebrand solubility parameter), it is called a *primary* plasticizer ("solvent" type). If the plasticizer used is not highly compatible with the given binder, it is called a secondary plasticizer ("non-solvent" type).

Generally, with increasing concentration of a plasticizer in a paint formulation, the film tensile strength will decrease, film elongation will increase, permeability will rise, and film toughness and adhesion will increase, reach an optimum, and decrease. Therefore, it is not only important to select the proper plasticizer but also the best part level to assure best results.

There are hundreds of different plasticizers on the market today; however, the phthalate plasticizers represent over half of the total volume used. Di-2 ethylhexyl phthalate (DOP) is by far the single largest volume plasticizer used. DOP is used with mostly PVC with some reported use in other binders. Also the more volatile dibutyl phthalate (DBP) is used extensively with nitrocellulose as well as other resin binders. Butylbenzyl phthalate (BBP) has been a preferred plasticizer for polymethyl meth-acrylate systems (lacquers).

DBP   DOP   BBP

Other plasticizers are also commonly used in paint formulations. Phosphates, such as triaryl phosphate as well as tris-β-chloroethyl phosphate (TCEP) are used in some cellulosic coatings. These plasticizers help reduce flammability properties of dry paint film to varying degrees. Also non-drying alkyds are used as plasticizers with other binders. For more detailed information concerning plasticizers and their use in polymers, refer to the previous chapter on the Plastics Industry.

Surfactants

These additives are very important to the paint industry, particularly in aqueous based coatings. These additives may also be referred to as surface-active agents. A surfactant molecule will generally contain both lipophilic (oil-loving) and hydrophilic (water-loving) groups as exemplified below.

$$CH_3(CH_2)_xCH_2-R$$
(Lipophilic)   (Hydrophilic)

By possessing both groups, these agents can "bridge the gap", so to speak, in order to function as emulsifying agents, dispersing agents, and wetting agents.

Emulsifying agents, are surfactants that reduce interfacial tension between different liquids, usually an oil and water. Through the use of the proper surfactant in an oil-water system, a dispersed emulsion of the oil in water can be formed and maintained or vice versa. Typically in the case of emulsifying an oil in water, the lipophilic (hydrophobic) groups of the surfactant molecules are attracted into the surface of the oil droplets. On the other hand, hydrophilic groups position themselves outwardly from the droplet. These hydrophilic groups form a "film" around the droplet to prevent coalescence. Also because like ionic charges repulse one another, the similarly charged droplets are kept from coalescing.

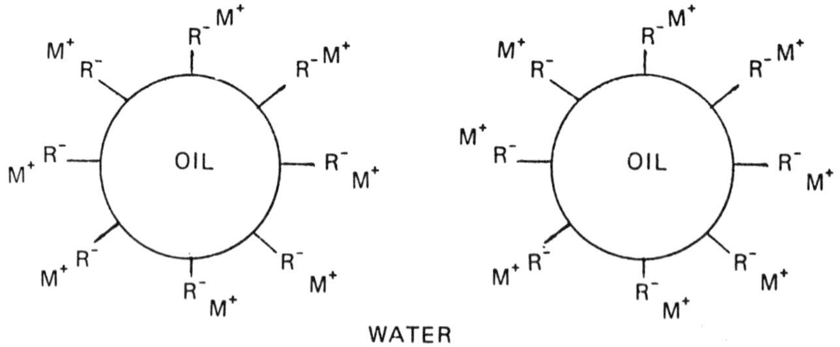

WATER

Surfactants are used as *dispersing agents* to wet and help keep solid pigment particles suspended in either an aqueous or non-aqueous medium. As discussed earlier, thickener additives can also help hold a suspension by increasing viscosity of the medium at rest. Surfactants, on the other hand, help "wet" the surface of the pigment particles and support a suspension through a different mechanism. In a similar manner to that just mentioned for emulsions, anionic surfactants will also impart a negative charge to small suspended pigment particles to help keep them apart in an aqueous medium. In fact, the addition of the proper surfactant to a pigment suspension will reduce the viscosity. Pigments do differ, however. Some have hydrophilic surfaces while others have hydrophobic (lipophilic) surfaces. Therefore different surfactants, or combinations of surfactants, will be required to wet and suspend different types of pigments.

Surfactants also can fulfill the role of *wetting agent* in aqueous based coatings. These agents reduce the surface tension of the water medium and reduce the interfacial tension between the coating and substrate. This enables the coating to wet and spread more easily over the substrate.

The following are the three basic types of surfactants that are used in the coatings industry.

(1) anionic

(2) cationic

(3) non-ionic

Anionic surfactants are surface-active agents which, upon ionization, have a lipophilic group in the anion (negatively charged ion). A common example of this is the negative stearate ion formed from sodium stearate or the negative lauryl sulfate ion from sodium lauryl sulfate.

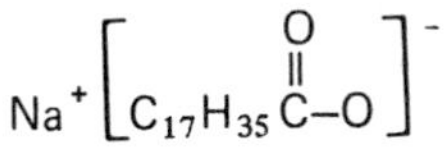

$$Na^{+}\left[C_{17}H_{35}\overset{\overset{\displaystyle O}{\displaystyle \|}}{C}-O\right]^{-} \qquad Na^{+}\left[C_{12}H_{25}-O-\overset{\overset{\displaystyle O}{\displaystyle \|}}{\underset{\underset{\displaystyle O}{\displaystyle \|}}{S}}-O\right]^{-}$$

Stearate Ion 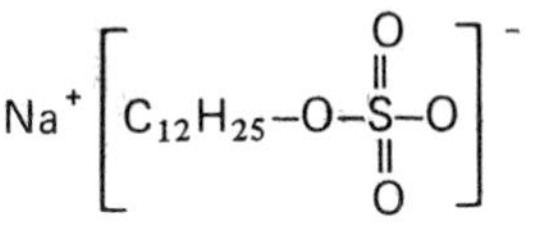   Lauryl Sulfate Ion

This class consists of carboxylates, alkyl sulfates, alkyl sulfonates, alkyl aryl sulfonates, phosphates, etc. Anionic surfactants as a class are the oldest agents used as surface-active agents. The following contains specific examples of commercial anionic surfactants used in industry.

sodium decylbenzenesulfonate

sodium dibutylnaphthalene sulfonate

sodium alkylbenzenesulfonate

sodium oleyl sulfate

sodium lignin sulfonate

sodium dioctyl sulfosuccinate

Cationic surfactants are surface-active agents which, upon ionization, have a lipophilic group(s) in the cation (positive charged ion). An example might be seen from the ionization of a dialkyl amine chloride.

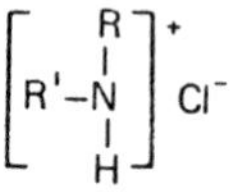

$$\left[R'-\underset{\underset{H}{|}}{\overset{\overset{R}{|}}{N}} \right]^{+} Cl^{-}$$

Cationic surfactants are seldom used in the coatings industry. They can not be used in latex paints because they cause the latex to coagulate. They do have very limited use reported in asphaltic emulsions and have occasionally been used as biocides. Some examples of cationic surfactants are given below.

> cetyldimethylethylammonium bromide
>
> alkylbenzyldimethylammonium chloride
>
> lauryltrimethylammonium chloride
>
> alkyltrimethylammonium chloride
>
> cetyltrimethylammonium bromide

Nonionic surfactants do not ionize but achieve their surface active properties by containing both lipophilic groups and hydrophilic groups (such as hydroxyl groups or ether linkages) on the same molecule. While these surfactants do not ionize, they do solubilize in water through the hydrogen bonding of hydroxyl or ether groups.

In the past thirty years, nonionic surfactants have grown in their diversity of applications. These agents are compatible with a wide range of substances and are sometimes effective where anionic surfactants are not. Some examples of commercially used nonionic surfactants are given below.

> tris(polyoxyethylene) sorbitan monolaurate
>
> polyoxyethylene Sorbitan monostearate
>
> Sorbitan monooleate
>
> Sorbitan monopalmitate
>
> Polyethylene glycol stearate
>
> Polyethylene glycol lauryl ether
>
> Diethylene glycol monostearate
>
> Sorbitan tristearate
>
> Tetraethylene glycol monooleate
>
> Polyoxyethylene cetyl alcohol

From the three classes discussed, there are over a thousand different commercially available surfactants used in industry. Many times two or more surfactants may be used together in paint formulation in order to disperse different pigments. Sometimes a nonionic surfactant is used with an anionic surfactant to maximize effect.

The selection of the proper surfactant or combination of surfactants which will be most cost effective involves a great deal of trial and error laboratory work. One method which helps take out some of the guess work is the use of HLB numbers developed by W. C. Griffin. HLB stands for *h*ydrophile-*l*ipophile *b*alance and attempts to measure the relative "strength" of the hydrophilic groups to the lipophilic groups on the same molecule. Surfactants with low HLB values of less than 4 have dominant lipophilic groups and do not disperse in water. On the other hand, surfactants with HLB values greater than 13 have dominant hydrophilic groups and are completely soluble in water. Many surfactants fall somewhere in between and form milky or translucent dispersions in water. Different pigments in a paint require different surfactants to be wetted and dispersed properly. Experimental work by the paint formulator to find the optimal HLB for color acceptance and minimum viscosity will aid greatly in selecting the final combination of surfactants for a particular paint.

Defoamers

These are important additives that are used to prevent foam. Technically, if the additive actually *prevents* foaming from occurring, it is an *anti-foaming* agent. On the other hand, if the additive destroys the foam after it forms, it is called a *defoaming* agent. In reality, however, the two terms are used interchangeably regardless of how the additive works.

Many aqueous paints containing surfactants or thickeners will foam upon agitation. Foam formation is undesirable because it can cause "cratering" during film formation. Also foaming can disrupt the paint manufacturing process.

While partially soluble solvents can be used as defoamers, usually special additives based on silicone oil or hydrophobic silica are used. Care should be taken in not using too high a concentration of these agents in that separation (fish eyes) can result.

Coalescing Agents

These are additives to latex paints which promote the coalescing of latex particles to form a continuous film during drying of the paint. These additives act as temporary plasticizers which are lost to volatilization after the film is formed. As discussed earlier, the latex particles in a paint must be soft enough to coalesce on drying at room temperature or lower. If the polymer is too hard or has too high a glass transition temperature (such as polyvinyl acetate homopolymer), then volatile plasticizers will aid in softening the latex particles to allow coalescing and film formation at a lower temperature. The plasticizer imparts "plastic flow" to the particles. After a few days, the film will harden from loss of these additives through volatilization. Examples of such coalescing agents include tributyl phosphate, hexylene glycol, dibutyl phthalate, or the volatile ether-alcohol plasticizer/solvents sold under the tradename of *Cellosolve* by Union Carbide Co.

Freeze-Thaw Stabilizers

These additives are used in latex paints to prevent freezing, especially in paints sold in the trade sale markets. In latex paints, as the water medium freezes at low temperatures, the protective layer of water formed by the dispersing agent around the particles is destroyed. This can lead to the particles coalescing. Coalescing agents, just discussed, can worsen this situation because they soften the colloidal particles which promotes coalescence. Freeze-thaw stabilizers simply lower the freezing point of water. The most commonly used stabilizers are the glycols. *Ethylene glycol* is the most common one used. Also *propylene glycol* is used. These agents are effective because they will solubilize in the water medium and suppress the freezing point without being solubilized in the latex particles. Also many nonionic surfactants act as freeze-thaw stabilizers.

Anti-Rust Agents

These additives should not be confused with corrosion inhibitors. Corrosion inhibitors are additives that protect a metallic substrate from corrosion after application. An anti-rust agent is an additive used in a water based latex paint to help protect the metallic container. One method to prevent can corrosion is to make the waterborne coating more alkaline, if possible. One of the most common anti-rust additives is *sodium benzoate*. Of course, the best protection against container corrosion is to use containers that are precoated for protection.

pH Buffers

These are additives to adjust the pH of a latex paint. The stability of a latex paint can be quite sensitive to pH. For anionic surfactants to be effective may require a specific pH range. Also thickeners are sensitive to pH. Some thickeners such as casein or carboxymethylcellulose, are only soluble in a specific alkaline range.

Usually a volatile pH adjuster is preferred, such as ammonium hydroxide. On drying, such a volatile adjuster will not hurt the water resistance of the dried film.

NEW DIRECTIONS

The paint and coatings industry in many ways is a "mature" industry. There are numerous coating producers in the marketplace and competition is keen. The growth rate in many markets is projected to be

2–3%/yr. through the 1980s. As a result, more paint producers have been decreasing their R and D efforts and relying on raw material suppliers for new technological innovations.

The quality of many of the finishes today is indeed excellent; however, there are still many areas which require technological development. One of the great challenges facing the industry is to achieve the same quality finish (achieved by conventional solvent based systems) by the use of coating systems which use little or no organic solvent. As discussed earlier, the reason for this need is the new tighter EPA solvent emission restrictions which are limiting the use of solvents. Because of this move away from solvents, water based latex paints have been used as substitutes wherever possible. However, there are limits to the number of applications for latex paints. As a result, there are several new emerging coating technologies which have evolved mainly to help reduce the use of organic solvents. These technologies are given below.

> High solids systems
>
> Two part catalyzed systems
>
> Water soluble binder systems
>
> Powder coatings
>
> Radiation curable systems.

These new technologies, which were discussed earlier, are growing at a very rapid rate. A number of markets will be utilizing these technologies in the late 1980's.

Another area of current improvement is in efficiency of application. New application methods, such as powder coating, electrostatic spraying, and electrodeposition, are improving efficiency in the transfer of paint to a substrate while also improving the uniformity of the film thickness. These techniques will reduce "overspray" and waste (paint sludge). A side effect of these improved efficiencies may be a further reduction in the future growth rate for paints.

Lastly, improvements in raw materials may also occur. For example, raw materials that will disperse more easily in the new "high solids" systems, which have higher viscosities, will be sought. Also new environmental concerns regarding dust in the workplace will force new techniques to be developed that will reduce or eliminate dust in the factory. More and more pigments may be used in predispersed forms.

SUGGESTED READING

Banov, Abel, *Paints & Coatings Handbook*, Structure Publishing Co., Farmington, Mich, 1973

Boxall, J., Fraunhofer, J. A., *Paint Formulations Principles & Practices*, Industrial Press, 1981

Chatfield, H. W., *The Science of Surface Coatings*, Van Nostrand, N.Y., 1962

DeRenzo, D. J., *Handbook of Corrosion Resistant Coatings*, Noyes Publications, Park Ridge, N.J., 1986

Federation Series on Coatings Technology, Units 1 through 27, Federation of Societies for Paint Technology, 1315 Walnut Street, Philadelphia, Pa., 19107

Flick, E. W., *Contemporary Industrial Coatings*, Noyes Publications, Park Ridge, N.J., 1985

Flick, E. W., *Handbook of Paint Raw Materials*, Noyes Publications, Park Ridge, N.J., 1982

Fuller, Wayne R., *Formation & Structure of Paint Films*, Federation of Societies for Coatings Technology, 1965

Martens, Charles R., *Emulsion and Water-Soluble Paints and Coatings*, Reinhold Publishing Co., N.Y., 1964

Martens, Charles R., *Technology of Paints, Varnishes and Lacquers*, Reinhold Book Corp. 1968

Martens, Charles R., *Waterborne Coatings*, Van Nostrand Reinhold, N.Y., 1981

Morgans, W. M., *Outlines of Paint Technology*, Charles Griffin & Co., 1982

Myers, Raymond R., Long, J. S., *Treatise on Coatings, Formulations Part I*, Marcel Dekker, N.Y., 1975

Myers, Raymond R., Long, J. S., *Treatise on Coatings, Volume 2, Characterization of Coatings: Physical Techniques, Part II*, 1976

Nylen, Paul, Sunderland, Edward, *Modern Surface Coatings*, Interscience Publishers, N.Y., 1965

Patton, Temple C., *Paint Flow & Pigment Dispersion (A Rheological Approach to Coatings)*, 2nd Ed., John Wiley & Son, N.Y., 1979

Solomon, D. H., *The Chemistry of Organic Film Formers*, Robert E. Krieger Publishing Co., Huntington, N.Y., 1977

Surface Coatings, A Complete Handbook of Paint Technology, By the Oil and Colour Chemists' Association, Australia, (in conjunction with the Australian Paint Manufacturer's Federation), New South Wales University Press, 1974

Sward, G. G., *Paint Testing Manual, Physical and Chemical Examination of Paints, Varnishes, Lacquers, and Colors*, (ASTM Special Technical Publication 500, 13th Ed.), ASTM Philadelphia, 1972

Taylor, C. J. A., Marks, S., *Part One, Non-Convertible Coatings*, Oil and Colour Chemists' Association, Chapman & Hall, London, 1969

Taylor, C. J. A., Marks, S., *Part Two, Solvents, Oils, Resins, and Driers*, 2nd Ed., Oil & Colour Chemists' Association, Chapman & Hall, London, 1969

Test Methods and Techniques for the Surface Coatings Industry, 2nd Ed., Shell Chemical Co., Industrial Chemicals Div., N.Y., 1967

Turner, G. P. A., *Introduction to Paint Chemistry*, Chapman & Hall, London, 1967

Weismantel, Guy E., *Paint Handbook*, McGraw-Hill, N.Y., 1981

Index

Ballotine microspheres - 62
Barium-cadmium heat stabi-
    lizers - 84
Barium ferrite - 65
Barium metaborate - 266,
    270
Barrier corrosion inhibitors -
    268
Basic principles - 11, 168,
    215
BBTS - 132
Bentones - 263
Bentonite clay - 263
Binders - 228
Biocides - 266
Black pigments - 89, 256
Blocked isocyanate poly-
    urethane - 242
Blocked polymer - 162, 183
Blowing agents - 156
Blow molding - 6
Blue pigments - 258
Boron flame retardants - 53
BR - 121
Bromobutyl rubber - 126
Bronze powder - 67
Butyl benzothiazole sulfen-
    amide (BBTS) - 132
Butyl rubber (IIR) - 125, 187

CAB - 39
Calcined clay - 259
Calcium carbonate - 63, 144,
    259, 266, 270
Calcium sulfate - 64, 259
Calcium-zinc heat stabilizers -
    85
Calendering - 8, 114
Carbon black - 64, 89, 140,
    256
Carbon fibers - 62
Carboxylic elastomers - 188
Casein - 189
Casting - 7

Catalytic type hardeners -
    201
Catalyzed moisture cured
    polyurethane - 242
Catalyzed two-part system -
    218
Cationic surfactant - 274
CBTS - 132
Cellulose acetate - 37
Cellulose acetate butyrate
    (CAB) - 39
Cellulose binders - 248
Cellulose derivatives - 190
Cellulose propionate - 39
Chain extenders (poly-
    urethane) - 159
Chemical plasticizers - 148
Chlorinated paraffin flame
    retardants - 155
Chlorinated polyvinyl
    chloride - 28
Chlorobutyl rubber - 126
Chloroprene (CR) - 124, 188
Chlorosulfonated polyethyl-
    ene - 125
Chromate corrosion inhibitors -
    269
Chromium type coupling
    agents - 70
Clay - 63, 143, 259
Coalescing agents - 275
Coal, powdered - 64
Cobalt driers - 260
Cohesive energy density - 171
CO, hydrin - 126
Color - 214
Colorants - 98, 155, 253
Components (for formulat-
    ing) - 220
Compounding, plastics - 14
Compounding, rubber - 116
Compression molding - 5
Convertible coatings - 216
Corrosion inhibitors - 268

DATE DUE

MAX 0 1			

Demco, Inc. 38-293